Plant Tissue Culture

– Emerging Trends –

Plant Tissue Culture
– Emerging Trends –

Editor
T. Pullaiah
Department of Botany
Sri Krishnadevaraya University
Anantapur – 515 003
Andhra Pradesh

2011
REGENCY PUBLICATIONS
Delhi - 110 002

© 2011, AUTHOR
ISBN 9789351241522

Published by	:	**REGENCY PUBLICATIONS**
		(*Division of Daya Publishing House*)
		4760-61/23, Ansari Road, Darya Ganj,
		New Delhi - 110 002
		Phone: 23245578, 23244987
		Fax: (011) 23260116
		e-mail : info@regency-books.com
		website : www.dayabooks.com
		website : www.regencybooks.com
Laser Typesetting	:	**Classic Computer Services**
		Delhi - 110 035
Printed at	:	**Chawla Offset Printers**
		Delhi - 110 052

PRINTED IN INDIA

Preface

The book "Plant Tissue culture – Emerging Trends" is a contribution of various scientists from different universities in India. This book is edited with the objective of analyzing the Recent advances in Plant Tissue Culture.

I am thankful to all the leading scientists who have contributed chapters clearly elucidating recent developments in Plant Tissue culture.

We hope that this book will be useful to students, researchers and teachers in Agriculture, Botany and Biotechnology.

Prof. T. Pullaiah

Contents

Abbreviations

―――――――

ABA :	Abscisic acid
AC :	Activated charcoal
BAP :	6-Benzyl amino purine
CH :	Casein hydrolysate
CM :	Coconut milk
2,4-D :	2,4-Dichlorophenoxy acetic acid
DKW medium :	Driver and Kuniyuki medium
GA_3 :	Gibberellic acid
IAA :	Indole acetic acid
IBA :	Indole-3-butyric acid
2-iP :	2-isopentenyl adenine
Kn :	6-furfurile aminopurine
K_2SO_4 :	Potassium sulphate
LS medium :	Linsmaier and Skoog medium
$MgSO_4$:	Magnesium sulphate
MS medium :	Murashige and Skoog medium
NAA :	Naphthalene-3-acetic acid
PG :	Phloroglucinol
TDZ :	Thidiazuron
WPM :	Woody plant medium
Z :	Zeatin

Chapter 1

Tissue Culture of Tree Legumes: A Review

T. Pullaiah, M. Sowghandika, P. Kalavathi and M. Anuradha

Department of Botany, Sri Krishnadevarya University, Anantapur – 515 003

Plant cell and tissue culture has blossomed over the years into a science and technology without parallel with hopes of production of newer plants with novel gene combinations of agronomic value. The progress has been overwhelming. Plant cell culture is rapidly developing technology which holds promise of restructuring agricultural and forestry practices. The technique greatly reduces the labor and space requirements for producing new varieties and can also markedly enhance propagation rates. During the last couple of decades plant cell, tissue and organ culture has made considerable advances not only in the realms of fundamental botany but also in the field of agriculture, horticulture, plant breeding, forestry, somatic cell genetics, physiopathology, industrial production of plant metabolites etc. With the far reaching implications of plant tissue culture as an experimental tool an ever-growing number of plant tissue biologists are now keen on exploiting this technology.

The first successful tissue culture experiments were performed with a leguminous tree *Robina pseudoacacia* (Gautheret *et al.* 1934) by employing explants and observed white healthy callus masses on its surface. Though it has been successful, *in vitro* culture and micropropagation of legume trees have not been widely used and long time stagnation was there before it flourished. Legume trees are fast growing, hardy, robust trees that enrich the soil with nitrogenous substance. There are several unattempted elite, endemic and endangered species with problem in conventional methods of propagation, such trees can be propagated rapidly by tissue culture technology.

In the following pages a review of the tissue culture work in the family Leguminosae (including all the three subfamilies – Papilionoideae, Caesalpinioideae and Mimosoideae) is given. Table 1.1 gives the list of species and the details of tissue culture work done. The review is arranged genera wise, and species wise within the genera in alphabetical sequence.

Acacia Species

Acacias are one of the short-rotation and fast growing tree species. Acacias are prolific seed setters and are generally propagated from seed. They also coppice well and several species spread by root suckers. The trees are heavy seeders but being outbreeders there is a large degree of variation in seedling progeny. There is a need to develop propagation methods for rapid clonal multiplication of elite individuals.

Beck and Dunlop (2001) gave a review on micropropagation of *Acacia* species. The first report on tissue culture of *Acacia* was from Prof. Bonner's laboratory as long ago as 1942 (Bonner, 1942). *In vitro* propagation methods have been reported for many species of *Acacia* (Barakat and Saligna, 1992; Beck and Dunlop, 2001; Beck *et al.*, 2000; Darus, 1992; Gassama, 1989; Yousoff *et al.*, 1999). *In vitro* clonal propagation has also been practiced in large scale for the hybrid clones of *A. auriculiformis* X *A. mangium* (Yousoff *et al.*, 1999; Banik and Islam, 1996).

To overcome the maturation effect on propagation coppice shoots were used in *A. auriculiformis* X *A. mangium* (Banik and Islam, 1996) and *A. mearnsii* (Beck and Dunlop, 1999; Beck *et al.*, 1998), whereas sprouting buds from stem cuttings were used in *A. auriculiformis*. From 4-year-old hybrid (*A. auriculiformis* X *A. mangium*) ortets maintained as hedge, axillary buds were collected from 1-year old ramets and rejuvenated *in vitro*, where cultures restored the seedling traits. The culture-derived shoots were found morphologically, identical to the seedling-derived Acacia plants with bipinnate leaves (Yasodha *et al.*, 2002). The *in vitro* micrografting technique has also been a useful tool for vegetative, non-destructive propagation of mature selected *A. mangium* ortets (Monteuui, 1995, 1996). Apical meristem was found more susceptible to rejuvenation during certain stages of development (Monteuuis, 1988). Apical meristems obtained from micropropagated plantlets and mature trees were used as source material for *A. mearnsii* (Beck *et al.*, 2000) and complete plantlets were produced. The explants obtained from mature trees of *A. mangium* (Bhaskar and Subhash, 1996; Zhang *et al.*, 1995), *A. auriculiformis* (Zhang *et al.*, 1995; Reddy *et al.*, 1995), *A. catechu* (Kaur *et al.*, 1998) and *A. senegal* (Badji *et al.*, 1993) were also micropropagated successfully.

Seedlings of hybrid Acacia (Darus, 1992), *A. mangium* (Darus, 1991), *A. stenophylla* (Crawford and Hartney, 1987), *A. seyal* (Al-Wasel, 2000), *A. mearnsii* (Beck *et al.*, 1998; Marguerite *et al.*, 2001), *A. catechu* (Kaur and Kant, 2000) and *A. auriculiformis* (Ide *et al.*, 1994; Watanabe *et al.*, 1994) were multiplied *in vitro* through axillary bud proliferation. Cotyledonary nodes of *A. mangium* (Douglas and McNamara, 2000) and hypocotyls of *A. sinuata* (Vengadesan *et al.*, 2000) were also used to regenerate

Table 1.1: Review of Leguminous Tree Tissue culture

Plant	Explant	Result	Authors
Acacia albida	Cotyledonary buds	Plantlets	Duhoux and Davies,1985
A. arabica	Zytotic embryo	Plantlets	Nanda and Rout, 2003; Das, 2011
A. auriculiformis	Axillary buds	Plantlets	Mittal *et al.*, 1989
A. auriculiformis	Seedling explants	*In vitro* propagation	Yang *et al.*, 1989; Ide *et al.*, 1994; Watanabe *et al.*, 1994
A. auriculiformis	*In vitro* nodules assessment	*In vitro* nodulation	Galiana *et al.*, 1990
A. auriculiformis	Hypocotyl, callus	Callus, plantlets	Rao and Prasad., 1991
A. auriculiformis	Cotyledons	Plantlets	Das *et al.*, 1993
A. auriculiformis	Mature trees	Micropropagation	Zhang *et al.*, 1995; Reddy *et al.*, 1995
A. auriculiformis X *A. mangium*	Coppice shoots	Micropropagation	Beck and Dunlop, 1999; Beck *et al.*, 1998
A. baileyana	-	Rapid propagation	Zhang, 2007a
A. bivenosa	Seedling explants	*In vitro* propagation	Jones *et al.*, 1990
A. catechu	Callus	Somatic embryogenesis and regeneration of plantlets	Rout *et al.*, 1995
A. catechu	Nodal segments	Plantlets	Sahni and Gupta, 2002
A. catechu	Mature nodal segments	Plantlets	Kaur *et al.*, 1998
A. catechu	Shoot apex (15 days old *in vitro* grown seedlings)	Multiple shoots, roots and plantlets	Kaur and Kant, 2000
A. catechu	Cotyledons, node from mature tree	Plantlets	Thakur *et al.*, 2002
A. catechu	Zygotic embryo	Plantlets	Das, 2011
A. chundra	Shoot tip and node	Plantlets	Rout *et al.*, 2008
A. cincinnata	Shoot cuttings	Micropropagation	Li *et al.*, 1996
A. crassicarpa	Phyllode	Adventitious shoots	Yang *et al.*, 2006
A. farnesiana	Immature zygotic embryos and somatic embryos	Plantlets	Ortiz *et al.*, 2000
A. floribunda		Rapid propagation	Zhang, 2007 b
A. georginae	Callus derived from leaves	Production of flouroacetate	Bennett *et al.*, 1983
A. harpophylla	Seedling explants	Callus	Mc Clean *et al.*, 1966
A. holosericea	Seedling explants	*In vitro* propagation	Jones *et al.*, 1990; Reddy *et al.*, 1995

Contd...

Table 1.1–Contd...

Plant	Explant	Result	Authors
A. implexa		Plantlet regeneration	Zhang and Liu, 2004
A. koa	Somatic callus	Gray colour callus, plantlets	Skoleman and Mapes, 1976, 1978, 1986
A. mangium	Nodal segments	Plantlets	Crawford and Hartney, 1986
A. mangium	Nodal segments	Micropropagation	Yang *et al.*, 1989; Ahmad, 1989; Sascha, 1998
A. mangium	Seedling explants	*In vitro* assessment of N_2 fixation	Galiana *et al.*, 1990a,b; 1991a,b; Douglas and John, 2000
A. mangium	Seedlings	Micropropagation	Darus, 1992
A. mangium	Axillary buds	*In vitro* propagation	Saito *et al.*, 1993
A. mangium	Mature trees	Micropropagation	Zhang *et al.*, 1995
A. mangium	Nodal segments	Micropropagation	Bhaskar and Subhash, 1996; Bon *et al.*, 1998; Monteuuis and Olivier, 2004
A. mangium	Immature zygotic embryos	Callus, somatic embryogenesis	Xie and Hong, 2000
A. mangium	Cotyledonary node	Micropropagation	Douglas and McNamara, 2000
A. mearnsii	Shoot tip	Seedling	Huang *et al.*, 1994
A. mearnsii	Hypocotyls, cotyledons	Micropropagation	Quoirin *et al.*, 1998
A. mearnsii	Meristems of seedlings and coppices	Micropropagation	Beck *et al.*, 2000
A. mearnsii	Hypocotyl	Callus	Zeijlemaker, 1972; Gong *et al.*, 1991
A. melonoxylon	Roots	Growth	Bonner,1942
A. melonoxylon	Seedling explants	Plantlets	Meyer and Vanstaden,1987
A. melanoxylon	Root sprout	Micropropagation	Su and Zhang, 2008
A. nilotica	Mature stem, Cotyledons	Adventitious buds, plantlets, embryogenesis	Mathur and Chandra,1983
A. nilotica	Cotyledonary nodes	Plantlets	Dewan *et al.*, 1992
A. nilotica	Endosperm culture	Somatic embryogenesis and regeneration of triploid plantlets	Garg *et al.*, 1996
A. nilotica		Plantlets	Dhabhai *et al.* (2010)
A. salicina	Cotyledonary nodes	Plantlets	Jones *et al.*, 1990
A. saligna	Shoot tip	Plantlets, multiple shoots	Jones *et al.*, 1990; Barakat and El-Lakany, 1992

Contd...

Table 1.1–Contd...

Plant	Explant	Result	Authors
A. schaffneri	Immature zygotic embryos and somatic embryos	Plantlets	Ortiz *et al.*, 2000
A. sclerosperma	Shoot tip	Plantlets	Jones *et al.*, 1990
A. senegal	Cotyledonary callus	Callus, roots	Kothjo and Tewari.,1979
A. senegal	Stem culture	Bud break, plantlets	Dave *et al.*,1980
A. senegal	Seedling explants	Callus formation and cell suspension	Hustache,1986
A. senegal	Mature tree adventitious buds	Multiple shoots	Gupta *et al.*,1994
A. senegal	Cotyledonary node	Plantlets	Kaur *et al.*, 1996
A. senegal	Callus tissues	Organogenetic differentiation	Kaur *et al.*, 1997
A. senegal	Seedling explants	Rapid method for producing plantlets	Badji *et al.*,1993
A. senegal	Cotyledonary node and node of mature tree	Micropropagation	Khalafalla and Dafalla, 2008
A. seyal	Seedling shoot tip	*In vitro* propagation	Al-Wasel 2000
A. sinuata	Callus derived from seedling leaf explants	Callus, somatic embryos plantlets	Vengadesan 2002; Vengadesan *et al.*, 2004
A. sinuata	Hypocotyl	Micropropagation	Vengadesan *et al.*, 2000
A. stenophylla	Cotyledonary node	Plantlets	Crawford and Hartney,1986
Acacia tortilis ssp. *raddiana*	Axillary buds of mature trees & spices from seedlings	Micropropagation	Detrez, 1994; Nandwani, 1995
Albizia amara	Hypocotyl	Callus, adventitious buds, plantlets, embryogenesis	Huss Danell,1982
A. amara	Hypocotyl	Callus, shoots, plantlets	Tomar and Gupta, 1988a,b
A. amrara	Leaf	Shoot bud regeneration	Ramamurthy and Savithramma, 2003
A. amara	Seedling explants	*In vitro* propagation	Indravathi and Pullaiah, 2004
A. chinensis	Petiolar and distal cotyledonary segments	Multiple shoot, rooting and regeneration	Sinha *et al.*, 2000
A. guachapele	Roots	Plantlets	Cerdas Lisette Valvarde *et al.*,1998
A. julibrissin	Hypocotyl	Shoot regeneration	Sankhla *et al.*,1993
A. julibrissin	Roots	Adventitious shoots	Sankhla *et al.*,1994,1996
A. julibrissin	Embryo culture	Somatic embryos	Burns and Wetzstein,1998

Contd...

Table 1.1–Contd...

Plant	Explant	Result	Authors
A. lebbeck	Hypocotyl	Embryogenesis, plantlets	Gharyal and Maheshwari, 1981
A. lebbeck	Hypocotyl	Shoots,Callus	Bhargava et al.,1983
A. lebbeck	Anthers	Plantlets	Gharyal and Maheshwari, 1983a
A. lebbeck	Seedling	Callus, plantlets organogensis	Gharyal and Maheshwari, 1983b
A. lebbeck	Anthers	Callus	De and Rao,1983
A. lebbeck	Seedling	Adventetious buds, plantlets in soil	Upadhyay and Chandra, 1983
A. lebbeck	Seedling	Callus, adventitious buds plantlets in soil	Rao and De,1987
A. lebbeck	Seedling	Callus, plantlets	Verghese and Amarjeet,1988
A. lebbeck	Stems and petioles	Plantlets	Gharyal and Maheshwari, 1990, 1993
A. lebbeck	Seedling explants	Multiple shoots	Philomina et al.,1992
A. lebbeck	Cotyledon, nodal segments of seedling & nodal segments of mature trees	Micropropagation	Mtin et al., 2004
A. lucida	Hypocotyl	Callus, shoots	Tomar and Gupta.,1988a
A. lucida	Axillary buds and hypocotyl	Plantlets	Gupta and Agarwal 1992
A. lucida	Mature stem and petiole	Plantlets	Paramjit et al., 1990
A. odoratissima	Seedling explants	In vitro regeneration	Phukan and Mitra,1982; Rajeswari and Paliwal, 2004, 2007
A. procera	Seedling	Callus, adventitious buds, plantlets	Anand and Bir,1986
A. procera	Seedling	Callus, plantlets	Datta,1987
A. procera	Axillary buds and Hypocotyl	Plantlets	Gupta and Agarwal 1992
A. procera	Leaflets	Shoot buds	Kumar et al., 1998
A. procera	Epicotyl and Hypocotyl (15± old mature +15 days old juvenile plants)	Callus, regeneration and multiplication	Swamy, 2004
A. richardiana	Axillary buds and hypocotyl	Callus, shoot buds, somatic embryogenesis, plantlets	Tomar and Gupta,1988a
A. richardiana	Callus culture	Organogenesis, somatic embryogenesis	Tomar and Gupta,1988b

Contd...

Table 1.1–Contd...

Plant	Explant	Result	Authors
A. richardiana	Hypocotyl	Somatic embryoids	Gupta and Agarwal, 1992
A. thompsonii	Hypocotyl	Plants	Anuradha and Pullaiah, 2001
Bauhinia blaceana	Seedling explants	Large scale tissue culture propagation	Rao and Lee, 1982
B. purpurea	Seedling nodes	Bud break, plantlets	Anand and Bir,1983
B. purpurea	Mature tree	Plantlets	Kumar,1992
B. vareigata	Nodes explants of mature trees	Plantlets	Mathur and Mukuntha kumar,1992
Caesalpinia pulcherrima	Nodes	Callus,shoots, plantlets to soil	Rahman et al.,1993
Calliandra tweedii	Internode and petiole	Somatic embryogenesis	Kumar et al., 2002
Cassia alata	Seedling	Callus for anthracene	Rai and Shok,1982
C. acutifolia	Cotyledon, hypocotyl and roots	Anthraquinon accumulation	Rady and Nazif,1997
C. angustifolia	Cotyledons	Somatic embryos	Agarwal and Sardar, 2006
C. angustifolia		In vitro propagation	Siddique and Anis, 2007
C. angustifolia	Root explants	Shoot regeneration	Praveen and Shahzad, 2010
C. fistula	Stem and petiole	Green meristemoids	Gharyal and Maheshwari, 1990
C. fistula	Anthers	Callus, embryogenesis	Bajaj and Dhanju,1983
C. fistula	Seedling	Callus culture for anthraquinones	Rai and Shok,1982
C. fistula	Seedling	Callus culture	Ahuja et al., 1991
C. nodosa	Seedling	Callus culture for anthraquinones	Rai and Shok,1982
C. podocarpa	Callus culture	Callus culture for anthraquinones	Rai and Shok,1982
C. roxburghii		Micropropagation	Babber et al., 1996
C. senna var. senna	Cotyledonary leaves of seedlings	Callus, shoot induction	Shrivastava et al., 2006
C. siamea	Cotyledonary node	Shoot regeneration	Praveen et al., 2010
C. siamea	Anthers	Callus, embryogenesis, plantlets	Gharyal et al., 1983a
C. torosa	Seedling explant	Produced Germichrysone	Noguchi and Sankawa, 1982
Ceratonia siliqua	Shoot tips, cotyledon cotyledonary nodes	In vitro propagation	Mohammed-Yassen et al., 1992

Contd...

Table 1.1–Contd...

Plant	Explant	Result	Authors
Cladrastis lutea		Somatic embryogenesis	Weaver and Trigiano, 1991
Dalbergia lanceolaria	Seedling	Callus, adventitious buds, plantlets	Anand and Bir,1984
D. lanceolaria	Cotyledon, hypocotyl	Callus, plantlets	Dwari and Chand,1996
D. latifolia	Roots, nodes (5 years tree)	Adventitious buds, plantlets in soil	Mascarenhas *et al.*,1982
D. latifolia	Seedling	Bud break, plantlets	Nataraj and Sudha Devi,1985
D. latifolia	Mature stem	Callus, adventitious buds, plantlets	Lakshmi Sita *et al.*, 1986
D. latifolia	Seedling	Callus, adventitious buds, plantlets	Sankara Rao,1986
D. lanceolaria	Cotyledon, hypocotyl	Callus, plantlets	Dwari and Chand,1996
D. latifolia	Explants from adult trees and immature embryos	Somatic embryogeny	Muralidhar Rao and Lakshmi Sita,1996
D. latifolia	Shoot and leaf segments	Buds and rooted plants	Rao *et al.*, 1985
D. latifolia	Roots and Nodes (5 years old tree)	Adventitious buds, plantlets in soil	Mascarenhas *et al.*,1982
D. latifolia	Seedling explants	Bud break, plantlets	Natraja and Sudha Devi, 1985; Ravishankar Rai and Jagadish Chandra, 1989
D. latifolia	Somatic callus tissue	Plantlets	Shankara Rao,1986
D. latifolia	Shoot callus	Regeneration	Lakshmi Sitha *et al.*, 1986
D. latifolia	Stem segments	Plantlets	Sudha Devi and Nataraja, 1987
D. latifolia	Mature tree shoot callus	Plantlets	Ravishankar Rai and Jagadish Chandra, 1988
D. latifolia	Nodal and Inter nodal explants	Multiple shoots, callus, plantlets	Lakshmi Sita and Raghava Swamy, 1992
D. latifolia	Nodal explants of mature tree	Multiple shoots, callus, plantlets	Raghava Swamy *et al.*,1992
D. latifolia	Leaf disc	Plantlets, control of necrosis and leaf abscission	Lakshmi Sita and Raghava swamy, 1993
D. latifolia	Callus	Multiple shoots and rooted plants	Lakshmi Sita and Chattopadhyay, 1986
D. latifolia	Cotyledon	Somatic embryogenesis	Muralidhar Rao and Lakshmi Sita,1996
D. latifolia	Immature zygotic embryos	Somatic embryogenesis	Lakshmi Sita, 1996

Contd...

Table 1.1–Contd...

Plant	Explant	Result	Authors
D. miscolobium	Nodes	Plantlets	Pereira-de-sa and Caldas, 1991
D. paniculata	Cotyledonary nodes	Multiple shoots	Sreedevi and Pullaiah,1999
D. retusa	Hypocotyl	Multiple shoot, roots and plantlets	Cerdas *et al.*, 2004
D. sissoo	Seedling explant and roots	Adventitious buds, plantlets	Mukhopadhyay and Mohan Ram, 1981
D. sissoo	Nodal explants	Plantlets	Swapna *et al.*, 1982
D. sissoo	Seedling explants	Clonal multiplication	Datta *et al.*, 1983
D. sissoo	Nodal explants, mature tree	Auxin induced regeneration	Datta and Datta,1983
D. sissoo	Nodal explant	Callus	Kumar, 1989
D. sissoo	Callus culture	Effect of growth regulator on callusing	Bhattacharya *et al.*, 1990
D. sissoo	Cambial callus of mature tree	Morphogenic response	Kumar *et al.*, 1991
D. sissoo	Hypocotyl and shoot tip	*In vitro* propagation	Ramulu., 1993
D. sissoo	Mature tree axillary tree	Plantlets	Chauhan *et al.*, 1996
D. sissoo	Hypocotyl	Plantlets	Sharma and Chandra,1988
D. sissoo	Nodal explants	Micropropagation	Gulati and Jaiwal, 1996; Joshi *et al.*, 2003
D. sissoo	Immature zygotic embryos	Somatic embryogenesis	Das *et al.*, 1997; Das 2011
D. sissoo	Stem callus of adult plant	Clonal multiplaction and plant regeneration	Mukhopadyay and Bhojwani, 1986
D. sissoo	Stem callus of adult plant	Shoot buds and rooted plants	Lakshmi Sita and Chatto-padyay, 1986
D. sissoo	Seedling explants	Clonal propagation	Dawra *et al.*, 1984
D. sissoo	Hypocotyl	Plantlets	Pattnaik, 2000
D. sissoo	Semi matured cotyledon	Callus, somatic embryos and plantlets	Singh and Chand, 2003; Singh *et al.*, 2001
D. sissoo	Nodal explants	Synthatic seeds and Plantlets	Chand and Singh, 2004; Singh and Chand, 2010
D. sissoo	Zygotic embryos	Plantlets	Das (2011)
Delonix elata	Leaf	Callus, adventitious buds	Sudharshana and Shan-thamma, 1986
D. regia	Aseptic seedlings	Multiple shoots and rooted plants	Hoque *et al.*, 1992
D. regia	Shoots tips, nodes, cotyledons	Plantlets	Yaseen *et al.*, 1992
D. regia	Seedlings	*In vitro* propagation	Gupta *et al.*, 1996

Contd...

Table 1.1–Contd...

Plant	Explant	Result	Authors
Erythrina berteroana	Vegetative apices, cotyledonary nodes	Plantlets	Berrios *et al.*, 1996
E. costaricensis	Vegetative apices, cotyledonary nodes	Plantlets	Berrios *et al.*,1991
E. fusca	Vegetative apices, cotyledonary nodes	Plantlets	Berrios *et al.*, 1991
E. poeppigiana	Vegetative apices, cotyledonary nodes	Plantlets	Berrios *et al.*, 1991
E. variegata	Cotyleon and leaf of seedling	Plantlets	Shasthree *et al.*, 2009
Faidherbia albida (= *Acacia albida*)	Root	Micropropagation	Ahee and Duhoux, 1994
Gynmocladus dioicus	Bud culture	Plantlets	Geneve and Kester, 1987
G. diocus	Axillary bud and shoot tip	Plantlets, *in vitro* rescue of mature male genotype	Smith and Obeidy, 1991
G. dioicus	Shoot tip	*In vitro* rescue of mature male genotype	Veneve *et al.*, 1990
Hardwickia binata	Mesocotyl	Multiple shoots	Anuradha and Pullaiah, 1996; Anuradha *et al.*, 2000
H. binata	Zygotic embryos	Somatic embryogenesis, Plantlets	Das *et al.*, 1995; Das, 2011
Leucaena glauca	Shoot tip	Multiple shoots and callus	Bandyopadyay and Ghosh, 1984
L. leucocephala	Seedling explants	Multiple shoots	Stokes,1981
L. leucocephala	Seedling explants	Mass propagation for improvement in forest tree biomass	Glovak and Greatbatch, 1992
L. leucocephala	Axial buds and embryos	Plantlets	Venkateswaran and Gandhi, 1982
L. leucocephala	Hypocotyl, cotyledonary segment	Morphogenic response	Nagamani and Venkatgeswaran, 1983a,b
L. leucocephala	Callus culture	Plantlet regeneration	Nagamani and Venkatgeswaran, 1983b,1987
L. leucocephala	Seedling explants	Plantlets	Ravishankar *et al.*, 1983
L. leucocephala	Single node and shoot tip	Reduction in cost	Dhawan and Bhojwani, 1984
L. leucocephala	Mature node, cotyledonary node, nodal explants	Bud break, plantlets in soil	Dhawan and Bhojwani,1985
L. leucocephala	Mature node	Plantlets	Datta and Datta,1985; Goyal *et al.*, 1985

Contd...

Table 1.1–Contd...

Plant	Explant	Result	Authors
L. leucocephala	Mature node	Bud break, plantlets in soil	Dawan and Bhojwani 1986
L. leucocephala	Seedling nodulation	*In vitro* nodulation	Dhawan and Bhojwani1987
L. leucocephala	Axillary bud, hypocotyls, leaflet	Plantlets and somatic embryos	Gupta and Agarwal,1992
L. leucocephala	Cotyledonary nodes	Plantlets	Toruan Mathius, 1992
L. leucocephala	Hypocotyl	Shoot buds	Ramulu, 1993
L. leucocephala	Aseptic seedling explants	Plantlets	Hoque *et al.,* 1992
L. leucocephala	Seedling nodulation	Invetronodulation	Dhawan and Bhojwani, 1988
L. leucocephala	Lateral buds	Plantlets	Goyal *et al.,*1985
L. leucocephala	Seedling nodulation	*In vitro* nodulation	Dhawan and Bhojwani, 1987
L. leucocephala	Lateral buds	Plantlets	Goyal *et al.,* 1984
L. leucocephala	Seedling nodulation	*In vitro* nodulation	Bhojwani *et al.,*1988
L. leucocephala	Nodal explant of mature tree, cotyledonary node	*In vitro* regeneration	Rastogi *et al.,* 2008
L. pallida	Cotyledonary node	Plantlets	Toruan Mathius, 1992
Mundulea sericea	Seedling explants	*In vitro* propagation	Kalavathi *et al*, 2004
Paraserianthus falcataria	Node of six day old seedling	Axillary shoot proliferation	Sasmitamihardja *et al.,* 2001
P. falcataria	Axillary buds and cotyledonary nodes	Plantlets	Sukarutiningsih *et al.,* 2002
P. falcataria	Node of seedlings	Micropropagation	Bon *et al.,* 1998
P. falcataria (=*Albizia falcataria*)	Cotyledon	Adventitious shoots	Sinha and Mallick,1993
P. falcataria	Nodal explants	Micropropagation	Bon *et al.,* 1998
Parkinsonia aculeata	Nodal explants of mature tree	Plantlets	Mathur and Mukuntha Mumar, 1992
P. aculeata	Seedling explants	Plantlets	Mathur *et al.,* 1991
Peltophorum dubium	Apical and nodal segments	Micropropagation	Bassan *et al.,* 2006
P. pterocarpum	Anthers	Haploid plants	Rao and De,1987
P. pterocarpum	Anthers	Haploid plants	De and Rao, 1983
P. pterocarpum	Seedling explants	Multiple shoots, callus	Lakshmi Devi *et al.,* 1996; Salah Uddin *et al.,* 2005
Pithecellobium dulce	Protoplast	Plantlets	Saxena and Gill, 1987
P. jiringa	Cotyledon shoot tip, nodal explants	Plantlets	Rahadyrau *et al.,* 1989

Contd...

Table 1.1–Contd...

Plant	Explant	Result	Authors
P. saman	Hypocotyl	Plantlets	Cerdas et al., 1997
Poinciana regia	Pollen	Embryoids	Bajaj and Dhanju, 1983
Pongamia glabra	Nodes	Callus	Sreelakshmi and Janardhana Reddy, 2004
P. pinnata	Axillary meristems from mature tree	Multiple shoots, rooting and plantlets	Sujatha and Sulekha, 2007; Sujatha et al., 2008
P. pinnata	Node	Micropropagation	Sugla et al., 2007; Shrivastava and Kant, 2010
Prosopis alba	Seedling explants	Shoot multiplication	Tobone et al., 1986
P. alba	Shoot tip	Embryogenesis plantlets	Jordan et al., 1985
P. alba	Axillary bud	Effect of N source on recalcitrant nature	Green et al., 1990
P. chilensis	Nodal explants	Plantlets	Jordan and Balboa,1985
P. chilensis	Shoot tip	Embryogenesis plantlets	Jordan,1987; Jordan et al., 1985b
P. chilensis	Nodal explants	Plantlets	Arce and Balboa,1991
P. chilensis	Nodal explants	Regeneration	Dun-Yi et al.1989
P. chilensis		Callus	Batchelor et al., 1990
P. chilensis	Nodal segments of seedlings and mature plants	Micropropagation	Caro et al., 2002
P. cineraria	Hypocotyl, seedling explants	Plantlets	Goyal and Arya,1981
P. cineraria	Vegetative shoot buds	Plantlets	Goyal and Arya 1983
P. cineraria	Lateral buds	Multiple shoots, plantlets	Goyal and Arya 1984a,b
P. cineraria	Gall callus tissue	Effect of auxin on growth	Kant and Vidya, 1986
P. cineraria	Cotyledon and hypocotyls	Shoot differentiation	Solanki and Kackar, 1992
P. cineraria	Axillary buds of juvenile explants, mature tree nodes	Callus, plantlets	Nandwani and Ramawat, 1993
P. cineraria	Nodal segments	Plantlets	Kackar et al., 1991
P. cineraria	Root	Plantlets	Kackar et al., 1992
P. glandulosa var. torreyana	Cotyledons, hypocotyl	Multiple shoots	Rubluo et al., 2002
P. juliflora	Seedling explants	Micropopagation	Wainright and England,1987
P. juliflora	Nodal explants of mature tree	Callus, bud break, plantlets	Nandwani and Ramawat, 1991
P. juliflora	Nodal explants	Regeneration	Dun-Yi et al., 1985

Contd...

Table 1.1–Contd...

Plant	Explant	Result	Authors
P. laevigata	Cotyledonary nodes	Multiple shoots	Buendia-Gonzalez *et al.,* 2007
P. tamarugo	Shoot tip	Plantlets	Jordan *et al.,* 1985
P. tamarugo	Nodal explants	Plantlets	Jordan and Balboa, 1985
P. tamarugo	Plumule, hypocotyl, plumule with cotyledon	Shoot buds, multiple shoots	Nandwani and Ramawat, 1992a, b
Pterocarpus marsupium	Shoot tip and nodal explants	Multiple shoots	Anuradha and Pullaiah,1999
P. marsupium	Cotyledonary nodes (20 days old axenic seedling)	Multiple shoots, rooting and plantlets	Suresh Chend and Ajay Kumar Singh, 2005
P. marsupium	Nodal segments	*In vitro* propagation	Tiwari *et al.,* 2004
P. marsupium	Seedling explants	*In vitro* plantlet regeneration	Anis *et al.,* 2005
P. marsupium	Nodal segments	*In vitro* propagation	Hussain *et al.*(2008)
P. santalinus	Cotyledons, nodal segment, shoot tips	Callus, adventitious shoots, plantlets	Sarita *et al.,* 1998; Patri *et al.,* 1988
P. santalinus	Shoot tips	Plantlets	Saritha *et al.,* 1998; Lakshmi Sita *et al.,* 1992
P. santalinus	Shoot tips of mature tree	Elongation, rooted plantlets	Kesava reddy and Srivasuki, 1990,1992
P. santalinus	Seedling explants	Tree improvement	Mithila and Srivasuki, 1992
P. santalinus	Mesocotyl, node and shoot tips	Multiple shoots, callus	Anuradha and Pullaiah, 1999a,b
P. santalinus	Seedling explants	*In vitro* propagation	Arockiasamy *et al.,* 2000
P. santalinus	Detached cotyledons from *in vitro* seedling	Plantlets	Arockiasamy *et al.,* 2000
P. santalinus	Mature nodes of 10 years old tree	Bud break, shoot sprouting, rooting and plantlets	Prakash *et al.,* 2006
Robinia ambigua var. *idahoensis*	Node	Axillary branching	Guo *et al.,* 2006
R. hispida	Anthers	Calus, plantlets	Cao *et al.,* 2003
R. neomexicana	Shoots buds	Plantlets	Lin and Wagner, 1995
R. pseudoacacia		Callus	Gautheret *et al.,* 1934
R. pseudoacacia	Root explants	Callus, shoot sprouds	Brown and Sommer, 1982
R. pseudoacacia	Nodal segments	Multiple shoots and rooted plantlets	Chalupa,1983
R. pseudoacacia	Roots explants	Shoot sprouts	Seelinger, 1959
R. pseudoacacia	Leaf disc explants	Shoots, rooted plantlets	Davis and Keathley, 1985

Contd...

Table 1.1–Contd...

Plant	Explant	Result	Authors
R. pseudoacacia	Mature tree leaves	Callus, shoots, rooted plantlets	Hu and Han, 1985 ; Wang *et al.*, 1985
R. pseudoacacia	Axillary buds	Micropropagation	Barghchi, 1987
R. pseudoacacia	Mature tree nodal explants	Micropropagation	Davis and Keathley, 1987a
R. pseudoacacia	Mature tree nodal explants during winter dormancy	Plantlets	Davis and Keathley, 1987b
R. pseudoacacia	Buds from procumbent trees	Micropropagation	Davis and Keathley, 1992
R. pseudoacacia	Mature tree explants	Somatic embryogenesis	Davis and Keathley, 1992 Merkle and Wiecku, 1989
R. pseudoacacia	Mature tree explants	Plantlets	Davis and Keathley, 1992
R. pseudoacacia	Cambial tissue culture	Callus, protoplasts culture	Han *et al.*, 1993
R. pseudoacacia	Cotyledon and leaf	Plantlets	Arrillaga and Merkle 1993
R. pseudoacacia	Seedling explants	Plantlets	Merkle,1995
R. pseudoacacia	Nodal segments and shoot tip	Plantlets	Chalupa,1984
R. pseudoacacia	Nodal explants	Micropropagation	Vertesy, 1991
R. pseudoacacia	Cotyledon, leaf	Plantlets	Arrillaga and Merkle, 1993
R. pseudoacacia	Staminate flower part	Somatic embryos	Merkle, 1995
R. pseudoacacia	Cambial tissue	Plantlets and effect of culture medium	Han *et al.*,1993
R. pseudoacacia	Hypocotyl	Shoots and transgenic plants	Han *et al.*, 1993
R. pseudoacacia	–	Artificial seeds	Arrilliga *et al.*, 1994
R. pseudoacacia	Suspension culture	Regeneration	Kanwar *et al.*, 2008
R. pseudoacacia f. *decaisneana*	Shoot segments	*In vitro* propagation	Luan and Luo, 2002
Samanea saman	Root, hypocotyl	Plantlets, callus	Vinolya Kumari.2000
Saraca asoca	Nodal segments	Multiple shoots	Rama Subba *et al.*, 2007
Sesbania aculeata	Hypocotyl	Callus, plantlets	Bansal and Pandey, 1993
S. aculeata	Seedling explants	Plantlets	Srivastava and Aggarwal, 2007
S. bispinosa	Cotyledon, hypocotyl and root	Plantlets	Kapoor and Gupta, 1984, 1986; Zhao *et al.*, 1993
S. bispinosa	Cotyledon and leaf	Plantlets	Yan Xiu *et al.*,1995
S. bispinosa	Cotyledons	Callus and shoot buds	Sinha and Mallick,1991

Contd...

Table 1.1–Contd...

Plant	Explant	Result	Authors
S. cannabina	Seedling explants; root, hypocotyls and cotyledon	Regeneration	Xu *et al.*, 1984; Zhao *et al.*, 1993
S. cannabina	Shoot tip explants	Shoot proliferation	Hussain *et al.*, 1990
S. drummondii	Nodal segments	Multiple shoots	Cheepala *et al.*, 2004
S. formosa	Root, hypocotyl, cotyledon	Plantlets	Zhao *et al.*, 1993
S. glauca	Root, hypocotyl, cotyledon	Regeneration	Bandyopadhyaya and Ghosh, 1984
S. grandiflora	Hypocotyl and cotyledons	Plantlets	Khattar and Mohan Ram, 1983
S. grandiflora	Hypocotyl and cotyledons	Callus, plantlets	Shankar and Mohan Ram, 1990b
S. grandiflora	Hypocotyl and cotyledons	Callus and somaclonal variation	Shankar and Mohan Ram, 1990b
S. grandiflora	Aspectic seedling	Plantlets	Hoque *et al.*, 1992
S. grandiflora	Cotyledon	Plantlets	Detrez *et al.*, 1994
S. rostrata	Shoot, leaf and root	Callus, regeneration	Pellegrineschi and Tepfer, 1993
S. rostrata	Cotyledons, hypocotyl, immature embryo	Callus, plantlets, studies with *in vitro* tumor induction	Vlachova *et al.*, 1987
S. rostrata	Crown gall callus	Increased frequency regeneration	Mathews *et al.*,1988
S. rostrata	Hypocotyl (12-15 days old seedling)	Multiple shoots, rooting and plantlets	Jha *et al.*, 2002
S. rostrata	Cotyledonary node	Multiple shoots	Jha *et al.*, 2004b
S. sesban	Cotyledon, hypocotyls, root	Callus, shoot buds	Khattar and Mohan Ram, 1982 ; Zhao *et al.*, 1993
S. sesban	Hypocotyl	Adventitious shoots	Shankar and Mohan Ram, 1987
S. sesban	Cotyledon, hypocotyl	Callus and somaclonal variation	Shankar and Mohan Ram, 1990a
S. sesban var. *bicolor*	Nodal and internodal explants of mature tree	Multiple shoots, micropropagation	Jha *et al.*, 2004
S. sesban	Embryo, adventitious buds	Plantlets, propagation	Subhan *et al.*, 1998
Sophora flavescens	Callus culture	Production of flavanones, Sophora flavanone-G and Lehmannin	Yamamoto *et al.*,1995

Contd...

Table 1.1–Contd...

Plant	Explant	Result	Authors
S. flavescens		Micropropagation	Zhao et al., 2003
S. macrocarpa	Node	Micropropagation	Gupta and Agarwal, 1992
S. secundiflora	Node	Shoot proliferation	Froberg et al., 1985
S. toromiro	Node, internode, rachis, callus	Plantlets	Zhao et al., 1993
S. toromiro	Node, internode, rachis	Plantlets	Iturriga et al., 1994
Swartzia madagascarinsis	Nodes, shoot apex	Plantlets	Berger and Schaffner, 1995
Tamarindus indica	Anthers	Embryoids	De and Rao, 1983
T. indica	Cell and tissue culture	Plantlets	Mascarenhas et al., 1987
T. indica	Seedling explants, shoot tips	Plantlets	Kopp and Nataraja, 1990
T. indica	Entire cotyledons	Plantlets to soil	Jaiwal and Gulati, 1991
T. indica	Nodes	Plantlets	Kopp and Nataraja, 1992
T. indica	Shoot tip and nodes	Plantlets	Jaiwal and Gulati, 1992
T. indica	Importance of tissue culture	Plantlets	Rao and Lee, 1982
T. indica	Comparison to normally grown plant	Plantlets	Mascarenhas, 1989
T. indica	Hypocotyl	Direct organogenesis, multiple shoots, roots and plantlets	Jaiwal et al., 1998

adventitious shoots. In *ex vitro* rooting, shoots more than 0.5 cm long gave high rooting success. In the case of *A. auriculiformis* X *A. mangium* hybrids, the rooting ability declined with increase in number of passages (Darus, 1991, 1992). Further the growth of micropropagated *A. mangium* plants was found to better than seedling raised plants (Setiawan *et al.*, 1991). The field performance of tissue cultured hybrid Acacia was also found better than the seedling and rooted cuttings (Darus, 1992).

The root culture of *Acacia melanoxylon* was maintained in liquid for over a year. The callus from seedling stem was obtained by McLean *et al.* (1966) from *Acacia harpophylla* and in *A. mearnsii* by Zeijlemaker (1972). Root differentiation from cotyledonary callus of *Acacia senegal* was achieved by Kathjo and Tewari (1979). Later Agarwal and Prasad obtained complete plantlets from the cultured stem tissue of *A. senegal*. Reports on successful propagation and transplantation of *A. koa* to the field was by Skoleman and Mapes (1976, 1978). Skoleman and Mapes (1976) successfully cultured shoot tips of *Acacia koa* and reported rooted plantlets mediated with callus.

In vitro somatic embryogenesis and subsequent plant regeneration was achieved by Nanda and Rout (2003) in callus cultures derived from immature zygotic embryos of *Acacia arabica* on semi-solid MS salts and vitamins supplemented with 8.88 μM BA,

6.78 µM 2,4-D and 30 g/l (w/v) sucrose. Somatic embryos proliferated rapidly by secondary somatic embryogenesis after transfer to MS medium supplemented with 6.66 µM BA, 6.78 µM 2,4-D. The maximum number of somatic embryos per g callus was 72.6 after 8 weeks of culture on medium containing 6.66 µM BA and 6.78 µM 2,4-D. The isolated somatic embryos germinated on half strength basal MS salts and vitamins supplemented with 0.04 µM BA, 0.94 µM ABA and 2& (w/v) sucrose. The embryo derived plantlets were acclimatized in the greenhouse (Nanda and Rout, 2003).

Multiple shoots have developed from axillary buds excised from *in vitro* grown seedlings of *Acacia auriculiformis* on B_5 medium supplemented with coconut milk (5 or 10 per cent) and BAP (10^{-6} M) (Mittal *et al.*, 1989). These shoots were transferred to IAA (10^{-7} M) or NAA (10^{-6} or 10^{-7} M) augmented B_5 medium produced roots at base.

Li *et al.* (1996) described methods for micropropagation of *Acacia cincinnata*. Shoot cuttings derived from a single plus tree were excised and cultured on MS medium supplemented with BA 0.5 mg/l and Kn 0.5 mg/l. Adventitious buds rooted on half-strength MS medium containing IBA 2 mg/l

A number of investigators later succeeded in propagating other *Acacia* species namely *A. nilotica* (Mathur and Chandra, 1983), *A. mangium* and *A. stenophylla* (Crawford and Hartney, 1986, Yang *et al.*, 1989). *Acacia mangium* and *Acacia stenophylla* have been regenerated using seedling explants. Shoot multiplication was achieved on MS (half-strength basal) medium with one micromole each of BAP and NAA. *In vitro* rooting in a moist vermiculite/perlite mixture was followed by hardening in plastic pots. The plants were later successfully transferred to the field. Ahmad (1989) cultured nodal explants from seedlings of *Acacia mangium* on MS medium supplemented with BAP and Kn at different concentration levels. MS medium with 0.5 mg/l BAP was the best combination to induce higher shoot multiplication with an average of 25.4 shoots per explant. For root formation excised shoots treated with Seradix 3 (commercial rooting powder) produced 85 per cent rooting.

Root culturing of *Faidherbia = Acacia albida* as a source of explant for shoot regeneration was studied by Ahee and Duhoux (1994). Root culture can be maintained over several months by successively subculturing excised roots in modified Bonner and Deverian medium which was found to be the best of the three media compositions tested. Adding auxins had an inhibiting effect on root growth characteristics. Mesoinositol at 0.05 mM slightly enhanced the overall elongation rate, and sucrose at 59 mM significantly increased root elongation. The effect of sucrose could not be replaced by glucose. Shoots were regenerated *in vitro* from root segments beginning with the first passage on 1/5 strength MS medium.

A method for *in vitro* micropropagation through shoot apices of *Acacia catechu* was described by Kaur and Kant (2000). Explants were excised from 15 days old *in vitro* grown seedlings. Shoot bud induction from shoot apex explants was observed on MS medium containing various growth regulators. A maximum of 12 shoots was obtained on MS medium supplemented with 1.5 mg/l BAP and 1.5 mg/l Kn. Well developed shoots were rooted on ¼ strength MS medium with 3.0 mg/l IAA and sucrose 1.5 per cent. *In vitro* regenerated plantlets were transferred to field conditions.

Efficient plant regeneration system via somatic embryogenesis have been developed by Ortiz *et al.* (2000) for *Acacia farnesiana* and *A. schaffneri*. The protocol used in this study consisted of placing immature zygotic embryos of these species in MS medium supplemented with 9.05 µM 2,4-D and 4.65 µM Kn to induce callus. Some parts of the callus were used for direct embryo differentiation and others for establishment of cell suspension culture. In the first case, somatic embryos were produced on semi-solid differention media without growth regulators or with abscisic acid (ABA). The highest number of somatic embryos, 345 and 198 embryos per g callus *A. farnesiana* and *A. schaffneri*, respectively was obtained in media without growth regulators, but adding ABA increased the percent of embryos that reached more advanced differentiation stages. The production of somatic embryos was achieved starting from cell suspensions only when these suspensions were plated into the semi-solid differentiation medium. Somatic embryos germinated on medium containing 217 µM adenine sulfate with efficiencies of 69 per cent in *A. farnesiana* and 47 per cent in *A. schaffneri*. Some somatic embryos that developed into plantlets were acclimatized in the greenhouse and they grew into normal plantlets (Ortiz *et al.*, 2000).

Rout *et al.* (1995) observed regeneration via somatic embryogenesis from callus derived from immature cotyledons of *Acacia catechu*. On WPM supplemented with 13.9 mm Kn and 2.7 mm NAA. The addition of 0.9-3.5mm L- proline to the medium influenced development of somatic embryos and also promoted secondary embryogenesis. The light – green somatic embryos germinated on half MS medium supplemented with 2 per cent (W/V) sucrose. Somatic embryos germinated into plantlets that were acclimatized in the green house and subsequently transferred to the filed.

Kaur (1998) developed a protocol for plantlet regeneration of *Acacia catechu* through callus tissue. At the time of media preparation adjuvents like adenine sulphate, ascorbic acid, glutamine and MgSO$_4$ along with cytokinin were added for multiple shoot induction. Ascorbic acid prevents browning of callus cultures due to leaching of phenolics. Adenine sulphate was found to re-inforce the effect of other cytokinins, glutamine stoped leaf fall in multiple shoots. Additional amount of MgSO$_4$ retain green colour of leaves.

Beck *et al.* (2000) cultured apical meristems of *in vitro* grown plants and coppices of *Acacia mearnsii* on MS medium supplemented with 2 mg/l BAP. They were able to regenerate the plants successfully.

Kaur and Kant (2000) cultured shoot tip of *Acacia catechu* for the production of plantlets. Maximum number of shoots was obtained on MS medium + 1.5 mg/l BAP and 1.5 mg/l Kn. The *in vitro* regenerated plants were rooted on MS medium + 3 mg/ l IAA and 1.5 per cent sucrose and were successfully transferred to field conditions.

In *Acacia catechu* Thakur *et al.* (2002) inoculated cotyledons of *in vitro* grown seedlings on MS medium + 0.25mg/1 2, 4-D and 0.25mg/1 NAA for callus induction. Plantlets were regenerated from the callus and mature nodal explants on MS medium + 2 mg/1 BAP and 2 mg/1 Kn and further multiplied on the same medium. Addition of 25 mg/1 adenine sulphate, 20 mg/1 ascorbic acid and 150 mg/1 glutamine enhanced axillary branching. Regenerated shoots were sub cultured on same medium repeatedly. For rooting microshoots of 2–2.5 cm length were dipped in 10mg/1 IAA solution for 24 hours followed by transfer to half MS medium + 0.20 per cent AC.

Das (2010) achieved somatic embryogenesis in *Acacia catechu, Acacia arabica, Hardwickia binata* and *Dalbergia sissoo* from immediate zygotic embryos inoculated on MS medium, 0.25- 1.0 mg/1 Kn + 2.0-3.0 mg/1 2,4-D or NAA and 3 per cent sucrose. MS medium + 2.0 mg/1 2,4-D + 1.0-1.5 mg/1 Kn was most effective in inducing friable embryogenic callus. Maximum number of somatic embryos was obtained in MS medium + 1.5 – 2.0 mg/1 2,4-D of NAA and 1.0-1.5 mg/1 Kn. Embryogenic callus was proliferated in cultures having 1.0-2.0 mg/1 2,4-D, 1.0-1.5 mg/1 Kn and 400-600 mg/l proline. Shoots developed in MS medium + 0.1 mg/l IAA + 0.25mg/l BA. Rooting was achieved in ½ MS medium supplemented with 0.1 mg/l IBA or IAA.

Rout *et al.* (2008) standardized a protocol for *in vitro* propagation of *Acacia chundra*. Shoot tip and node derived from *in vitro* grown plants were used as explants. High frequency of multiple shoot induction was observed on MS medium augmented with 1.5 mg/l BA, 0.01 to 0.05 mg/l IAA and 50mg/l Ads. *In vitro* shoots were rooted on half MS medium supplemented with 0.25 mg/l IBA or IAA and 20 g/l sucrose after 10-12 days.

Yang *et al.* (2006) reported the establishment of *Acacia crassicarpa* regeneration through organogenesis. They used phyllode (leaf) explant excised from 60 days old *in vitro* seedlings. Green compact nodules and adventitious shoots were induced in 10 and 40 days, respectively, on medium containing a combination of 0.5 mg/1 TDZ and 0.05 mg/1 NAA. Efficient shoot elongation was achieved by transferring the clusters of adventitious shoots to medium containing 0.1 mg/1 TDZ with in 2 months. The elongated shoots were rooted at the rate of 96.5 per cent on half MS medium + 0.5 mg/1 IBA in 1 month. Rooted plants were hardened and successfully established in soil with an 80 per cent survival rate. This is the first report describing a detailed protocol for regeneration through organogenesis using phyllodes as explant.

A micropropagation system was developed for *Acacia mearnsii* by Huang *et al.* (1994). Shoot tips, 5 mm long from 3-week old seedlings germinated *in vitro* were cultured on three-fourth-strength MS medium supplemented with combinations of auxins (IBA and NAA) and cytokinins (Kn and BAP). Multiple shoot formation was promoted by BAP at 2 mg/l and higher combined with or without 0.01 mg/l IBA. *In vitro* produced shoots were induced to root on a range of NAA concentrations supplemented to half- or full-strength MS medium. The highest frequency of root proliferation was on half-strength MS medium supplemented with 0.6mg/l NAA. Plantlets survived in potting soil (Huang *et al.*, 1994).

Nitrogen fixing potential of micropropagated clones were observed in *Acacia mangium* by Galiana *et al.* (1991a,b) by collecting seedlings of different shoot lengths and micropropagated after two weeks. 3 specific *Bradyrhizobium* spp. strains were inoculated in 15 combinations. After 5 months most efficient nodulation was observed, but no interaction was observed between two. It indicates that these two can be selected separately. Douglas and John (2000) developed shoot regeneration from cotyledonary nodes of seedling explants, explants were cultured on DKW medium with N-6 benzyladenine and either thidiazuron of N-(2-chloro 4-piridyl)-N-urea. They observed adventitious buds arising from wound tissue of the cotyledons and cotyledonary node. Monteuuis and Olivier (2004) developed shoot regeneration and rooting in the same plant by using mature and juvenile genotypes. Shoot regeneration occurred in BAP, roots in 0.068 mg/1 IAA, IBA or NAA. In rooting medium the proportion of rooted microshoots significantly increased. Juvenile explants responded better than mature explants. Xie and Hong (2000) observed somatic embryos and whole plant regeneration in callus cultures derived from immature zygotic embryos of *Acacia mangium*. Embryogenic callus was induced on MS medium + 1.2 mg/1 TDZ + 0.25 – 2.0 mg/1 IAA and a mixture of amino acids. Globular embryos developed on embryogenic callus cultured on the induction medium. Nearly 42 per cent of embroygenic cultures with globular embryos produced torpedo and cotyledonary-stage embryos by a two steps maturation phase. First stage occurred on half MS medium + 30 g/1 sucrose + 5mg/1 GA$_3$ followed by the second stage somatic embryos, 11 per cent germinated into seedlings that could be successfully transferred to pots. Bon *et al.* (1998) studied the influence of five different macronutrient formulations and various growth regulators on micropropagation of single node explants of *Acacia mangium*.

Saito *et al.* (1993) described methods for *in vitro* propagation from axillary buds of *Acacia mangium*. Nodal segments from 6-month old seedlings were cultured on MS medium supplemented with different concentrations of BAP. 10 µM BAP was effective for inducing shoots from axillary buds. Shoots were rooted in half strength MS medium supplemented with 10µM IAA. The plantlets were acclimatized and transferred to potting mixture.

Plant regeneration of *Acacia mangium* was achieved through organogenesis in callus cultures by Xie and Hong (2001). Calli were induced from five types of explants (embryo axes and cotyledons of mature zygotic embryos as well as leaflets, petioles and stems of seedlings) on MS medium containing 9.05 µM 2,4-D and 13.95 µM Kn. 22 per cent of the nodules formed adventious shoots on MS medium containing 0.045 µM TDZ. Shoots were elongated on MS medium containing 0.045 µM TDZ supplemented with 7.22 µM gibberellic acid. The medium containing 10.75 µM NAA and 2.33 µM Kn promoted rooting of 10 per cent of the elongated shoots.

Acacia mearnsii "principal source of the world's tanbark" was micropropagated by Hung *et al.* (1994). Shoot tips of *in vitro* seedlings were cultured on NAA, Kn and BAP. Higher levels of BAP promoted more multiple shoots and shoot elongation. These shoots were sub cultured on different concentration of NAA (0.0-0.8mg/1) +

half MS or MS medium. The highest frequency of root proliferation was on half MS media + 0.6 mg/1 NAA. Plantlets survived in potting soil and exhibited normal growth under green house conditions. Sascha (1998) inoculated nodes of 3 and 9 months old green house grown *Acacia mearnsii* plants on MS medium + 2 mg/1 BA for multiple shoot induction. High frequency of roots was observed on MS medium + 1 mg/1 IBA. Plants were acclimatized with 90 per cent success rate under green house conditions.

Quoirin *et al.* (1998) investigated the effect of growth regulators on indirect organogenesis of *Acacia mearnsii* tissues cultured *in vitro*. Sections of hypocotyls and cotyledons of young seedlings were cultivated on MS medium supplemented with Kn, NAA and TDZ. The best results for bud regeneration were obtained with cotyledons, when cultured in the presence of 2.33 to 9.30 μM Kn or 1.82 μM TDZ and 2.69 μM NAA.

A method of combined the root sprout and tissue culture were applied in the rapid propagation of clones of *Acacia melanoxylon* by Su and Zhand (2008). High frequency of sterilization with 80.9 per cent was found when 10-20 cm root sprout were propagated as explant. The results showed that good performance with rhizogenesis as explants were root sprouts. Two key factors, the prevention of yellow leaves and the promotion of adventitious buds stratch were noticed in the proliferation. Decreasing the sugar content of the medium was an effective approach. In this way, the number of aseptic sprouts were average of 23 each bottle with the three-step-gap culture.

Garg *et al.* (1996) inoculated immature endosperm of *Acacia nilotica* on MD medium + 2, 4-D, BAP and CH. Callus differentiated into somatic embroyos in third passage. Embryos germinated on MS medium only after 15 days pretreatment on modified MS medium in which major salts were replaced by those of major salts of B_5 medium and supplemented with glutamine, CH and CW. Triploid nature of the somatic embryos was confirmed by Fuelgen cytophotometry.

Cotyledonary node explants of *Acacia nilotica* ssp. *indica*, differentiated multiple shoots on B_5 medium supplemented with cytokinins like BAP, Kn and Zeatin. Of the four BA supported maximum multiple shoot differentiation, the highest average number of shoots (6l3) per explant was 1.5mg/l. Rooting was obtained in a medium containing 2mg/l IAA. They were successfully transferred to the field (Dewan *et al.,* 1992).

An efficient regeneration protocol was developed by Dhabhai *et al.* (2010) for *in vitro* propagation of *Acacia nilotica* through direct regeneration. *In vitro* nodal segments cultured on MS medium supplemented with NAA (0.6 mg/l) and Kn (1.0 mg/l) for shoot proliferation. NAA was found to be more effective than Kn for shoot multiplication. The highest number of shoots (4.6) were achieved on MS medium augmented with NAA (0.6 mg/l). Excised shoots were rooted on half strength MS medium supplemented with IBA (0.5 mg/l) after 15-20 days of culture. The microalgapropagated plants were hardened, acclimatized and transferred to natural conditions.

Barakat and El-Lakany (1992) obtained multiple shoots successfully from shoot tips of *Acacia saligna* by placing explants into solidified MS medium supplemented with 5.0 to 9.0 mg/l BAP. Sequential culture treatment was highly effective for shoot elongation using MS medium containing 0.3 mg/l BAP and 0.2 mg/l IAA. The shoots rooted best on medium supplemented with 2.0 mg/l IBA.

In *Acacia senegal* Gupta *et al.* (1994) observed leaf fall prevention in the regenerated shoots by the addition of glutamine to the medium along with cytokinins and auxins. Kaur *et al.* (1996) developed plantlets of same plant by inoculating cotyledonary node on different concentrations and combinations of MS medium.

Khalafalla and Dafalla (2008) developed a protocol for *in vitro* micropropagation and micrografting for *Acacia senegal*. Multiple shoots were regenerated from cotyledonary node derived from 7-days old *in vitro* raised seedling and nodal segment derived from 12-months old plant growting in a greenhouse. The maximum number of shoots per cotyledonary node (8.3) and nodal (5.3) explant were obtained on MS medium supplemented with 1.0 mg/l BA after 4 weeks of culture. *In vitro* regenerated shoots were either rooted *in vitro* on MS medium supplemented with IBA or micrografted on *in vitro* induced root stock. The *in vitro* rooted shoots and successful grafts were transplanted to plastic pots hardened off and transferred to greenhouse (Khalafalla and Dafalla, 2008).

In vitro propagation of *Acacia seyal* was achieved using seedling shoot tip explants on MS medium supplemented with BA or TDZ with NAA (Al-Wassel, 2000). The best result was obtained with BA in the presence of NAA. The greatest shoot multiplication with long shoots was observed on media containing 2 mg/l BA with 0.1 or 0.5 mg/l NAA and 4.0 mg/l BA with 0.1 mg/l NAA, with 6.4 and 6.7 mean number of shoots, respectively. TDZ also induced multiple shoots but most of the shots were stunted. Microshoots were rooted better on half-MS salts supplemented with IBA (4.0 mg/l). The plantlets successfully survived acclimatization *ex vitro*.

In *Acacia sinuata* Vengadesan *et al.* (2002) developed somatic embryos in cell suspension culture. It is initiated from calli derived from leaf explant of *in vitro* grown seedling. Callus was induced on MS (liquid) medium + 0.099 mg/l 2, 4-D + 10 per cent CW, it resulted in high frequency of somatic embryos. Different stages of somatic embryos occurred on auxin free MS medium + 2.97 mg/l sucrose + 4.79 mg/l glutamine, cytokinin leads to recallusing of embryos. 8-10 per cent of embryos were converted into plantlets. Vengadesan *et al.* (2003) standardized protocol for high frequency plant regeneration from cotyledonary callus. Cotyledon was inoculated on MS medium + 0.8 per cent agar or 0.15 per cent phytogel, 0.145 mg/l NAA and 0.048 mg/l BAP to produce calli. Callus cultured on MS medium + 10 per cent CM + 0.29 mg/l BAP, 0.052 mg/l Zeatin resulted in high frequency regeneration of adventitious buds. Optimum concentration of shoot bud induction was 2.97 mg/l sucrose, 0.057 mg/l favored shoot elongation. *In vitro* raised shoots were rooted on half MS medium + 0.14 mg/1 IBA. Plantlets were transferred to soil, within 3 months cotyledon yielded 40 plantlets.

Vengadesan *et al.* (2004) standardized a protocol for callus induction from seedling hypocotyls explants of *Acacia sinuata* from 10 days old seedling and inoculated on MS medium + 3 per cent sucrose + 0.8 per cent Agar + 6.78 µM 2,4-D and 2.22 µM BAP. Regeneration of adventitious buds from callus was observed when callus was cultured on MS medium + 10 per cent coconut water + 13.2 µM BAP and 3.42 µM IAA. Addition of Gibberellic acid favoured shoot elongation. Rooting was observed on MS medium + 7.36 µM IBA.

Detrez (1994) studied shoot propagation from mature tree explants of *Acacia tortilis* ssp. *raddiana*, a tropical tree legume. Apices from seedlings, axillary buds from young lignified branches of mature trees, and axillary buds obtained through *in vitro* axillary branching from microcuttings isolated from mature trees were all successfully micrografted on seedling rooted stocks.

A procedure was described by Nandwani (1995) for micropropagation of *Acacia tortilis* subsp. *raddiana* using embryonic explants. Shoot bud formation from cotyledonary nodes was observed on MS medium containing various doses of cytokinins (2.5-5.0 mg/l), *viz.*, BA and Kn, with or without incorporation of auxin (0.1-0.2 mg/l), *viz.* NAA and IAA. Multiple shoot regeneration (13-15) was achieved on MS medium supplemented with NAA (0.1 mg/l) and BA (5.0mg/l). Incorporation of auxin, *e.g.*, IAA and NAA promoted callusing in the explants. Shoot bud formation was restricted to the cotyledonary node and proximal region of the cotyledons. Half-strength MS medium containing IBA (3.0mg/l) was found suitable for rooting on regenerated shoots. *In vitro* developed plantlets were transferred into soil.

The study on 8 species of ornamental *Acacia* was carried out by Ruffoni *et al.* (1995) to define a protocol of micropropagation to be used industrially in order to program plant production. Specific features of the woody species, such as ethylene production, phenols release, loss in juvenility, slow growth, were overcome by increasing iron-chelates or by utilizing antioxidant agents or by modifications in salt composition. Shoot proliferation can be obtained using BAP, IAA and GA_3 0.5 ppm each, roots induction and development occurred in media containing 1 ppm of IAA. Plants were acclimatized and then hardened and grown in the field (Ruffoni *et al.* 1995).

Somatic embryogenesis has been reported in *Acacia koa* by Skoleman (1995b).

Albizia Species

Tomar and Gupta (1988a,b) observed maximum number of shoots in *Albizia amara, A. lucida, A. richardiana* on B_5 + 0.002 mg/1 2, 4-D and B_5 + 0.22 mg/1 BAP. BAP enhanced shoots differentiation. After sub culturing on B_5 + 0.017 mg/1 IAA for best rooting, plantlets were successfully transferred to soil.

Somatic embryogenesis has been reported in *Albizia amara* by Tomar and Gupta (1986), in *Albizia lebbeck* by Gharyal and Maheshwari (1981), *Albizia lucida* by Tomar and Gupta (1986) and *Albizia richardiana* by Tomar and Gupta (1988) all from hypocotyls segments.

In *Albizia chinensis* Sinha *et al.* (2000) cultured proximal and distal cotytledonary segments on MS medium and induced to form adventitious shoot buds in the presence of BAP, Kn, and TDZ. Higher concentration of BAP inducing shoot bud and differentiation was observed. Proximal cotyledonary segments was more morphogenic to shoot bud differentiation than distal cotyledonary segments. TDZ was highly effective in inducing shoot buds, but arrested shoot growth, while Kn produced more callus during differentiation of shoots. Rapid and high rate of shoot multiplication per explant was achieved through subculture in MS medium containing 1 mg/l BAP and 0.5 mg/l IAA. BAP at low concentration was required to enhance shoot multiplication and elongation. Successful rooting of regenerated shoots was carried out in a two-step culture procedure in MS media with 2 mg/l IBA and subcultured in MS medium.

Regeneration of adventious buds in *Albizia guachapele* was achieved from hypocotyls explants by Cerdas *et al.* (1998) from epicotyl explants. Four explants were cultured in each petridish on half strength modified MS medium and five concentrations of BA were studied. Cerdas *et al.* (2008) also gave *in vitro* propagation of *Albizia guachapele*.

In vitro regeneration of silktree (*Albizzia julibrissin*) from excised roots was described by Sankhla *et al.* (1996). Root segments (1 cm long) were excised from 15-20 day old seedlings grown on B_5 medium. About 50 per cent of the control (no growth regulators added) root explants formed shoot buds within 15 days after placement on the culture medium. After 30 days, there were about 4 shoots per control explant. Addition of BA, Zeatin and TDZ to the culture medium increased both the percentage of explants that formed shoots and the number of shoots per explant. TDZ was highly effective in stimulating shoot formation at low concentration. At 0.05 μM TDZ, 95 per cent of the explants produced shoots and about 10 shoots were formed per explant. Upon excision and transfer to B5 medium, regenerated shoots developed into normal rooted plantlets. Sankhla *et al.* (1994) also described TDZ induced *in vitro* shoot formation from roots of intact seedlings of *Albizzia julibrissin* via callus formation.

In *Albizia julibrissin* Burns and Wetzstein (1998) placed immature seeds, embryo cotyledons and embryo axes (cotyledons removed) on induction media with different concentration of 2,4-D. Two distinct embryogenic responses occured either proembryo masses or cotyledonary stage embryos. Twenty five percent of all embryo axes cultured on basal medium produced cotyledonary somatic embryos. Six percent of immature seed explants generated proembryo masses. These masses proliferated in liquid culture in the dark. Proembryos developed further when transferred to a basal semi solid media in the light. Somatic embryos derived from proembryo suspensions or cotyledonary embryo cultures on semisolid medium germinated to form plants. Plants grew vigorously when transfered to soil.

In vitro differentiation of plantlets from tissue culture of *Albizzia lebbeck* was given by Gharyal and Maheshwari (1983a). Attempts at inducing differentiation in various explants resulted in the production of shoot buds from the hypocotyls, root, cotyledon and leaflet explants, both directly and indirectly (*i.e.*, without and with the intervention of callus formation). Rooting was achieved on transfer of the shoots to

BM + 2 mg/l IAA after some growth. The plants could be successfully transferred to soil providing a method for mass propagation of this tree species.

Upadhyaya and Chandra (1983) studied shoot and plantlet formation in organ and callus cultures of *Albizia lebbeck*. Plantlets were produced *in vitro* from root and hypocotyl explants formed from seedlings. These explants formed shoots when cultured with 5 mg/l Kn and 1 mg/l IAA in MS medium. Shoots were also induced in large numbers from callus treated with BAP. About 20 per cent of the shoots rooted and were grown into plants.

Rao and De (1987) established callus cultures from hypocotyls, leaf and stem tissue of *Albizia lebbeck*. High frequency multiple buds and shoots were obtained on callus derived from somatic explants on MS medium with NAA and BAP/Kn combinations. Roots were induced in Bonner's salt solution with IBA. 30-40 plantlets were regenerated in an 8 week period from each explant.

Stem and petiole explants obtained from mature trees of *Albizia lebbeck* callused and differentiated shoot-buds and later shoots on B_5 medium supplemented with either 0.5mg/l IAA + 1 mg/l BAP or BM + 2 mg/l NAA + 0.5mg/l BAP (Gharyal and Maheshwari, 1990). Plantlets were rooted, acclimatized and successfully transferred to the field.

Matin *et al.* (2004) obtained plantlets by culturing the cotyledon, nodal segments of *in vitro* grown seedlings and nodal segments of field grown mature trees of *Albizia lebbeck*. Among all the hormonal supplements used BA-NAA combination with MS medium was proved best in all respect of callusing response. Among the explants the *in vitro* internodal segments were the best for callus induction. Among these three explants of *Albizia lebbeck* only calli derived from cotyledon regenerated best shoots in MSA media supplemented with BA singly or in combination with NAA and Kn. In respect of direct shoot regeneration nodal explants produced highest range of regenerated shoots (Matin *et al.*, 2004).

Mamun *et al.* (2004) cultured cotyledon, nodal segments of *in vitro* grown seedlings and nodal segments of field grown mature tree for micropropagation of *Albzia lebbeck*. Among all the hormonal supplements used BA-NAA combination with MS medium was proved best in all aspect of callusing response. Among the three explants only calli derived from cotyledon regenerated best shoots in MS medium supplemented with BA singly or in combination with NAA and Kn. In respect of direct shoot regeneration nodal explants produced highest range of regenerated shoots (Mamun *et al.*, 2004).

Rajeswari and Paliwal (2004) studied *in vitro* adventitious shoot organogenesis and plant regeneration from seedling explants of *Albizia odoratissima*. Epicotyl, petiole and cotyledon explants derived from 14 day old seedlings were cultured on MS basal medium supplemented with different concentrations of either BAP solely or in combination with 0.5μM NAA. The best response in terms of the percentage of shoot regeneration was obtained from epicotyls cultured horizontally on MS medium supplemented with 5 μM BAP, whereas the highest number of shoots per responding explant was recorded on medium containing 2.5 μM BAP and 0.5 μM NAA. Successful rooting was achieved by placing the microshoots on to MS medium containing 25 µM IBA for 24h first, then transferring to the same medium without IBA. Rajeswari and Paliwal (2007) cultured nodal explants from seedlings and two year old potted plants of *Albizia odoratissima* on MS medium supplemented with different growth regulators.

Kumar *et al.* (1998) reported direct regeneration of shoots from leaflet explants in *Albizia procera* by further addition of silver nitrate into the MS medium + 0.22 mg/1 BAP + 0.018 mg/1 NAA + 2.6g /1 phytogel. This medium enhanced adventitious buds regeneration..

Swamy *et al.* (2004) collected explants from 15 ± 2 years-old mature trees from 15 days old juvenile seedling of *Albizia procera* and were regenerated with exogenous application of different hormones. Epicotyl and hypocotyls explants excised from juvenile seedling showed higher callusing than axillary bud and shoot tip explants derived from mature trees. 100 per cent callusing in epicotyl and hypocotyls explants on half MS medium + 3 mg/1 BA. Callus derived from epicotyl and hypocotyl explants proliferated and formed *de novo* shoots and leaflets on a medium containing 3 mg/1 GA_3 and rooting was most successful on half MS medium + 6 mg/1 IBA alone. Sand or vermiculate supplemented with 4 ml of Yoshida solution proved as best hardening media, which recorded 70-80 per cent survival of plantlets.

In *Albizia thompsoni* Anuradha and Pullaiah (2001) observed effective seed germination on B_5 medium + 0.1mg/1 BAP. 5 minute scarification treatment with alcohol improved percentage of germination.

Bauhinia Species

Mathur and Mukuntha Kumar (1992) reported multiple shoots on MS medium + 0.048-0.68 mg/1 6-benzyladenine and rooting on MS medium + 0.049-0.29 mg/1 IBA in *Bauhinia variegata*. Axillary shoot proliferation was achieved from nodal explants from mature trees using MS medium supplemented with 2.22-31.1 µM of BAP. Subsequent rooting of the regenerated shoots was achieved on medium containing 2.46-14.8µM of IBA. Successful transfer of the regenerated shoots to soil has been accomplished (Mathur and Mukuntha Kumar, 1992).

Protocol for micropropagation of *Bauhinia purpurea* was given by Kumar (1992). Stem explants of mature tree were cultured on MS medium with various auxins and cytokinins. Plantlets were obtained via callus growth.

Calliandra Species

In vitro somatic embryogenesis was established from 1-1.2 cm long internodal and petiolar segments excised from 20-years-old woody legume *Calliandra tweedii* by Kumar *et al.* (2002). Within 6-7 weeks, globular, heart-shaped, torpedo-shaped and dicot normal embryos were obtained in 71 per cent of internodal segments. Petiolar explants responded best on 0.009 mg/1 NAA where in 30 per cent of the cultures developed normal embryos. Somatic embryoids developed shoots and roots when they were retained for 90 days in old desiccated MS medium.

Cassia Species

Stem and petiole explants obtained from mature trees of *Cassia fistula* and *C. siamea* callused and differentiated shoot-buds and later shoots on B_5 medium supplemented with either 0.5mg.l IAA + 1 mg/l BAP or BM + 2 mg/l NAA + 0.5mg/l BAP (Gharyal and Maheshwari, 1990). The stem explants were more responsive than the petiole explants. However in *C. fistula* the type of explants rather than the medium compostion had an overriding influence on shoot differentiation since those from petiole hardly responded in either medium. Plantlets were rooted, acclimatized and successfully transferred to the field.

In *Cassia angustifolia*, Siddique and Anis (2007) observed the highest rate of shoot multiplication on MS medium + 0.11 mg/1 TDZ + 0.017 mg/1 IAA. Regenerated shoots were followed by repeated subcultures on MS basal medium which increases the rate of shoot multiplication and shoot length by the end of fourth subculture passage. Rooting was achieved on the isolated shoots using MS medium + 1.2 mg/1 IBA 15 per cent AC for 1 week and subsequently transferring the shootlets to half MS liquid medium. Agarwal and Sardar (2006) observed dark compact callus on MS medium + 0.22 mg/1 BA + 0.022 mg/1 2,4-D within 10-15 days. Callus pieces were subcultured on MS medium with BA or Kn. Where as 0.11 mg/1 BA was optimum for eliciting morphogenic response in 83.33 and 70.83 per cent cultures with an average of 4.16 ± 0.47 and 3.70 ± 0.56 shoots in cotyledons and leaflet derived calli respectively. Regenerated shoots were subcultured on MS medium + 0.09 mg/1 NAA + 0.11 mg/1 BA further elevated the maximum average number of shoots to 12.08 +1.04 and 5.37 ± 0.52 for cotyledon and leaflet calli. Nearly 95 per cent shoots developed an average of 5.4 ± 0.41 roots on half MS medium + 0.2 mg/1 IBA.

In vitro regeneration through somatic embryogenesis as well as organogenesis using cotyledons *Cassia angustifolia* was reported by Agarwal and Sardar (2007). Maximum somatic embryos were observed from cotyledons of semi matured seeds in MS medium + 2.5 μM BA + 10 μM 2,4-D. Somatic embryos started germination on same media and developed into full plantlets only if transferred to MS basal media with 2 per cent sucrose.

Praveen and Shahzad (2010) cultured root explants taken from 30 day old aseptic seedlings of *Cassia angustifolia* on MS medium supplemented with different plant growth regulators: BAP, Kn, TDZ. Organogenic nodular calli obtained on MS + TDZ (1.0μM) were transferred to shoot regeneration medium supplemented with different cytokinins (BA, Kn of TDZ) either alone or in combination with auxins IAA, NAA. Maximum shoot regeneration frequency (90 per cent) was obtained on MS +BA (2.5 μM) + NAA (0.6 μM) wherein a maximum of 42.76 shoot buds per explant were induced with a maximum conversion rate of 35.63 shoots per explant and average shoot length of 5.43 cm. Elongated microshoots were successfully rooted under *ex vitro* conditions by pulse treatment in 200 μM of IBA for half an hour.

Shrivastava *et al.* (2006) investigated the biosynthetic potential of *in vitro* grown callus cells of *Cassia senna* var. *senna*. Cotyledonary leaves of aseptically grown seedlings were used as explants for callus induction and were inoculated on MS medium with different concentrations and combinations of growth regulators. In the medium supplemented with 16 µM BA and 10 per cent CM, along with the shoot-forming nodules, non-morphogenic compact callus mass was produced.

A protocol for rapid micropropagation of *Cassia siamea* was developed by Sreelatha *et al.* (2008) by using shoot tip, cotyledonary node, and nodal explants derived from seedlings grown *in vitro*. Maximum response was observed with nodal explants on MS macro salts +B_5 micro salts medium supplemented with TDZ 0.5 mg/l with shoot number of 6.70 and shoot length of 3.95 cm. Maximum number of shoots 20.15 was induced on MS macro salts + B_5 micro salts medium containing Kn 0.1 mg/l + TDZ 0.1 mg/l + 2 iP. The shoots were subcultured on medium containing 2-iP + GA_3 3 mg/l for elongation. High frequency of rooting was obtained on half strength MS medium, NAA 1 mg/l and IBA 0.25 mg/l. The plantlets were hardened and were successfully established in natural soil.

A method for rapid *in vitro* propagation of *Cassia siamea* using cotyledonary node explants, excised from 14-day old aseptic seedlings, has been established by Praveen *et al.* (2010). MS medium supplemented with different concentrations of BA, Kn and TDZ singly or in combination with auxins was used for regeneration studies. Among the single treatment of three cytokinins BA at 1.0 µM was found to be optimum for direct shoot regeneration as it induced an average of 8.20 shoots per explant. The regeneration frequency further enhanced with the application of auxin along with optimal BA concentration. The highest frequency of shoot regeneration (90 per cent) the maximum number of shoots per explant (12.20) and the maximum shoot length (6.40cm) were obtained on a medium consisted of MS + 1.0 µM BA + 0.5 µM NAA. Successful *in vitro* rooting was induced from cut end of the microshoots when placed on half strength MS + IBA (2.5 µM). The regenerated shoots with well developed root system were successfully acclimatized and established (Praveen *et al.*, 2010).

Bajaj and Dhanju (1983) successfully reported the induction of pollen embryogenesis in tree legume *Cassia fistula*.

Gharyal *et al.* (1983a) reported haploid chromosome number (n=24) in callus derived form anthers of *Cassia siamea* cultured on B_5 + CM (15 per cent ; v/v), 2.0 mg/1 2, 4-D and 0.5 mg/1 Kn.

A callus culture of *Cassia torosa* which produced germichrysone in high yield was established by Noguchi and Sankawa (1982) on MS medium containing IAA (3ppm) and BA (0.1 ppm). In six-week old callus culture the main pigment was pinselin and the germichrysone content was markedly decreased. The germichrysone production in relation to growth was investigated with the shake culture and the production of germichrysone was found to possess two maxima. The first maximum was observed in lag phase and the second coincided with active growth.

Ceratonia siliqua

A protocol for *in vitro* shoot regeneration was developed by Mohammed-Yasseen *et al.* (1992) from aseptically germinated seedlings of *Ceratonia siliqua*. Shoot tips, stem nodes, cotyledonary nodes and cotyledons were cultured on MS media supplementd with combination of TDZ and IAA and IBA or BA and NAA. Multiple shoots were obtained after four weeks culture under long day period.

Cladrastis Species

An emphasis has been made to propagate the legume trees with high quality of timber, which would be of immense use in forestry. *Cladrastis lutea* (yellow wood) is difficult to propagate by conventional methods was successfully multiplied by somatic embryogenesis (Weaver and Trigiano, 1991) using immature embryos.

Dalbergia Species

Dalbergia is an important timber-yielding genus. For timber, good tree forms are highly desirable but due to cross pollination, the seed progeny gives rise to heterogenous population. Thus there is a need to develop large scale cloning of desired genotypes. *Dalbergia lanceolaria* (Anand and Bir, 1984), *D. latifolia* (Mascarenhas *et al.*, 1982; Nataraj and Sudhadevi, 1985; Lakshmi Sita *et al.*, 1986; Sankara Rao, 1986) and *D. sissoo* (Mukhopadhyay and Mohan Ram, 1981) have been propagated from seedling explants as well as adult tissues.

Callus-mediated shoot bud formation was demonstrated in *Dalbergia latifolia* by Shankara Rao (1986). Cultures were raised from shoot explants of six-year old plants on MS medium supplemented with NAA and BA. A sequential treatment of callus with increasing BA levels and decreasing NAA ensured shoot bud induction. Rooting of shoots was achieved by three-step culture procedure involving 1) White's liquid medium containing IAA, NAA and IBA, 2) half-strength MS agar solidified medium with charcoal (0.25 per cent and 3) half-strength MS liquid medium.

Of these species, *Dalbergia sissoo* was the first to be regenerated, this being done from seedling root explants (Mukhopadhyay and Mohan Ram, 1981). Shoots regenerated directly on the explants or from callus were successfully rooted and transplanted. Sewal *et al.* (1099) achieved 10-15 fold multiplication in every 8 weeks, from cotyledonary node explants. The *in vitro* multiplied shoots could be treated as microcuttings and rooted under *in vivo* conditions with 85 per cent transplantation success. Datta *et al.* (1983) reported multiple shoot formation from nodal explants of 30 year old tree and were successful in inducing rooting, indicating the potential of the species to regenerate under culture conditions. Datta and Datta (1983) followed up their earlier work and reported callus and plantlet formation in same auxin combinations. Multiple shoots were obtained which were successfully rooted under *in vitro* conditions. Nataraja and Sudhadevi (1985) successfully established callus from seedling explants (hypocotyls, stem and root) of *Dalbergia latifolia* followed by shoot bud differentiation. The shoots could be rooted and entire plants were raised.

Induction of single and multiple shoots was obtained from nodal explants of *Dalbergia latifolia* trees on MS medium supplemented with BAP and NAA of IAA (Raghava Swamy *et al.*, 1992). Multiplication of shoots was obtained on MS (reduced major elements) or Woody Plants Medium supplemented with BAP and Kn. Excised shoots were rooted on half-strength MS with IBA (2 mg/l) to obtain complete plantlets. The regenerated plantlets have been acclimatized and successfully transferred to the soil.

Micropropagation of Indian rosewood (*Dalbergia latifolia*) by tissue culture was given by Ravishankar Rai and Jagadish Chandra (1988). Multiple shoots were induced on excised hypocotyl segments and shoot tips of *in vitro* germinated seedlings on MS medium supplemented with cytokinins and auxins. Roots were induced when individual shoots were treated first with half strength MS medium supplemented with NAA, IAA and IBA (1 mg/l each) and subsequently transferred to hormone-free half-strength MS medium. The plantlets were then transferred to pots and grown in greenhouse.

Muralidhar Rao and Lakshmi Sita (1995) developed a protocol for direct somatic embryogenesis from immature embryos of *Dalbergia latifolia*. Cultured on high concentration of 2, 4-D medium after four weeks of subculture on low 2, 4-D + high sucrose induce direct somatic embryos. Embryos transferred to MS medium + 0.5-1 mg/l BAP, embryos developed into plantlets. Direct regeneration of somatic embryos with out callus phase has direct application for genetic manipulation studies. Pradhan *et al.* (1998) observed high frequency plant regeneration from cell suspension cultures. Friable callus developed by inoculating hypocotyls on MS medium + 10.8Wm NAA + 2.2Wm BA. Calli were increased by sub-culturing on MS medium + 10 per cent CW. High frequency shoot bud differentiation on MS medium + 2.7Wm NAA + 13.3Wm BA. Regeneration frequency declined at the higher BA concentration. The organogenic potential of the cell suspensions were influenced by the age of the culture. Multiple shoot induction from cotyledonary node explant collected from 1-week-old axenic seedlings. High frequency shoot proliferation (99 per cent) and maximum number of shoots per explant (7.9 shoots) were recorded on MS medium + 0.19 mg/l BA. After repeated subculturing on growth medium 60-70 shoots were obtained in 3 months from a single cotyledonary node. 91 per cent of the shoots developed roots on half MS medium + 0.10 mg/l Indole-3-propionic acid.

In vitro plantlet regeneration from seedling explants was achieved by Sreedevi and Pullaiah (1999) in *Dalbergia paniculata* on MS medium supplemented with BAP. Cotyledonary node responded well with 83 per cent shoot bud regeneration. IBA was found to be more suitable for induction of rooting.

In *Dalbergia retusa* Cerdas (2004) reported more number of shoots on MS medium + 0.193 mg/l BA, and highest average number of roots obtained on MS medium + 20 g/l sucrose + 0.49 mg/l IBA. Plants were obtained via organogenesis from hypocotyls explants from *in vitro* germinated seedlings.

Swapna *et al.* (1982) observed multiple shoots from nodal explants of 30-years-old trees of *Dalbergia sissoo* on MS medium supplemented with auxins + cytokinin combinations. IAA alone promoted 15 per cent rooted shoot buds. A combination of IAA + Kn gave 100 per cent rooted shoot buds. Ascorbic acid in the medium prevented the death of callus and plantlets, which followed darkening of the medium.

Chand and Singh (2004) inoculated semi-mature zygotic embros of *Dalbergia sissoo* for high frequency callus induction in MS medium 0.198 mg/l 2, 4-D and 0.02 mg/l Kn. Regeneration occurred on MS medium + 0.19mg/l BAP and 0.02 mg/l NAA. The regenerated shoots were rooted on half strength MS medium + 0.02mg/ l IBA. A method has been developed by Singh and Chand (2010) for plant regeneration by encapsulation of somatic embryos obtained from callus cultures derived from semi-mature cotyledon explants of *Dalbergia sissoo*. Embryogenic callus was developed from cotyledon pieces on MS medium supplemented with 9.04 mM 2,4-D and 0.46 mM Kn. The somatic embryos were induced from embryogenic callus on hormone free ½ MS medium with 2 per cent sucrose. Cotyledonary stage somatic embryos were encapsulated using sodium alginate (2.5 per cent) and calcium chloride (75 mM) as gelling matrix. The highest frequency (43.3 per cent) for conversion of encapsulated somatic embryos into plantlets was achieved on ½ MS medium with 2 per cent sucrose. Plantlets with well developed shoots and roots were established in pots containing autoclaved mixture of peat moss and soil.

Micropropagation of *Dalbergia sissoo* has been developed by Gulati and Jaiwal (1996). Node of 20-25 year old plants produced more shoots inoculated on MS medium supplemented with 4.4×10^{-6} M BAP and 4.4×10^{-7} M NOA within 4 weeks, following this procedure 18-24 shoots were produced from single nodal segments with in 60 days. 80 per cent of the shoots directly produced roots when they were treated with MS medium containing 10^{-5} M IBA and subsequently transferred to half strength MS medium + activated charcoal followed by half strength MS basal medium.

In *Dalbergia sissoo* Chand and Singh (2004) reported plant regeneration from encapsulated nodal segments collected from basal sprouts of mature trees. Root induction treatment was given to nodal segments for 10 days. For synthetic seed production nodal segments were encapsulated with 3 per cent sodium alginate + 1.05 mg/1 $CaCl_2$ $2H_2O$. Rooting was carried out in half MS medium. Singh and Chand (2003) cultured cotyledon pieces of *Dalbergia sissoo* on MS medium + 0.19 mg/1 2, 4-D and 0.009 mg/1 Kn to obtain more amount of callus. Somatic embryos were obtained from callus sub-cultured on half MS medium. Half MS medium + 0.009 mg/1 L-glutamine enhanced somatic embryos. Half MS medium + 2 per cent sucrose enhanced the conversion of somatic embryos to plantlets. A method has been developed by Singh and Chand (2010) for plant regeneration by encapsulation of somatic embryos obtained from callus cultures derived from semi-mature cotyledons explants of *Dalbergia sissoo*.

A procedure was outlined by Singh *et al.* (2001a) to induce adventitious shoot organogenesis from semi-mature as well as mature cotyledons lacking the embryonic axis of *Dalbergia sissoo*. Shoot buds were induced in the proximal region of the semi-mature cotyledons on MS medium supplemented with 4.44 µM BAP and 0.26µM NAA. These buds elongated into shoots following transfer to similar medium containing half-strength macro-nutrients. Adventitious shoot bud formation was also induced in the mature cotyledons. Regenerated shoots derived from semi-mature and mature cotyledons rooted on half-strength MS medium containing 1.23µM and 4.92 µM IBA respectively.

Kumar *et al.* (1991) reported regeneration of plantlets from cell suspension derived calli of cambial origin from mature tree of *Dalbergia sissoo* cultured on MS medium + 2 mg/1 2, 4-D + 0.1 mg/1 BAP. Shoot bud differentiation from cells was observed in MS medium + 2 mg/1 BAP. Rooting was better on MS medium + low organic salts and auxins.

Das *et al.* (1997) developed a protocol for development of somatic embryos from callus derived from zygotic embryos of *Dalbergia sissoo*. Zygotic embryos cultured on MS medium + 0.005 mg/1 – 0.024 mg/1 Kn, 0.14 -1.19 mg/1 2, 4-D and 30g/1 sucrose. Secondary somatic embryos developed on half MS + 0.009 – 0.024 mg/1 Kn and 0.149 -0.198 mg/1 2,4-D with 2 per cent w/v sucrose. Light green somatic embryos germinated on half MS medium + 0.5 mg/1 ABA + 2 per cent w/v sucrose.

Micropropagation studies were undertaken by Joshi *et al.* (2003) using nodal explants collected from 60-year-old superior tree of *Dalbergia sissoo*. MS and B$_5$ media were used to find out the suitability of the medium. Bud break was achieved in both the media within 6-8 days under different media combinations supplemented with BAP (0.1-1.0 mg/l) alone as well as in combinations with IAA or NAA (0.1-0.5 mg/l). Maximum number of shoots per explant (8.04) was observed in the MS medium supplemented with 1.0 mg/l BAP + 0.25 mg/l NAA. Maximum number of roots per plantlet (4147) was observed in ½ MS supplemented with (1.0mg/l) IBA within 18 days. Plantlets were acclimatized and transplanted to pots.

Delonix regia

A protocol for *in vitro* shoot regeneration was developed by Mohammed-Yasseen *et al.* (1992) from aseptically germinated seedlings of *Delonix regia*. Shoot tips, stem nodes, cotyledonary nodes and cotyledons were cultured on MS media supplemented with combination of TDZ and IAA or IBA or BA and NAA. Multiple shoots were obtained after four weeks culture under long day period.

Erythrina Species

Multiple shoot production could be achieved when cotyledon and leaf from *in vitro* grown seedlings of *Erythrina variegata* were cultured on MS medium containing BAP, 2,4-D and NAA (Shasthree *et al.*, 2009). MS medium supplemented with 10 per cent, 15 per cent and 20 per cent of coconut milk in addition to TDZ triggered the induction of multiple shoots. MS medium with 1 mg/l BAP and 2 mg/l L-glutamic acid also favoured the induction of multiple shoots which ranged from 12-16 from cotyledon segments.

Hardwickia binata

An optimal *in vitro* propagation procedure for *Hardwickia* was achieved by Anuradha *et al.* (2000) using mesocotyls, shoot tips and axillary buds as source of explants. The highest frequency of 80 per cent shoot bud proliferation with 3 shoot buds was observed in cultures established on MS medium supplemted with 2 mg/l BAP and 2 mg/l Kn. The proliferated shoots readily rooted *in vitro* on MS medium supplemented with 4 mg/l IBA. The rooted plantlets were successfully transferred to soil.

Leucaena Species

Venkateswaran and Gandhi (1982) and Glovak and Greatbach (1992) succeeded in regenerating plantlets from seedling explants of *Leucaena leucocephala*. Similarly Ravishankar *et al.* (1983) obtained plantlets from shoot tips of 4 day old *in vitro* grown seedlings of *Leucaena leucocephala*. Nagamani and Venkateswaran (1983a,b) cultured seedling explants of three species of *Leucaena* of which only *L. diversifolia* formed immature plantlets through somatic embryogenesis. In the cultures of *L. leucocephala* and *L. retusa* only leaf shoots were formed.

A tissue culture method was described by Goyal *et al.* (1985) for clonal multiplication of *Leucaena leucocephala* K67 using single lateral bud explants from 2-3 tall greenhouse grown trees. BA (3.0 mg/l) and NAA (0.05 mg/l) in MS medium were found to be best suited for multiple shoot differentiation in 4-5 week old cultures. A shoot multiplication rate of 22 shoots per bud explant was obtained in 150 days on ½ strength MS medium with 3.0 mg/l BA and 0.05 mg/l NAA. Shoots developed adventitious roots within 15 days in ½ strength MS medium containing IBA (3.0 mg/l) and Kn (0.05 mg/l). Plantlets were transplanted in greenhouse.

Rastogi *et al.* (2008) developed *in vitro* regeneration system for *Leucaena leucocephala* from mature tree derived nodal explants as well as seedling derived cotyledonary node explants. Best shoot initiation and elongation was found in MS medium supplemented with 20.9 µM BAP and 5.37 µM NAA. Rooting was induced in half strength MS medium containing 14.76 µM IBA and 0.23 µM Kn. Rooted plantlets were subjected to hardening and successfully transferred to greenhouse. Somatic embryogenesis from nodal explants via an intermediate callus phase was also established.

Mundulea Species

Kalavathi *et al.* (2004) obtained shoot buds from cotyledonary segments of *Mundulea sericea* inoculated on MS medium + Dicamba in combination with IAA, IBA, and NAA. Shoot buds on transferring to the shoot induction medium containing 3 mg/l BAP + 0.5 mg/l NAA + 0.5 g/l GA_3 developed into 20-25 shoots. These shoots were best rooted on half MS medium + 2.5 mg/l NAA. Shoot tip explants cultured on MS medium + 5 mg/l Picloram + 1 mg/l NAA + 0.1mg/l BAP produced 29-30 translucent somatic embryos. Regenerating capacity was limited.

Paraserianthes falcataria

Bon *et al.* (1998) studied the influence of five different macronutrient formulations and various growth regulators on micropropagation of single node explants of *Paraserianthes falcataria*. Sasmitamihardja *et al.* (2001) achieved highest number of shoot multiplication in *Paraserianthes falcataria* (=*Albizia falcataria*) from 6-days old seedlings cultured on MS medium + 0.1µM IAA and 0.5 µM BA. After 8 weeks in culture, up to 23 shoots per explant were produced. Regenerated shoots were elongated on MS medium without growth regulators.

Sinha and Mallick (1993) obtained regeneration and multiplication of shoot in *Paraserianthes falcataria* (=*Albizia falcataria*). They cultured the cotyledons from 15 d old *in vitro* seedlings on MS medium containing BA.

Parkinsonia aculeata

Mathur and Mukuntha Kumar (1992) reported multiple shoots on MS medium + 0.048-0.68 mg/l 6-benzyladenine and rooting on MS medium + 0.049-0.29 mg/l IBA in *Parkinsonia aculeata*. Axillary shoot proliferation was achieved from nodal explants from mature trees using MS medium supplemented with 2.22-31.1 µM of BAP. Subsequent rooting of the regenerated shoots was achieved on medium containing 2.46-14.8 µM of IBA. Successful transfer of the regenerated shoots to soil has been accomplished (Mathur and Mukuntha Kumar, 1992).

Peltophorum Species

Bassan *et al.* (2006) investigated the phenolic exudation, type of explant and nutritive media in *Peltophorum dubium in vitro* establishment. MS medium was found to be more efficient than WPM medium in the *in vitro* establishment. Both apical segment and nodal segments responded better for culture establishment

Lakshmi Devi *et al.* (1996) gave the protocol for *in vitro* regeneration from seedling explants of *Peltophorum pterocarpum*. Mesocotyl explants cultured on MS medium supplemented with BAP 2 mg/l showed 90 per cent frequency with 2-3 shoots per explant.

In *Peltophorum pterocarpum* Rao and De (1987) observed haploid plants from anther culture. Pretreatment of flower buds at moderate temperature of 14°C for 8 days was most effective for callus production. The frequency of callus and shoot production was highest when anthers were cultured at mid or late – uninucleate stage and high sucrose concentration.

Salah Uddin *et al.* (2005) in *Peltophorum pterocarpum* reported the highest number of multiple shoots on MS medium + 2 mg/1 Kn + 0.5 mg/1 NAA. In some cases, BAP also showed better result. The regenerated shoots were transferred on MS medium + IBA for adventitious root initiation.

Pithecellobium Species

Cerdas *et al.* (1997) inoculated hypocotyl explants of *Pithecellobium saman* on a medium containing BA. Adventitious bud induction was affected by mineral salts and BA concentration, explant age, position of explant on the medium, and BA exposure time. Best results were seen on half MS medium + 0.58 mg/1 BA when 5 to 10 days old explants were placed horizontally and 7 days exposure period to BA. Proximal and intermediate section of hypocotyls showed the highest organogenic repose. Activated charcoal in the medium increased bud development and shoot elongation.

Pongamia pinnata

In *Pongamia pinnata* Sujatha and Sulekha (2007) observed a swelling developed at the axil at higher concentrations of plant growth regulators. Multiple shoot primordia appeared and differentiated from this swelling after culturing these explants on MS medium for six passages of 2 weeks each. Shoots were harvested and cultured on 0.009 mg/1 TDZ for further proliferation. Primary explants after harvesting of shoots were identified as "STUMP". Reculturing of stumps on 0.009 mg/1 TDZ produced, more shoots. This step was followed for six cycles to obtain additional shoots in each cycle. Shoots maintained on 0.009 mg/1 TDZ elongated and rooted on growth regulator free medium. Repeated proliferation of caulogenic buds from same origin may also find application in rescue of endangered germplasm.

A complete protocol for the micropropagation of *Pongamia pinnata* was given by Sugla *et al.* (2007). Multiple shoots were induced *in vitro* from nodal segments through forced axillary branching. MS medium supplemented with 7.5 μM BAP induced up to 6.8 shoots per node with an average shoot length of 0.67 cm in 12 d. Shoots formed *in vitro* were rooted on full strength MS medium supplemented with 1.0μM IBA. Plantlets were successfully acclimatized, established in soil and transferred to the nursery.

Sujatha *et al.* (2008) reported role to TDZ in inducing adventitious organogenesis in *Pongamia pinnata.* TDZ at 0.249 mg/1 concentration was optimum for the induction of shoots and rapid elongation. Shoots induced at higher concentrations elongated after several passages in growth regulator free medium. Exposer of the explant for 20 days yielded more number of buds than 10 days. Proximal segment of the cotyledon was more responsive. Contact of abaxial surface in the medium was more effective and generated more buds than the adaxial side. Buds differentiated and elongated on transfer to MS basal medium for 8-12 passages of 15 days each. Rooting and elongation of shoots was achieved in charcoal supplemented half MS medium.

Prosopis Species

Several attempts have been made in the micropropagation of *Prosopis* species as reviewed by Jordan (1987) and Ramawat and Nadwani (1991). Most of the work has been done using juvenile material, and good results have been achieved. However, although it is possible to propagate juvenile plants with relative ease, it is most important to develop the ability to propagate field material from adult plants that have tolerance to salinity and dryness, high productivity and a fast growth rate, among other characteristics. Important results using adult material as the source of explants have been obtained in *P. cineraria, P. juliflora* was well as in *P. chilensis* using juvenile material taken from adult trees.

In general juvenile material of most species is easier to propagate than adult material. This is especially true in *Prosopis,* where some species are recalcitrant (Jordan, 1987; Arce and Balboa, 1991) due to the high concentration of phenolic compounds in their tissues. Some of these compounds, intermediates in the lignin synthesis pathway, are known to be inhibitors of morphogenic responses in explants cultured *in vitro.* For this reason, in some cases it is preferable to work with very young plants. In *P. tamarugo* it has been necessary to use nodal and apical explants taken from very young plants or even from germinated seeds (hypocotyls, cotyledons and embryogenic axes).

Complete plants of *P. alba, P. alpataco, P. chilensis, P. cineraria, P. juliflora* and *P. tamarugo* have been obtained using juvenile explants that include meristematic tissues such as nodes, apical sections and lateral buds.

Some *Prosopis* species show higher regenerative capacity. For example it has been possible to micropropagate *P. chilensis* plants by different methods. Jordan *et al.* (1985b) used shoot tip explants on MS medium with 0.3 or 1 mg/l of NAA, 0.1mg/l

BA and 0.01 mg/l GA$_3$ for the induction of shoot development and subsequent rooting. The use of shoot tips as explants has the advantages of reduction of contamination and browning and the possible elimination of pathogens. Jordan *et al.* (1985) have evaluated the effect of NAA and the antioxidant cysteine on regeneration of complete *P. chilensis* plants. Nodal sections of 1-4-month-old plants were cultured on MS liquid medium plus 0-10 mg/l NAA and 0-10 mg/l cysteine (Arce and Balboa, 1991). Callus culture and cellular suspensions of *P. chilensis* was also established by Jordan *et al.* (1987). For callus induction, shoots with nodes were cultured in MS medium containing 5 mg/l NAA and 15 mg/l cysteine. Up on subculturing these calli in the same medium enriched with double concentration of vitamins and Mg sulfate, several whitish globular aggregates formed. When this cell suspension was planted on B$_5$ medium supplemented with 2 mg/l 2,4-D and 60 mg/l cysteine small dividing cells appeared, developing proembryonic structures, but no further development was observed. In an attempt to regenerate shoots from hypocotyls of *P. chilensis*, Batchelor *et al.* (1990) used a range of media with high cytokinin content. Callus and root regeneration was obtained in several media and one explant (of 800 tested) regenerated shoots.

Tabone *et al.* (1986) reported that in *Prosopis alba* clone B$_2$V$_{50}$ high concentrations of BAP are the key factors in stimulating shoot production. Juvenile explants of *P. alba* have also been micropropagatd by Jordan *et al.* (1985 a,b). They used shoot tips from 9-month old plants. These shoot tips initiated growth in MS medium with NAA (0.01-1 mg/l) and BA (0.01-0.5 mg/l).

Dun-Yi *et al.* (1989) cultured nodal explants from *Prosopis chilensis, P. cineraria* and *P.juliflora* on solidified MS medium with combination of cytokinins and auxins including Kn (0.05-15 mg/l), BAP 0.05-15 mg/l), 2,4-D (0.005- a mg/l), IAA (1-10 mg/l), IBA (1-15 mg/l), and NAA (1-15 mg/l). Shoot growth and leaf number of *P. chilensis* were greatest with Kn 0.05 mg/l, IBA 3 mg/l, explants developing a mean of 5 nodes per shoot. With *P. cineraria* shoot growth was greatest (3 nodes per shoot) with Kn 3 mg/l, NAA 3 mg/l, but BA 1 mg/l, IBA 3 mg/l gave reduced leaf abscission. Kn 0.05 mg/l, IBA 15 mg/l proved most effective for shoot growth of *P. juliflora*. BAP at 10 or 15 mg/l with NAA or IAA was most effective in inducing shoot proliferation of *P. chilensis* with BAP 15 mg/l, NAA 5 mg/l being the best treatment and giving an average of 4.3 shoots per node. Shoot proliferation of *P. cineraria* was also promoted by high levels of BAP (>3mg/l) and the best treatment (BAP 10 mg/l, NAA 5 mg/l) resulted in 1.8 shoots per node. Results of *P. juliflora* were less conclusive, in very few cases were multiple shoots obtained (Dun-Yi *et al.,* 1989).

Arce and Balboa (1991) collected nodal and apical segments of field grown and juvenile seedlings of *Prosopis chilensis.* Juvenile material gave 80 per cent regeneration rate of complete plantlets on MS media fortified with 5 mg/l NAA and 10 mg cysteine. In the same medium, the regenertative response of segments obtained from rooted cuttings was 60 per cent. Material collected from field showed no rooting response.

Caro *et al.* (2002) developed a protocol for micropropagation of *Prosopis chilensis* from young and mature plants. Nodal segments from 4 month-old plants grown in the green house and from adult trees grown in a natural environment were selected.

These cuttings were cultured on MS or BTMm and treated with 0.05 mg/l BA and 3 mg/l of either IAA, IBA or NAA. Culturing in BTMm resulted in significantly greater shoot and root biomass than culturing on MS.

Shekhawat *et al.* (1993) identified various factors like genotype, age of tree, nature of explants and size (length and diameter), season of explants collection, explants position on medium, plant growth regulators and certain additives (ascorbic and citric acids, adenine sulphate L-arginine, glutamine and ammonium citrate), incubation conditions, and sub-culturing period greatly influenced the *in vitro* clonal propagation of *Prosopis cineraria*.

In *Prosopis cineraria* Sharma *et al.* (1997) described a high frequency regeneration procedure by inoculating cotyledonary node on MS medium + 2 mg/l BAP for multiple shoot induction. Elongated shoots could be rooted on 2 mg/l IBA gelled with phytogel. Several longisections of the cultured nodes revealed vertical and sideways expansion of each of the two axillary meristems and subsequent differentiation of three vertically arranged shoot buds in each of the two axils of cotyledons. Secondary proliferation especially of the terminal buds lead to a large number of shoots buds.

Rubluo *et al.* (2002) cultured cotyledons with hypotyls of *Prosopis glandulosa* var. *torreyana* on MS medium enriched with NAA at 0-1 mg/l combined with Kn at 0-0.1 mg/l. All twenty of the treatments utilized produced vigorous shoots, even the control (no growth regulators added). The best response was attained at 0.05 mg/l Kn with an average of 8 shoots per explant.

Prosopis juliflora nodal explants were successfully initiated into aseptic culture with a two stage sterilization procedure (Wairight and England, 1987). This involved a 2 minute immersion in 70 per cent ethanol and then a 20 minute treatment with sodium hypochlorite (2 per cent available chlorine) and wetter Kn or BA were necessary in the medium for shoot extension of axillary buds. Auxins, either IAA or NAA were not required in the initiation medium, as their inclusion caused callus growth at the base of nodal explants and a reduction in shoot growth. The condition of the media plant was also shown to influence *in vitro* performance of the explants (Wainright and England, 1987).

Buendia-Gonzalez *et al.* (2007) developed a protocol for clonal propagation of *Prosopis laevigata* via cotyledonary nodes. The highest number (3.37 + 0.51) of multiple shoots was observed in MS medium + 0.19 mg/l 2, 4-D + 0.14 mg/l BA. The regenerated shoots were then transferred into rooting media. Best rooting efficiency of 44 per cent was obtained when 0.28 mg/l NAA and vermiculate were used. After rooting the cloned plantlets were successfully hardened to *ex vitro* conditions.

High frequency plantlets regeneration from seedling explants of *Prosopis tamarugo* was carried out by Nandwani and Ramawat (1992a,b). Callus cultures were established from hypocotyls and cotyledons on MS medium supplemented with 2 mg/l NAA and 0.2 mg/l BAP. Regeneration through various juvenile explants was obtained on hormone free and high cytokinin containing MS medium. Multiple shoot induction was observed from the embryonic axis on MS medium + 5 mg/l BAP or with out BAP. *In vitro* produced shoots were rooted on MS medium supplemented with IBA or NAA singly or in combination.

Pterocarpus Species

Anuradha and Pullaiah (1999b) cultured nodal and terminal buds on B_5 medium fortified with various concentrations of cytokinins and coconut milk to induce multiple shoots and also to enhance axillary branching. Nodal explants only responded with axillary branches. About 80 per cent of cultured nodal segments formed three to five shoots with about 20 axillary shoots in *P. marsupium* on 3 mg/l BAP containing medium. The regenerated axillary shoots were excised and transferred to White's medium fortified with 4mg/l IAA (Anuradha and Pullaiah, 1999b). Chand and Kumar (2004) developed a protocol for *in vitro* plant regeneration of *Pterocarpus marsupium* by inoculating cotyledonary node on different hormonal media. Kalimuthu and Lakshmanan (1994) could obtain three or four shoots from cultured nodal segments of *Pterocarpus marsupium*.

Tiwari *et al.* (2004) described methods for *in vitro* propagation of *Pterocarpus marsupium*. Nodal segments were inoculated on seven different media components each supplemented with 1.0 mg/l BAP and 0.5 mg/l NAA or 0.2 mg/l IBA. Regenerated plants were acclimatized and successfully transferred under field conditions.

In *Pterocarpus marsuspium* Husain *et al.* (2007) observed the highest shoot regeneration frequency (90 per cent) and maximum number (15.2 ± 0.20) of shoots per cotyledonary node explant of MS medium + 0.008 mg/l TDZ. Continuous exposure to TDZ inhibited shoot elongation. Maximum (90 per cent) shoot elongation with an average shoot length of 5.4 ± 0.06 cm was observed on MS medium + 0.11 mg/l BA as secondary medium. Regenerated shoots produced a maximum number (4.4 ± 0.2) of roots per shoot by a two-step culture procedure employing pulse treatment and subsequent transfer of treated shoots to low concentration of 0.004 mg/l IBA along with 0.04 mg/l Phloroglucinol.

Husain *et al.* (2008) standardized a protocol for *in vitro* propagation of *Pterocarpus marsupium*, nodal segments of *in vitro* derived seedling were used as explants. Highest number of multiple shoots and length was obtained from MS medium supplemented with 4.0 μM BA, 0.5μM IAA and 20μM Ads. Best rooting was induced in microshoots cultured on half strength MS semisolid basal medium, after dip treatment for 7 days in half MS liquid medium containing 100μM IBA and 15.84μMPG.

Saritha *et al.* (1988) reported differentiation of shoots from callus derived from shoot tip cultures of *Pterocarpus santalinus*. Lakshmi Sita *et al.* (1992) cultured shoot tips of aseptically raised seedlings and reported a maximum of eight shoots per explant. Anuradha and Pullaiah (1999a,b) reported healthy shoot formation in *Pterocarpus santalinus* cultured on B_5 medium by employing a combination of cytokinins, reported well expanded normal leaves in Red sanders (*Pterocarpus santalinus*). They extended their work for multiple shoot induction with a combination of cytokinins and obtained eight shoots per explant. Anuradha and Pullaiah (1999a, b) established efficient protocols for shoot multiplication *in vitro* for *Pterocarpus santalinus*. Factors responsible for rapid multiplication were screened at various levels. The highest shoot bud regeneration frequency (10-15 per cent) was achieved by culturing mesocotyl explants on B_5 medium fortified with 3mg/l BAP + 1mg/l NAA within a six week culture period. The regenerated shoots were rooted on half strength MS medium supplemented with IAA (Anuradha and Pullaiah, 1999a,b).

Prakash *et al.* (2006) developed a protocol for plantlets development of *Pterocarpus santalinus* by inoculating different explants on different concentrations of hormonal media. In the same species Arockiasamy (2000) inoculated detached cotyledons from *in vitro* germinated seedlings and cultured on MS medium + 0.1mg/1 NAA, 1 mg/1 Kn. These shoots were rooted on half MS medium +1 mg/1 IAA. Chand (2006) developed a protocol for *in vitro* clonal multiplication by using mature nodal explant of 10 years old tree. Rejeswari and Pailwal (2008) observed highest shoot multiplication rate and shoot length from cotyledonary node on MS medium with 0.05 mg/1 2-iP. For rooting microshoots were dipped in 0.085 mg/1 IAA solution.

Robinia Species

Successful shoot regeneration has been reported for black locust (*Robinia pseudoacacia*) from seedling-derived callus (Han and Keathley, 1989; Woo *et al.*, 1995a), callus derived from shoot cultures of mature trees (Han *et al.*, 1990), cambial tissues (Han *et al.*, 1993b, 1997), and from leaf disks (Davis and Keathley, 1985). Along with these, isolation and culture of protoplasts (Han and Keathley, 1988) and regeneration of black locust via somatic embryogenesis (Arriliga *et al.*, 1994; Merkle and Wiecko, 1989; Woo *et al.*, 1995b) have been reported. Micropropagation of black locust (reviewed in Davis and Keathley, 1992) has been achieved using axillary buds (Bargchi, 1987; Davis and Keathley, 1987) and both nodal segments and shoot tips (Bargchi, 1987; Chalupa, 1992).

Procedures have been developed by Barghchi (1987) for micropropagation of *Robinia pseudoacacia* using axillary bud explants. Axillary shoot multiplication was established in MS medium supplemented with 0.25-1.0 mg/1 BAP. The effect of NAA, 7-aza-indole (AZI), GA$_3$ and ABA on shoot growth and proliferation *in vitro* was also investigated. Rooting of shoots *in vitro* was achieved on half strength MS medium containing 0.5-1.0 mg/1 IBA and plantlets produced *in vitro* were established readily in soil compost.

Robinia ambigua var. *idahoensis* presumably originated from interspecific hybridization of *R. pseudoacacia* and *R. hispida*. Guo *et al.* (2006) developed an efficient protocol for micropropagation of *R. ambigua* by enhanced branching of axillary buds. The culture system consisted of sequential use of three media, namely the bud-induction medium (MS medium supplemented with 0.8-1.4 mg/1 BA, 0.05-0.08 mg/1 NAA and 0.07-0.1 mg/1 GA), elongation medium (MS medium added with 0.35-0.5 mg/1 BA, 0.05-0.08 mg/1 NAA and 0.07-0.1 mg/1 GA) and root induction medium (1/4 MS medium fortified with 1.7-2.5 mg/1 IAA and 0.05-0.5 mg/1 IBA). In addition they investigated the genetic stability (relative to the donor plant) of a sample of 41 morphologically normal plants randomly taken from ca. 13,000 micropropagated plants, by using the inter-simple sequence repeat (ISSR) marker with 32 selected primers. They found that of the 226 reproducible bands scored, 24 were polymorphic (10.62 per cent), thus pointing to the occurrence, though at a relatively low level compared with an earlier study on *R. pseudoacacia,* of genomic variation in those micropropagaed plants. Further sequencing on seven loci underlying the variation showed that two had significant homoogy to know or predicated plant genes. Chen *et al.* (1998) regenerated plantlets form root explants by suspension culture. Davis and Keathely (1987) excised buds from the stems of five dormant, mature (20-30 year old) *R. pseudoacacia* trees and placed on MS medium + 6-BAP in different concentrations. In all treatments, bud explants from two of the trees produced shoots which could be subcultured. Whole plants were obtained from cultures of these two trees. Explants from other two trees became vitrified or produced callus, respectively, when cultured on medium containing between 0.0007 mg/1 and 0.022 mg/1 6-BA, subculturable shoots were only obtained when the buds from those trees were cultured on medium containing 0.07 mg/1 BAP. No shoots cultures which could be subcultured were obtained from the tree used in these experiments.

The anthers of *Robinia hispida* were cultured by Cao *et al.* (2003) and the results showed that the highest induction rate of callus was 41.5 per cent when they were cultured on MS medium supplemented with 2,4-D 0.1 mg/1 and BA 3.0 mg/1 for 20 days. Green buds formed from the callus after 2 months subcultured on MS medium supplemented with BA 5.0 mg/1. When the shoots grew to 2-3 cm long, they were detached and transferred to MS medium supplemented with IBA 1.0 mg/1 for rooting. The shoots produced roots within 2 weeks of culture and formed complete plantlets.

Han *et al.* (1997) studied the *in vitro* response of cambial tissue and dormant vegetative buds obtained from top of epicormic branches of mature *Robinia pseudoacacia* trees. Cambial tissues isolated from epicormic branches produced more callus from cambial tissues isolated from top branches, whereas *in vitro* shoot cultures derived from buds excised from epicormic branches.

Naujoks *et al.* (2000) observed the survival and growth of micropropagated black locust (*Robinia pseudoacacia*) plantlets during transfer to soil was improved by an early inoculation of rooted shoots *in vitro* by different nodule forming bacteria strains isolated from older black locust tree.

Luan and Luo (2002) cultured shoot segments of *Robinia pseudoacacia* f. *decaisneana*. The best medium for initial culture was SH medium supplemented with 0.5mg/1 BA + 0.05 mg/1 NAA, with a propagation coefficient of 4.1 (per microcutting in a month) and for subculture it was B_5 + 0.5 mg/1 BA + 0.05 mg/1 NAA + 10 mg/1 Glu., with a propagation coefficient of 4.7. The best rooting medium was ½ MS medium + 0.5 mg/1 NAA + 10 mg/1 Glu., with a rooting rate of 84.4 per cent.

Kanwar *et al.* (2008) described a method for plant regeneration in *Robinia pseudoacacia* from cell suspension culture. Non regenerative friable callus from hypocotyls and cotyledon explants from *in vitro* raised seedling induced on solid medium supplemented with 0.05 mg dm^{-3} 2,4-D was used for initiation of cell suspension cultures on same MS medium but without agar. Single cells were isolated after 3 d and the optimum cell density was $1-3 \times 10^4$ cells for cm^3 of the liquid MS medium. Planting efficiency was 29.6 per cent and callus formed with in 4 weeks was subcultured and transferred to solid MS medium supplemented with 0.6 mg dm^{-3} BA along with 0.05 mg dm^{-3} NAA for the induction of adventitious bud primordial. The shoots developed were isolated and re-cultured on MS medium containing 0.6 mg^{-3} BA. These microshoots after dipping in 1-2 cm^3 of 10 mg dm^{-3} IBA for 24 h in dark were cultured on half strength solid MS medium supplemented with 0.05 per cent charcoal and showed 80-82 per cent of rooting within 4 weeks.

Samanea saman

In *Samanea saman* (Vinolya Kumari and Pullaiah, 2003) cotyledonary node inoculated on MS medium + 2 mg/1 BAP + 0.5mg/1 NAA highest number of multiple shoots were produced. The regenerated shoots rooted on MS medium + 2 mg/1 IBA or combination with 0.5mg/1 BAP, 60 per cent of plantlets survived in field.

Saraca asoca

Shahid *et al.* (2007) observed antibacterial activity of aerial parts as well as *in vitro* raised calli of the medicinal plant *Saraca asoca*. Alcoholic extracts of all the explants and calli showed better result. Extract derived from calli showed comparable results to the extracts from explants. This is the first report on antibacterial activity of *S. asoca*, especially through *in vitro* raised calli. Rama Subbu *et al.* (2008) observed highest frequency of shoot organogenesis from nodal segments treated with 0.5 mg/ 1 of BAP with a mean of 11.71 + 0.53 adventitious shoot. The microshoots were well rooted on MS medium supplemented with 4 mg/1 of IAA.

Sesbania Species

Srivastava and Aggarwal (2007) induced photoautotrophy in regenerants of *Sesbania aculeata*. Reentrants raised from seedling explants were sequentially transferred to liquid medium containing decreasing concentration of sucrose. On sucrose free medium the plantlets underwent an initial withering followed by axillary proliferation indicating the activation of autotrophic metabolism. It reduced infection rate during hardening.

Multiple shoots differentiated from hypocotyls explants of *Sesbania bispinosa* when cultured on Gamborg's medium alone or in combination with BAP (10^{-7} – 10^{-4}M) (Kapoor and Gupta, 1986). For cotyledonary explants BAP (10^{-6} – 10^{-4}M) was necessary. The shoots rooted when cultured on Gamborg's basal medium containing IBA (10^{-5} M). Plantlets thus formed were transferred to soil (Kapoor and Gupta, 1986).

Protoplasts were isolated from cotyledons of *Sesbania bispinosa* in a liquid-over-agar culture system with MS medium supplemented with 1 mg/1 2,4-D, 2mg/1 BA, 1mg/1 glutamine and 0.5 and formed callus. The first division occurred after 3-4 days. Callus formed from protoplasts differentiated shoots by organogenesis on MS medium with 1 mg/1 IBA and 1 mg/1 BA. These shoots developed into complete plantlets when excised and cultured on MS medium with 0.5mg/1 IBA.

Root, hypocotyls and cotyledon explants of *Sesbania bispinosa, Sesbania cannabina, Sesbania formosa* and *Sesbania sesban* were cultured on MS medium with BA (2.22, 4.44, 8.88 µM) in combination with 2,4-D (2.26, 4.52, 9.05 µM), IBA (0.25, 0.49, 4.92 µM) or NAA (2.69, 5.37, 10.74 µM) (Zhao *et al.*, 1993). Although all the explant types developed some callus, callus occurred earliest and continued to grow fastest with hypocotyls. Media including 2,4-D or NAA gave the fastest growing callus. Shoots regenerated readily from both hypocotyls or cotyledons but not from roots. Shoot organogenesis was most frequent with IBA (0.25-4.92 µM) in combination with BA (4.44 – 8.88 µM) and did not occur with 2,4-D. Shoots that differentiated were excised and cultured on MS medium without growth regulators or with IBA (2.46, 4.92, 9.84 µM). Roots developed after 3-8 days on appropriate rooting medium, often without IBA. Rooted plantlets were transplanted to pots in a green house (Zhao *et al.*, 1993).

In vitro regeneration of *Sesbania drummondii* was achieved by Cheepala *et al.* (2004). The nodal segments isolated from seedlings and young plants proliferated into multiple shoots on MS medium + 0.48mg/1 BA. MS medium + 0.048 and 0.099 mg/1 TDZ induced 5-6 shoots per node from 3-months – old plants. Callus induced on cotyledonary explant when subcultured on 0.048 mg/1 TDZ containing medium resulted in its mass proliferation having numerous embryiod like structures. 0.004 – 0.049 mg/1 IBA was found suitable for root induction.

In *Sesbaia grandiflora* Detrez *et al.* (1994) identified the age and the light conditions to the explants were critial factors for both bud induction and bud elongation. Additional sites of regeneration were obtained after wounding on the epidermal surface of explants, suggesting a large distribution of regenerative cells all along the explants.

Shoot formation in calli of *Sesbania rostrata* was achieved by Pellegrineschi and Tepfer (1993) by varying the nutrient composition of the medium, the photoperid and the amount and the source of plant hormones. Induction of adventitious shoots from the hypocotyls explants of 12-15 days old seedlings of *Sesbania rostrata* by Jha *et al.* (2002) on Nitsch medium + 1.0mg/1 BA. A maximum of 5.9 + 3.4 shoots per explant in 100 per cent of cultures were obtained. Sucrose at 3 per cent exhibited the development of the maximum of 3.5 + 0.9 shoots per explants with an average shoot length of 4.7 + 3.9 cm. The shoot development in all cases was accompanied by the development of moderate to profuse callus at the basal cut end of the explants. The *in vitro* regenerated shoots produced roots when transferred to half MS media + 1 mg/1 IBA + 3 per cent sucrose. Field transferred plants produced flowers, fruits and exhibited the development of prominent and organized stem nodules.

In *Sesbaia rostrata* Jha *et al.* (2004b) observed multiple shoots from the cotyledonary nodes cultured on Nitsch medium supplemented with 1mg dm^{-3}. BA proved to be the best, eliciting 5.8 ±1.0 shoots per explant in 100 per cent cultures. The elongation of shoots was best at 2 mg dm^{-3} BA. Following the repeated harvesting on same media of shoots an average of 33 shoots produced roots when transferred to half MS medium + 1mg dm^{-3} IBA.

Subhan *et al.* (1998) carried out investigations using the vesicular arbuscular mycorrhizal fungus *Glomus fasciculatum* to improve the success in transplanting micropropagated plantlets of *Sesbania sesban*. Plantlets were developed from somatic embryos and/or adventitious buds (induced from various explants on Gamborg's medium supplemented with BAP), in the presence of 10^{-7} M NAA and 5×10^{-6} M GA$_3$. The observations showed that mycorrhizal association helped to increase the potential of micropropagated plantlets to successfully withstand transplantation shock.

The nodal and internodal explants from the orthotropic shoots of *Sesbania sesban* var. *bicolor* elicited the development of shoots directly from the explants as well as via an intervening callus phase on Nitsch medium (Jha *et al.*, 2004a). Nodal explants on Nitsch medium supplemented with 1.5 mg dm⁻³ Kn developed an average of 12.5 shoots per explant in 100 per cent cultures, while internodal explants induced an average of 11.6 shoots per explant in 75 per cent explants at 0.5 mg dm⁻³ Kn. The *in vitro* regenerated shoots developed roots when implanted on Nitsch medium supplemented with 2 mg dm⁻³ IAA, after 30 days of inoculation (Jha *et al.*, 2004a).

Sophora Species

Zhao *et al.* (2003) developed a micropropagating system based on young stem node segments of *Sophora flavescens*. MS basal medium supplemented with 0.88 µM BA plus 2.69 µM NAA and that with only 5.37 µM NAA were found the best in promoting proliferation of shoots and induction of root respectively. The segments of the regenerated shoots could be continuously induced to reproduce new shoots through subculture on the same medium in 30-d intervals and still kept this activity after being subcultured for 6 generations.

Iturriage *et al.* (1994) developed a protocol for plantlet formation of an endangered species *Sophora toromiro* from 3-4 months old seedlings. A range of NAA and BA concentration induced root formation in nodal segment explant, developing plantlets, and also promoted axillary bud development. In subculture, nodal sections derived from axillary growth initiated multiple shoot formation and roots in a liquid medium leading to plantlet formation. Jordan *et al.* (2001) cultured different parts of 20 years old tree explants. Embryonic shoot tips were the only explants capable of regenerating plants. They developed rapidly *in vitro* in the presence of NAA and BA while in subculture roots were induced at the proximal end in the presence of 0.009 mg/1 IBA with in 40-60 days. Isolated cotyledons cultured produced callus only, while axillary buds and leaves did not show any response in the presence of several growth regulators assayed. Inoculation of seedling with various strains of rhizobia under *in vitro* conditions resulted in root outgrowths, but not in nodules that are typical of rhizobia infection.

Swartzia madagascariensis

Shoot induction frequency for the leguminous tree *Swartzia madagascariensis* was higher on MS and WP media than on B₅ (Berger and Schaffner, 1995). Explants incubated on media solidified with agar produced more shoots with a lower tendency to hyperhydricity than explants on agarose or Gelrite media. Maximum shoot production was obtained with an agar-solidified MS medium containing 2.2 µM BA (37 shoots/explant). Shoots rooted after transfer to half-strength MS medium supplemented with 26.8 µM NAA.

Tamarindus indica

Optimal culture conditions for high frequency plant regenerations from excised cotyledons of *Tamarindus indica* were established by Jaiwal and Gulati (1991). Maximum shoot bud differentiation (100 per cent) occurred when the adaxial surface of the entire cotyledon (excised from 12-d old seedlings) was in contact with medium containing 5 x 10⁻⁶ M BAP. Roots were formed on MS basal medium. Shoot or root formation was confined to nodal tissue at the top of the notch present on the adaxial surface at the proximal end of the cotyledon. 35-95 shoots were regenerated in a 4 month period from individual cotyledons. Shoots were rooted on MS medium + 5.7 x 10^{-6} m IAA. IAA (5.7×10^{-7} m) alone induced complete plant formation. Regenerated plants were established in the soil with 70 per cent success.

In *Tamarindus indica* Jaiwal *et al.* (1998) observed direct regeneration of shoot buds from hypocotyls segments of 12-d-old *in vitro* grown seedlings. Highest shoot regeneration (66.6 per cent) and the maximum number of shoots (3-4) per explant were obtained on MS medium + BAP (5×10^{-6} M). Best rooting was observed on IBA (5×10^{-6} M). Plantlets were hardened and transferred to soil.

In vitro induction of multiple shoots from axillary buds of the tamarind genotype 'Urigam' was given by Balakrishnamurthy and Ganga (1997). Culturing of axillary buds in MS medium supplemented with 5.0 mg/l BA and 0.5 mg/l GA recorded the highest per cent bud break. The regenerated microshoots were rooted on half strength MS medium 0.5 mg/l each of IAA and IBA (Ganga and Balakrishnamurthy, 1997).

Germination of tamarind seeds in medium containing TDZ resulted in induction of nodular protrusions in and around cotyledonary node meristem (Mehta *et al.*, 2005). The structures developed radially in well defined circles and subsequently spread towards the cotyledonary bridge and also in the proximal part of the hypocotyls. The structures developed into shoots on transfer to medium devoid of growth regulators.

A protocol for *in vitro* shoot regeneration was developed by Mohammed-Yasseen *et al.* (1992) from aseptically germinated seedlings of *Tamarindus indica*. Shoot tips, stem nodes, cotyledonary nodes and cotyledons were cultured on MS media supplemented with combination of TDZ and IAA and IBA or BA and NAA. Multiple shoots were obtained after four weeks culture under long day period.

References

Agarwal, V. and Sardar, P.R. 2006. Propagation of *Cassia angustifolia* through leaflet and cotyledon derived calli. Biologia Plantarum. 50: 233-238. DOI: 10.1007/s10535-005-0084-8

Agarwal, V. and Sardar, P.R. 2007. *In vitro* regeneration through somatic embryogenesis and organogenesis using cotyledons of *Cassia angustifolia* Vahl. *In vitro* Cell Dev. Biol. – Plant 43(6): 585-592. DOI: 10.1007/s11627-007-9058-1.

Ahmad, D.H. 1989. Micropropagation of *Acacia mangium* from aseptically germinated seedlings. J. Trop. For. Sci. 3(3): 204-208.

Ahee, J. and Duhoux. 1994. Root culturing of *Faidherbia = Acacia albida* as a source of shoot regeneration. Plant Cell Tiss. Org. Cult. 36: 219-225. DOI: 10.1007/BF00037723.

Ahuja, A., Samyal, M. and Kausik, J. P. 1991. Regulation of anthraquinone production by nutritional and hormonal factors in *Cassia fistula* callus cultures. Fitoterapia. 62: 205-214.

Ahuja, M.R. 1987. Biotechnology of forest trees. Plant Res. Development 33: 106-120.

Ahuja, M.R. 1993. Micropropagation of woody plants. Springer, pp. 507.

Al-Wasel, A.S. 2000. Micropropagation of *Acacia seyal* Del. *in vitro*. J. Arid Environ 46(4): 425-431.

Anand, M. and Bir, S.S. 1984. Organogenetic differentation in tissue cultures of *Dalbergia lanceolaria*. Curr.Sci. 53: 1305-1307.

Anand, M. and Bir, S.S. 1986. *In vitro* regeneration of plantlets from seedling explants of *Albizia procera*. J. Plant Sci. 2:25-28.

Anis, M., Husain, M.K. and Shahzad, A. 2006. *In vitro* control of shoot tip necrosis (STN) in *Pterocarpus marsupium* Roxb. – a leguminous tree. Physiol. Mol. Biol. Plants 12: 259-262.

Anis, M., Husain, M.K. and Shahzad, A. 2007. *In vitro* propagation of Indian kino (*Pterocarpus marsupium* Roxb.) using thiadiazuron. *In vitro* Cell Dev. Biol. Plant 43: 59-64.

Anis, M., Husain, M.K. and Shahzad, A. 2008. *In vitro* propagation of a multipurpose leguminous tree (*Pterocarpus marsupium* Roxb.) using nodal explants. Acta Physiologiae Plantarum 30: 353-359.

Anis, M., Kashif, H.M. and Anwar, S. 2005. *In vitro* plantlet regeneration of *Pterocarpus marsupium* Roxb., an endangered leguminous tree. Curr. Sci. 88: 861-863.

Anis, M., Shahzad, A. and Ahmad, N. 2006. An improved method for organogenesis from cotyledon callus of *Acacia sinuata* (Lour.) Merr. using thidiazuron. J. Plant Biotech. 8: 1-5.

Anis, M. and Siddiqui, I. 2007a. High frequency multiple shoot regeneration and plantlet formation in *Cassia angustifolia* Vahl using thiadiazuron. Medicinal and Aromatic Plant Science and Biotechnology 1: 282-284.

Anis, M. and Siddiqui, I. 2007b. *In vitro* shoot multiplication and plantlet regeneration from nodal explants of *Cassia angustifolia* Vahl. – a medicinal plant. Acta Physiologiae Plantarum 29: 333-338.

Anuradha, M. 1995. Investigations on *in vitro* cultures of *Pterocarpus santalinus* L.f., *P. marsupium* Roxb. and *Hardwickia binata* Roxb. Ph.D. thesis, S.K.University, Anantapur.

Anuradha, M. and Pullaiah, T. 1996. *In vitro* shoot regeneration in *Hardwickia binata*. In: Irfan A. Khan (ed.) Frontiers in Plant Science pp 273-277.

Anuradha, M. and Pullaiah, T. 1999a. Propagation studies of Red Sanders (*Pterocarpus santalinus* L.f.) *in vitro* – An endangered taxon of Andhra Pradesh, India. Taiwania 44: 311-324.

Anuradha, M. and Pullaiah, T. 1999b. *In vitro* seed culture and induction of enhanced axillary branching in *Pterocarpus santalinus* and *P. marsupium* a method for rapid multiplication. Phytomorphology 49:157-163.

Anuradha, M., Kavi Kishor, P. and Pullaiah, T. 2000. *In vitro* propagation of *Hardwickia binata*. Roxb. J. Indian Bot. Soc. 79: 127-131.

Anuradha, T. and Pullaiah, T. 2001. *In vitro* germination studies on endangered tree taxon *Albizia thompsonii* Brandis. Proc. of A.P. Academy of Sciences. 5:137-140.

Arce, P. and Balboa, O. 1991. Seasonality in rooting of *Prosopis chilensis* cutting and *in vitro* micropropagation. For. Ecol. Manage. 40: 163-174.

Arockiasamy, S., Ignacimuthu, S. and Melchias, G., 2000. Influence of growth regulators and explant type on *in vitro* shoot propagation and rooting of red sandal wood (*Pterocarpus santalinus* L.). Indian J. Exp. Biol. 38: 1270 –1273.

Arrillaga, I. and Merkle, S.A. 1993. Regeneration plants from *in vitro* culture of black locust cotyledon and leaf explants. *Hort. Science.* 28: 942-943.

Arrillaga, I., Tobolski, J.J. and Merkle, S.A. 1994. Advances in somatic embryogenesis and plant production of black locust (*Robinia pseudoacacia* L.). Plant Cell Rep. 13: 171-175.

Arya, H.C. and Shekhawat, N.S. 1986. Clonal multiplication of tree species in the Thar desert through tissue culture. For. Ecol. Manage. 16: 201-208.

Babber, S. Kiran, Vinod, S. and Varghese, T. M. 1996. Micropropagation of *Cassia roxburghii* DC through cultural technique. J. Indian Bot. Soc. 75: 263-266.

Badji, S., Marione, Y., Ndiaye, I., Merline, G., Danthu, P., Neville, P. and Colonna, J.P. 1993. *In vitro* propagation of the gum arabic tree (*Acacia senegal* (L.) Willd.). Developing a rapid method for producing plants. Plant Cell Rep. 12: 629-633.

Bajaj, Y.P.S. 1997. High-tech and Micropropagation. In: Biotechnology in Agriculture and Forestry. Vol. 389. p. 395

Bajaj, Y.P.S. and Dhanju. 1983. Pollen embryogenesis in three ornamental trees – *Cassia fistula, Jacaranda acutifolia* and *Poinciana regia.* J. Tree. Sci. 2: 16-19.

Balakrishnamurthy, G. and Ganga, M. 1997. *In vitro* induction of multiple shoots from axillary buds of the tamarind genotype 'Urigm'. In: Proceedings of National Symposium on *Tamarindus indica* L. A.P.Forest Department, India pp.110-112.

Bandyopadhyay, S. and Ghosh, P.D. 1984. Tissue culture studies of leguminous trees with particular reference to *Leucaena glauca*. Appl. Biotech. Med. Aromatic Timber Plants. 203-210.

Banik, R.L. and Islam, S.A.M.N. 1996. *In vitro* clonal propagation of hybrid *Acacia* (*A. auriculiformis* X *A. mangium*). Bangladesh J. For. Sci. 25: 1-7.

Bansal, Y.K. and Pandey, S. 1993. Micropropagation of *Sesbania aculeata* Pers. by adventitious organogenesis. Plant Cell. Tiss. Org. Cult. 32: 351-355.

Barakat, M.N. and El-Lakany, M.H. 1992. Clonal propagation of *Acacia saligna* by shoot tip culture. Euphytica 59(2-3): 103-107. DOI: 10.1007/BF 00041260.

Barghchi, M. 1987. Mass clonal propagation *in vitro* of *Robinia pseudoacacia* (black locust) Cv. Jaszkiseri. Plant Sci. 53:183-189. DOI: 10.1016/0168-9452(87) 90129-4

Bari, M.A., Ferdaus, K.M.K.B., Hossain, M.J. 2008. Callus induction and plantlet regeneration from *in vitro* nodal and internodal segments and shoot tip of *Dalbergia sisssoo* Roxb. J. Bio-Sci. 16: 41-48. DOI: 10.3329/jbs.v.16i).3740.

Bassan, J.S., Reiniger, L.R.S., Rocha, B.H.G., Severo, C.R.P. and Flores, A.V. 2006. Phenolic oxidation, type of explant and nutritive media in *Peltophorum dubium* (Spreng.) Taub. *in vitro* establishment. Ciencia Florestal 16(4): 381-390.

Batchelor, C.A., Yao, D., Koehler, M. J. and Harris, P.J.C. 1989. Propagation of *Prosopis* species (*P. cineraria* and *P. juliflora*). Ann. Sci. 46(suppl): 110-112.

Beck, S.L. and Dunlop, R. 1999. Vegetative propagation of black wattle (*Acacia mearnsii* de Willd.). ICFR Bull. 8: 15.

Beck, S.L. and Dunlop, R. 2001. Micropropagation of *Acacia* species – A review. *In Vitro* Cell Dev. Biol. Plant 37(5): 531-538.

Beck, S.L., Dunlop, R. and Van Staden, J. 1998a. Micropropagation of *Acacia mearnsii* from *ex vitro* material. Plant Growth Reg. 26: 143-148.

Beck, S.L., Dunlop, R. and Van Staden, J. 1998b. Rejuvenation and micropropagation of adult *Acacia mearnsii* using coppice material. Plant Growth Reg. 26: 149-153.

Beck, S.L., Dunlop, R. and Van Staden, J. 2000. Meristem culture of *Acacia mearnsii*. Plant Growth Reg. 32(1): 49-58. DOI: 10.1023/A: 1006304826910.

Benneth, L.W., Miller, G.W., Yu, M.H. and Lykin, R.I. 1983. Production of flouroacetate by callus tissue from leaves of *Acacia georginae*. Flouride. 16: 111-117.

Berger, K. and Schnaffner, W. 1995. *In vitro* propagation of leguminous tree *Swartzia madagascariensis*. Plant Cell Tiss. Org. Cult. 40(3): 289-291. DOI: 10.1007/BF00048136.

Berrios, A., Sandoval, J. and Muller, L.E. 1991. Clonal *in vitro* propagation of different species of *Erythrina*. Turrialba. 41: 607-614.

Bhargava, S., Upadhyaya, S., Garg, K. and Chandra, N. 1983. Differentiation of shoot buds in hypocotyl explants and callus cultures of some legumes. In: S.K. Sen and K.L. Giles, (eds.) Plant cell culture in crop improvements Plenum Press, New York. 22: 431-433.

Bhaskar, P. and Subhash, K. 1996. Micropropagation of *Acacia mangium* Willd. through nodal bud culture. Indian. J. Exp. Biol. 34: 590-591.

Bhattacharya, K., Senguta, L.K. and Shara, N.C. 1990. Effect of some growth regulators on callusing in *Dalbergia sissoo*. Bionature. 10:19-22.

Bhojwani, S.S., Benerjee, M. and Mukhopadhyay. 1988. Legumes improvement through tissue culture. Plant breeding. Gent. Engg. 233-268.

Blaydes, D.F. 1965. Interaction of kinetin and various inhibitors in the growth of soyabean tissue. Physiol. Plant. 19: 748-753.

Bon, M.C., Bonal, D., Goh, D. K. and Monteuuis, O. 1998. Influence of different macronutrient solutions and growth regulators on micropropagation of juvenile shoot *Acacia mangium* and *Paraserianthus falcataria* explants. Plant Cell Tiss. Org. Cult. 53: 171-177.

Bonga, J.M. and Von Aderkas, P. 1992. *In vitro* culture of Trees, volume 38. Kluwer Academic Publishers, Dordrecht.

Bonner, J. 1942. Culture of isolated roots of *Acacia melanoxylon*. Bull. Torrey. Bot. Cl. 69: 130-133.

Borges Junior, N., Sobrosa, R.C. and Martins-Corder, M.P. 2004. Multiplicacao *in vitro* de gemas axilares de acacia-negra (*Acacia mearnsii* de Willd.). Revista Arvore Vicosa 28(4): 493-498.

Buendia-Gonzalez, L., Orozco-Villafuerte, J., Cruz-Sosa, F., Chavez-Avila, V.M and Vorno-Carter, E.J. 2007. Clonal propagation of mesquite tree (*Prosopis laevigata* Humb. & Bonpl. ex Willd. M. C. Johnston). via cotyledonary nodes. *In vitro* Cell. Dev. Biol.- Plant. 43(3): 260-266. doi: 10.1007/s11627-007-9027-8

Burns, J.A. and A. Wetzstein. 1998. Embryogenic cultures of the leguminous tree *Albizia julibrissin* and recovery of plants. Plant cell tiss. Org. Cult. 54: 55-59.

Butcher, D.N. and Street, H.E. 1964. Excised root cultures. Bot. Rev. 30:513-586.

Cao, K.-Y., Wang, Z.-Z. and Zhao, Y.-P. 2003. Anther culture and plant regeneration of *Robinia hispida*. Acta Botanica Boreali-Occidentalia Sinica. DOI: cnki:ISSN: 1000-4025.0.2003-03-021.

Caro, L.A., Polci, P.A., Lindstorm, L.I., Echenique, C.V. and Hernandez, L.F. 2002. Micropropagation of *Prosopis chilensis* (Mol.) Stuntz. from young and mature plants. Biocell 26(1): 25-33..

Castillo de Meier, G. and Bovo, O.A. 2000. Plant regeneration from single-nodal stem explants of legume tree *Prosopis alba* Griseb. Biocell 24: 89-95.

Cerdas, L.V., Dufour, M. and Villalobos, V.M. 1997. *In vitro* propagation of *Pithecellobium saman* (Jacq.) Benth. Rain tree. *In vitro* Cell. Dev. Biol. Plant. 33:38-42.

Cerdas L. V., Dufour – Magali, Villalobos – Victor. 1998. *In vitro* organogenesis in *Albizia guachapele, Cedrela obovata* and *Swietenia macrophylla* (Fabaceae, Meliaceae), Revista-de Biological Tropical. 46:225-228.

Cerdas, L.V., Dufour, M. and Villalobos, V.M. 2004. *In vitro* organogenesis in *Dalbergia retusa* (Papilionaceae). I. Rev. Biol. Trop. 52: 41-46.

Cerdas, L.V., Rojas-Vargas, A. and Hine-Gomez, A. 2008. *In vitro* propagation of *Albizia guachapele, Cedrela odorata, Platymiscium pinnatum* and *Guaiacum sanctum*. Plant Tiss. Cult. Biotech.18(2): 151-156. DOI:10.3329/ptcb.v.18i2. 3397.

Chalupa, V. 1983. *In vitro* propagation of Willows (*Salix* spp.) europena moutain – ash (*Sorbus aucuparia* L.) and black locust (*Robinia pseudoscacia*). Biol. Plant. 25:305-307.

Chalupa, V. 1984. *In vitro* propagation of broad leaves forest trees. Plant cell Tiss. Org. Cult. Appl. Crop. Impr. Meet. 545-546.

Chalupa, V. 1992. Tissue culture propagation of black locust. In: Black Locust: Biology, Culture and Utilization. Proc. International Conference on Black Locust, June

17-21, 1991, East Lansing, MI, USA, pp. 115-125. (eds.) J.W.Hanover, K.Miller and S.Plesko). Michigan State University.

Chand, S. and Singh, A.K. 2004. Plant regeneration from encapsulated nodal segments of *Dalbergia sissoo* Roxb., a timber-yielding leguminous tree species. J. Plant Physiol. 161: 237-243.

Chauhan, V.A., Josekutty, P.C., Jasrai, Y.T. and Prathapasena, G. 1996. Rapid *in vitro* multiplication of *Dalbergia sissoo* Roxb. J. Plant Biochem. Biotechnol. 5: 117-118.

Cheepala, S.B., Sharma, N. C. and Sahi, S.V. 2004. Rapid *in vitro* regeneration of *Sesbania drummondii*. Biologia Plantarum. 48: 13-18. DOI: 10.1023/B: BIOP. 0000024269.72171.42.

Chen, R., Gyokusen, K. and Saito, A. 1998. Plantlets regeneration from root explants of *Robinia pseudoacacia* L. by suspension culture. Journal of Forest Research. 3(3): 161-165.

Correia, D. and Graca, M.E.C. 1995. *In vitro* propagation of black wattle (*Acacia mearnsii* de Willd.). IPEF Piracicaba 48/49: 117-125.

Crawford, D.F. and Hartney, V.J. 1986. Micropropagation of *Acacia mangium* and *A. stenophylla*. In: *Acacia* in developing countries. ACIAR Proc. No. 16. Proc. Workshop, Queenland, Australia, 64-65.

Darus, H.A. 1989. Micropropagation of *Acacia mangium* by stem cuttings from aseptically germinated seedlings. J. Trop. For. Sci. 3: 204-208.

Darus, H.A. 1991. Multiplication of *Acacia mangium* by stem cuttings and tissue culture techniques. Advances in Tropical *Acacia* Res. J.W. Turnbull (ed.). ACIAR Proceedings Int. Workshop, Bangkok, Thailand pp. 32-35.

Darus, H.A. 1992. Micropropagation techniques for *Acacia mangium* X *Acacia auriculiformis*. In: L.T.Carron and K.M.Aken (eds.). Proc. Int. Workshop on breeding technologies for tropical Acacias, held on 1-4 July, ACIAR, Malaysia 37: 119-121.

Das, A.B., Rout, G.R. and Das, P. 1995. *In vitro* somatic embryogenesis from callus cultures of the timber yielding tree *Hardwickia binata* Roxb. Plant Cell Rep. 15: 147-149.

Das, P. 2011. Somatic embryogenes in four legumes. Biotechnology Research International (in press). doi: 10.4061/2011/737636.

Das, P., Samantaray, S., Roberts, A.V. and Rout, G.R. 1997. *In vitro* somatic embryogenesis of *Dalbergia sissoo* Roxb. A multipurpose timber yielding tree. Plant. Cell. Rep. 16: 578-582. DOI: 10.1007/BF).1142327.

Das, P., Samantray, S. and Rout, G.R. 1996. *In vitro* propagation of *Acacia catechu*, a xerophilous tree. Pl. Tissue Cult. 6: 117-126.

Das, P.K., Chakravarthi, V. and Maity, S. 1993. Plantlet formation in tissue culture from cotyledon of *Acacia auriculiformis* A. Cunn. ex. Benth. Indian J. For. 16: 189-192.

Das, T. and Chatterjee, A. 1993. *In vitro* studies of *Pterocarpus marsupium* – An endangered tree. Indian J. Plant physiol. 36: 269-272.

Datta, K.and Datta, S.K. 1983. Auxin induced regeneration of forest tree *Dalbergia sissoo* Roxb. Curr. Sci. 52: 434-436.

Datta, S.K.., Datta, K. and Pramanik, T. 1983. *In vitro* clonal multiplication of mature trees of *Dalbergia sissoo* Roxb. Plant Cell Tissue Organ Cult. 2: 15-20.

Datta, K. and Datta, S. K. 1985. Auxin + KNO_3 induced regeneration of leguminous tree *Leucaena leucocephala* through tissue culture. Curr. Sci. 54: 248 – 250.

Dave, Y.S., Goyal, Y., Vaishnawa, G.R., Surana, N.M. and Arya, H.C. 1980. Clonal propagation of desert plants through tissue culture. Plantlet formation in *Acacia senegal* stem culture. 3rd All India Botanical Conference, Lucknow, 57-59 (Abstr.).

Davis, J.M. and Keathley, D.E. 1985. Regeneration of shoots from leaf disc explants of black locust, *Robinia pseudoacacia* L. In: 4th North Central Tree improvement conference, East lansing, M.I. 29-34.

Davis, J.M. and Keathley, D.E. 1987a. Differential response to *in vitro* culture in mature *Robinia pseudoacacia* (Black loust). Plant Cell Rep. 6: 431-434.

Davis, J.M. and Keathley, D.E. 1987b. Towards efficient propagation of mature locust trees using tissue culture. Nitrogen fixing Trees. Res. Rep. 5: 57-58.

Davis, J.M. and Keathley, D.E. 1988. *In vitro* propagation of black locust tree with an unusual phenotype. Nitrogen fixing tree res. Rep. 6: 57-67.

Davis, J.M. and Keathley, D.E. 1992. Micropropagation of black locust (*Robinia pseudoacacia* L.). In: Y.P.S. Bajaj (ed.), Biotechnology in agriculture and forestry. High-Tech and micropropagation 11 Springer – Verlag, Berlin Heidelberg. 18: 24-39.

Dawara, S., Sharma, D.R. and Choudary, J.V. 1984. Clonal propagation of *Dalbergia sissoo* Roxb. through tissue culture. Curr. Sci. 53: 807-809.

De, D.N. and Rao, P.V.L. 1983. Androgenic haploid callus of tropical leguminous trees. In: Sen, S.K. and Giles, K.L. (eds) Plant cell culture in crop improvement pp. 469 –474.

Detrez, C. 1994. Shoot production through cutting culture and micrografting from mature tree explants in *Acacia tortilis* (Forsk.) Hayne subsp. *raddiana* (Savi) Brenan. Agroforestry systems 25(3): 171-179. DOI: 10.1007/BF00707458.

Detrez, C., Ndiaye, S. and Dreyfus, B. 1994. *In vitro* micropropagation of the tropical multipurpose leguminous tree *Sesbania grandiflora* from cotyledonary explants. Plant Cell. Rep. 14: 87-93.

Dewan, A., Nanda, K. and Gupta, S.C. 1992. *In vitro* micropropagation of *Acacia nilotica* sub sp. *indica* via cotyledonary node. Plant Cell Rep. 12:18-21.

Dhabhai, K., Sharma, M.M. and Batra, A. 2010. *In vitro* clonal propagation of *Acacia nilotica* L. – A nitrogen fixing tree. Researcher 2(3): 7-11

Dhawan, V. 1993. Micropropagation of N$_2$ fixing trees. In: M.R.Ahuja (ed.) Micropropagation of Woody plants. Kluwer Acad. Publishers, Dordrecht.

Dhawan, V. and Bhojwani, S.S. 1984. Reduction in cost of tissue culture of *Leucaena leucocephla* (Lam.) De Wit replacing AR grade sucrose by sugar cubes. Curr. Sci. 53: 1159-1161.

Dhawan, V. and Bhojwani, S.S. 1985. *In vitro* vegetative propagation of *Leucaena leucocephala* (Lam.) De Wit. Plant. Cell. Rep. 4:315-318.

Dhawan, V. and Bhojwani, S.S. 1986. Transplantation of micropropagated plants of *Leucaena leucocephala* (Lam.) De Wit. Int. Congr. Plant. Tiss Cell. Cult. 6 meet. Minnesota, 113.

Dhawan, V. and Bhojwani, S.S. 1987. Transplantation of micropropagated plants of *Leucaena leucocephala*. Proc. Indian Natl. Sci. Acad. 53: 351.

Dhawan, V. and Bhojwani, S.S. 1988. *In vitro* nodulation of micropropagated plants of *Leucaena leucocephala* with *Rhizobium*. Basic Life Sci. 4: 464.

Di, Michele, M.N. and Bray, L. 1994. Multiplication vegetative *in vitro* d'*Acacia flava* syn. *Ehrenbergiana*. In: Quel avenir pour l'amelioration des plantes. AUPELF-URF, pp 195-204. Paris. John Libby Eurotext, Paris.

Douglas, G.C. and McNamara, J. 2000. Shoot regeneration from seedling explants of *Acacia mangium* Willd. *In vitro* Cell Dev. Biol. Plant 36: 412-415.

Duhoux, E. and Davies, D. 1985. Shoot production from cotyledonary buds of *Acacia albida* (Synonym *Fadherbia albida*) and influence of sucrose on rhizogenesis. J. Plant Physiol. 121: 175-180.

Duhoux, E., Galiana, A., Ahee, J. and Franche, C. 1998. Applications des cultures *in vitro* daus le genere *Acacia*. In: L'acacia au Senegal Claudine Campa. Claude Grigon Mamalo Guaye Serge. Hammon ed. Paris.

Dun-Yi, Y., Batchelor, C.A., Koehler, M.J. and Harris, P.J.C. 1989. *In vitro* regeneration of *Prosopis* species (*P. chilensis*, *P. cineraria* and *P. juliflora*) from nodal explants. Chinese J. Bot. 1(2): 89-97.

Dwari, M. and Chand, P.K. 1996. Evaluation of explants growth regulators and culture passage for enhanced callus induction, proliferation and plant regeneration in the tree legume, *Dalbergia lanceolaria* L., Phytomorphology. 46: 123-131.

Galiana, A., Alabarce, J. and Duhoux, E. 1990a. *In vitro* nodulation *Acacia mangium* Willd. (Leguminosae). Ann. Sci. For. 47: 451-460.

Galiana, A., Chaumont, J., Diem, H.G. and Dommergues, Y.R. 1990b. *In vitro* assessment of N$_2$ fixation potential of *Acacia mangium* and *Acacia auriculiformis*. Biol. Fertil. Soils, 9: 261-267.

Galiana, A., Goh, D., Chevallier, M.H., Gidiman, J., Moo, H., Hattah, M. and Japarudin, Y. 2002. Micropropagation of *Acacia mangium* X *A. auriculiformis* hybrids in Sabah. Bois Forets Tropiques. pp. 77-82.

Galiana, A., Tibok, A. and Duhoux, E. 1991a. *In vitro* propagation of nitrogen fixing tree legume, *Acacia mangium* Willd. Plant and Soil. 135: 151-160.

Galiana, A., Tibok, A. and Duhoux, E. 1991b. N_2 fixing potential of micropropagated clones of *Acacia mangium* with different *Bradyrhizobium* spp., strains. Plant and Soil. 135: 161-166.

Ganga, M. and Balakrishnamurthy, G. 1997. *In vitro* rhizogenesis of axillary bud derived microshoots of the tamarind genotype 'Urigm'. In: Proceedings of National Symposium on *Tamarindus indica* L. A.P.Forest Department, India pp.107-109.

Garg, L., Bhandari, N.N., Rani, V. and Bhojwani, S.S. 1996. Somatic embryogenesis and regeneration of triploid plants in endosperm cultures of *Acacia nilotica*. Plant Cell. Rep. 15: 55-58.

Gassama, Y.K. 1989. *In vitro* culture and improvement of symbiosis using mature *Acacia albida*. In: Proc. Regional seminar on Trees for development in sub-Saharan Africa held on 20-25 Feb, 1989. ICRAF House, Nairobi, Kenya, pp. 286-290.

Gassama, Y.K. and Duhoux, E. 1986-87. Micropropagation de l'*Acacia albida* Del. (Leguminosae) adulte. Bull. I.F. A.N. 46: 314-320.

Gassam-Dia, Y.K. and Duhoux, E. 1992. Regeneration de bourgeons apartir de culture de raciones d'*Acacia albida*. In: International Plants – Microorganisms IFS (International Foundation for Science). Dakar.

Gautheret, R.J. 1934. Culture du tissue Cambial. C.R. Hebd. Seances Acad. Sci. 198: 2195-2196.

Gharyal, P.K. and Maheshwari, S.C. 1981. *In vitro* differentiation of somatic embryoids in a leguminous tree, *Albizia lebbeck* L. Naturwissenschaften. 8: 379-380.

Gharyal, P.K. and Maheshwari, S.C. 1983a. *In vitro* differentiation of plantlets from tissue culture of *Albizia lebbeck* L. Plant Cell Tiss. Org. Cult. 2: 49-53. DOI: 10.1007/BF00033552.

Gharyal, P.K. and Maheshwari, S.C. 1983b. *In vitro* differentiation in explants from mature leguminous trees. Plant Cell Rep. 8: 550-553.

Gharyal, P.K. and Maheshwari, S.C. 1990. Differentiation in explants from mature leguminous trees. Plant Cell Rep. 8: 550-553.

Gharyal, P.K., Rashid, A. and Maheshwari, S.C. 1983a. Production of haploid plantlets in anther cultures of *Albizia lebbeck* L. Plant Cell Rep. 2: 308-309.

Gharyal, P.K., Rashid, A. and Maheshwari, S.C. 1983a. Androgenic response from cultured anthers of leguminous tree. *Cassia siamea* Lam. Protoplasma. 118: 91.

Glovak, L. and Greatbatch, W. 1992. Successful tissue culture of *Leucaena*. *Leucaena* Res. Rep. 3: 81-82.

Gong, Z., Khayri, A.I. and Huang, F.H. 1991. *Acacia mearnsii* tissue culture, callus inducton from hypocotyl. Tree cult. Prot. Util. 105.

Goyal, Y. and Arya, H.C. 1981. Differentiation in cultures of *Prosopis cineraria* Linn. Curr. Sci. 50: 468-469.

Goyal, Y. and Arya, H.C. 1983. Tissue cultures of desert trees – 1. Clonal multiplication of *Prosopis cineraria* by bud cultures. J. Plant Physiol. 115: 183-189.

Goyal, Y. and Arya, H.C. 1984. Effects of sugars, nitrogen, amino acids and vitamins on shoot differentiation from single bud *in vitro* culture of *Prosopis cineraria*. Indian J. Exp. Biol. 22: 592-595.

Goyal, Y., Bingham, R.L. and Felker, P. 1985. Propagation of tropical tree *Leucaena leucocephala* K 67 by *in vitro* bud cultures. Plant Cell. Tissl. Org. Cult. 4:3-10.

Goyal, Y., Felker, P. and Kleberg, C. 1984. Exploitating unique germplasm resources of *Leucaena leucocephala* K67 by single bud *in vitro* culture. *In vitro* Cell Dev. Biol. 20: 278-279.

Green, B., Tabone, T. and Felker, P. 1990. A comparison of amide and uriede nitrogen sources in tissue culture of tree legume *Prosopis alba* clone B$_2$ V$_{50}$. Plant Cell. Tiss. Org. Cult. 21: 83-96.

Gulati, A. and Jaiwal, P.K. 1996. Micropropagation of *Dalbergia sissoo* from nodal explants of mature trees. Biologia Plantarum 38(2): 169-175. DOI: 10.1007/BF02873840.

Guo, W., Li, Y., Gong, L., Li, F. Dong, Y. and Liu, B. 2006. Efficient micropropagation of *Robinia ambigua* var. *idahoensis* (Idaho Locust) and detection of genomic variation by ISSR markers. Plant Cell Tiss. Org. Cult. 84: 343-351.

Gupta, P, Patni, V., Kant, U. and Arya, H.C. 1994. *In vitro* multiple shoot formation from mature trees of *Acacia senegal* (L.) Willd. J. Indian Bot. Soc. 73:331-332.

Gupta, N., Jain, S.K. and Srivastava, P.S. 1996. Micropropagation of a multipurpose leguminous tree *Delonix regia*. Phytomorphology. 46: 267-276.

Hamzah, M.B., Alang, Z.C. and Salekan, J. 1989. *In vitro* propagation of *Acacia mangium* from young seedlings. In: A.N.Rao and Aziah Mohe, Y. (eds.) Proc. of the seminar on Tissue Culture of forest species.

Han, K.H. and Keathley, D.E. 1988. Isolation and culture of protoplasts from callus tissue of black locust (*Robinia pseudoacacia* L.). Nitrogen fixing Tree Res. Rep. 6: 68-70.

Han, K.H. and Keathley, D.E. 1989. Regeneration of whole plants from seedling-derived callus of black locust. Nitrogen fixing Tree Res. Rep., 7: 112-114.

Han, K.H., Davis, J.M. and Keathley, D.E. 1990. Differential response persist in shoot explants regenerated from callus of two mature black locust trees. Tree Physiol. 6; 235-240.

Han, K.H., Keathley, D.E., Davis, J.M. and Gordon, M.P., 1993a. Regeneration of transgenic woody legume (*Robinia pseudoacacia* L. black locust) and morphological alterations induced by *Agrobacterium rhizogenes*-mediated transformation. Plant Sci. 88: 149-157.

Han, K.H., Keathley, D.E. and Gordon, M.P. 1993b. Cambial tissue culture and subsequent shoot regeneration from mature black locust (*Robinia pseudoacacia* L.). Plant cell. Rep. 12: 185-188.

Han, K.-H. and Park, Y.G. 1999. Somatic embryogenesis in Black Locust (*Robinia pseudoacacia* L.). In: S.M.Jain, P.K.Gupta and R.J.Newton (eds.) Somatic embryogenesis in woody plants, Volume 5 pp 149-161.

Han, K.-.H., Shin, D.-I. and Keathley, D.E.1997. Tissue culture responses of explants taken from branch sources with different degrees of juvenility in mature black locusts (*Robinia pseudoacacia* L.). Tree Physiol 17: 671-675.

Hu, Q.J. and Han, Y.F. 1985. A study on induction of plantlets from mature leaves of *Robinia pseudoacacia*. Hereditas. 7: 20-21.

Huang, F.H., Khayri, J.M. and Gbur, E.E., 1994. Micropropagation of *Acacia mearnsii*. *In vitro* Dev. Biol. Plant. 30: 70-74.

Husain, M.K., Anis, M. and Shahzad, A. 2007. *In vitro* propagation of Indian Kino (*Pterocarpus marsupium* Roxb.) using Thidiazuron. *In vitro* Cell. Dev. Biol.-Plant. 43: 59-64.

Husain, M.K., Anis, M. and Shahzad, A. 2008. *In vitro* propagation of a multipurpose leguminous tree (*Pterocarpus marsupium* Roxb.) using nodal explants. Acta Physiologia Plantatum 30(3): 353-359.

Hussain, M., Ahmed, G, and Hussain, S.N. 1990. Rapid clonal propagation in *Sesbania cannabina* through *in vitro* shoot proliferation from shoot tip explants. Bangladesh J. Bot. 19: 183-188.

Hustache, G., Barnoud, F. and Joseleau, J.P. 1986. Callus formation and induction of cell suspension culture from *Acacia senegal*. Plant Cell Rep. 5: 365-367.

Ide, Y., Watanabe, Y., and Ikeda, H. 1994. Tissue culture of *Acacia auriculiformis* using the aseptically germinated seedlings. J. Japan. Forestry Soc. 76: 576-583.

Indravathi, G. 2005. *In vitro* propagation studies of *Albizia amara* Boiv. M.Phil. dissertation, S.K.University, Anantapur.

Indravathi, G. and Pullaiah, T. 2004. Micropropagation of *Albizia amara* National symposium on recent trends in plant sciences souvenir and abstracts poster presentation, Department of Botany, Sri Krishnadevaraya University, Anantapur. 29-31, October.

Iturriga, L., Jordon, M., Roveraro, C. and Goreu, A. 1994. *In vitro* culture of *Sophora toromiro* (Papilionaceae), an endangered species. Plant cell Tiss. Org. Cult. 37: 201-204.

Jain, S.M. and Haggman, H. 2007. Protocols for micropropagation of woody trees and fruits. Springer.

Jain, S.M. and Ishi, K. 2003. Micropropagation of woody trees and fruits. p. 840.

Jaiwal, P.K. and Gulati, A. 1991. *In vitro* high frequency plant regeneration of a tree legume, *Tamarindus indica* L. Plant Cell Rep. 10: 569-573.

Jaiwal, P.K. and Gulati, A. 1992. Micropropagation of *Tamarindus indica* L. from shoot tip and nodal explants. National Acad. Sci. Letters, India. 15: 63-67.

Jaiwal, P.K. and Gulati, A. and Dahiya, S. 1998. Direct organogenesis in hypocotyls cultures of *Tamarindus indica*. Biologia plantarum 41: 331-337. DOI: 10.1023/A:1001873.

Jha, A.K., Prakash, S., Jain, N., Nanda, K. and Gupta, S.C. 2002. Production of adventitious shoots and plantlets from the hypocotyls explants of *Sesbania rostrata* (Bremek. & Obrem). *In vitro* Cell. Dev. Biol.-Plant. 38: 430-434.

Jha, A.K., Prakash, S., Jain, N., Nanda, K. and Gupta, S.C. 2004a. Micropropagation of *Sesbania sesban* from the mature tree derived explants. Biologia Plantarum 47(1): 121-124. DOI: 10.1023/A: 1027349419134.

Jha, A.K., Prakash, S., Jain, N., Nanda, K. and Gupta, S.C. 2004b. Micropropagation of *Sesbania rostrata* from the cotyledonary node. Biologia plantarum 4: 289-292.

Jones, C. 1986. Getting started in micropropagation of Tasmanian Blackwood (*Acacia melanoxylon*). Combined proceedings Intern. Plant Propagators Soc. 36: 477-481.

Jones, C. and Smith, D. 1988. Effect of 6-benzylamonopurine and 1-naphthyl acetic acid on *in vitro* axillary bud development of *Acacia melanoxylon*. Combined Proc. Int. Plant Propagators Soc. 38: 389-393.

Jones, C., Smith, D., Gifford, F. and Nicholas, I. 1991. Field trial results of *Acacia melanoxylon* from tissue culture. Combined Proceedings Intern. Plant Propagators Soc. 41: 116-119.

Jones, T.C., Batchelor, C.A. and Harris, P.J.C. 1990. *In vitro* culture and propagation of *Acacia* sps., *Acacia bivenosa*, *A. holosericea*, *A.saligna* and *A. sclerosperma*. Int. Tree Crops J. 6: 183-192.

Jordan, M. 1987. *In vitro* cultures of *Prosopis* species. In: Bonga, J.M. and Durzan, D.J. (eds.) Cell and Tissue culture in Forestry. Vol.III.General principles of Biotechnology Martinus, Nijhoff, Dordrecht. pp.370-384.

Jordan, M. 1996. Metodes de propagacion biotecnologicos y convencionales de legumiosas de usos multiples para zonas aridas. In: J. Izquierdo, y G. Palomino (eds.). Tecnicas convencionales y biotecnologicas para la propagacion de plantas de zonas aridas. Oficina Regional de la FAO para la America Latina y el Caribe, Santiago, Chile, pp 111-150.

Jordan, M. 1998. *In vitro* regeneration of some leguminous multipurpose trees from seedling explants. Phyton 63: 249-256.

Jordan, M. and Balboa, O. 1985. *In vitro* regeneration of *Prosopis tamarugo* and *Prosopis chilensis* (Mol.) Stunz from nodal section. Garttenbauesenschaft. 50: 138-142.

Jordan, M., Pedraza, J. and Goreux, A. 1985. *In vitro* propagation studies of three *Prosopis* species (*P.alba, P.chilensis* and *P.tamaurgo*) through shoot tip culture. Garttenbausenschaft. 50: 265-267.

Jordan, M., Lorrain, M., Topia, A. and Roveraro, C. 2001. Regeneration of *Sophora toromiro* from seedling explants. Plant cell Tiss. Org. Cult. 66: 89-95.

Joshi, I., Bisht, P., Sharma, V.K. and Uniyal, D.P. 2003. Studies on effect of nutrient media for clonal propagation of superior phenotypes of *Dalbergia sissoo* Roxb. through tissue culture. Silvae Genetica 52 (3-4): 143-147

Kackar, N.L., Solnki, K.R., Sing, M. and Vyas, S.C. 1991. Micropropagation of *Prosopis cineraria*. Indian J. Exp. Biol. 29: 65-67.

Kackar, N.L., Vyas, S.C., Sing, M. and Solanki, K.R. 1992. *In vitro* regeneration of *Prosopis cineraria* (L.) Bruce using root explant. Indian J. Exp. Biol. 30: 429-430.

Kalavathi, P. 2003. *In vitro* propagation, phytochemical and antimicrobial studies of *Mundulea sericea* (Willd.) A. Cheval. Ph.D. thesis, S.K. University, Anantapur.

Kalavathi, P., Chakradhar, T. and Pullaiah, T. 2004. Shoot bud induction and somatic embryogenesis in callus cultures of *Mundulea sericea* (Willd) A. Cheval. Plant Cell Biotech. Mol. Biol.. 5: 149-154.

Kalimuthu, K. and Lakshmanan, K.K. 1994. Preliminary investigation on micropropagation of *Pterocarpus marsupium* Roxb. Indian J. For. 19: 192-195.

Kant, U. and Vidya, R. 1986. Effects of auxin on growth of insect induced gall tissues of *Prospis cineraria* (Linn.) Druce. Intl. Congr. Plant Tiss. Cell Cult. 6th Meet, 300.

Kanwar, K., Kaushal, B., Abrol, S. and Deepika, R. 2008. Plant regeneration in *Robinia pseudoacacia* from cell suspension cultures. Biologia Plantarum 52(1): 187-190. DOI: 10.1007/s10535-008-0042-3.

Kapoor, S. and Gupta, S. 1986. Rapid *in vitro* differentiation *of Sesbania bispinosa* plant, a leguminous shrub. Plant Cell Tiss. Org. Cult. 7: 263-268.

Kathjo, S. and Tewari, M.N. 1979. Development of the root from the cotyledonary callus of *Acacia senagal* Labd. J. Sci. Tech. P.B. Life Sci. 11: 84-85.

Kaur, K. 1997. Plantlet regeneration of *Acacia catechu* Willd. through tissue culture. J. Indian Bot. Soc. 76: 291-292.

Kaur, K., Gupta, P. and Kant, U. 1996. Micropropagation of *Acacia sengal* (L.) Willd. via cotyledonary nodes. J. Indian. Bot. Soc. 75: 175-178.

Kaur, K., Gupta, P., Verma, J.K. and Kant, U. 1998. *In vitro* propagation of *Acacia senegal* (L.) Willd. from mature nodal explants. Adv. Plant Sci. 11. 229-233.

Kaur, K. and Kant, U. 2000. Clonal propagation of *Acacia catechu* Willd. by shoot tip culture. Plant growth Regulation. 31: 143-145. DOI: 10.1023/A:1006362318265.

Kaur, K., Kant, U. and Arya, H.C. 1997. Organogenic differentiation in callus tissues of *Acacia senegal* (L.) Willd. J. Indian Bot. Soc. 76 : 55-57.

Kaur, K., Verma, B.B. and Kant, U. 1998. Plants obtained from the khair tree *Acacia catechu* Willd. using mature nodal segments. Plant Cell. Rep. 17:427-429.

Kesava Reddy, K. and Srivasuki, K.P. 1990. Vegetative propagation of Red sanders *Pterocarpus santalinus* L.f. Indian For. 116: 536-539.

Kesava Reddy, K. and Srivasuki, K.P. 1992. Biotechnological approach for tree improvement in Red Sanders. In: K. Kesava Reddy (ed.) Vegetative propagation and Biotechnologies for tree improvement. Natraj Publishers, Dehradun, pp. 141-146.

Khalifalla, M.M. and Dafalla, H.M. 2008. *In vitro* micropropagation and micrografting of gum Arabic tree [*Acacia senegal* (L.) Willd.]. Int. J. Sust. Crop. Prod. 3(1): 19-27.

Khatter, S. and Mohan Ram, H.Y. 1982. Organogenesis in the cultured tissues of *Sesbania sesban*, a leguminous shrub. Indian J. Exp. Biol. 20: 216-219.

Khattar, S. and Mohan Ram, H.Y. 1983. Organogenesis and plantlet formation in *Sesbania grandiflora* (L.) Pers. Indian J. Exp. Biol. 21: 251-256.

Kiran, S. 2000. Study on micropropagation of *Cassia roxburghii* DC through *in vitro* culture technique. M.Sc. thesis, CCS Haryana Agricultural University, Hissar.

Kirst, M. and Sepel, L.M.N. 1996. Micropropagacao de *Peltophorum dubium* (Sprengel) Taubert a partir de apices caulinares de plantulas. In: Simposio Sobre Ecossistemas Naturais du Mercosul. 1. Santa Maria pp. 141-146.

Kitani, S. and Yasutani, I. 1997. Callus cultures of *Acacia* species, initiation, growth optimization and organogenesis. Proc. Intern. Workshop BIO-REFOR, IUFRO/SPDC. Brisbane.

Kogal, F., Haagen Smit, A.J. and Erxleben, M. 1934. Uber cinneues Auxin ("Hetero auxin") aus Harn. Hoppe-Seyler-S. Z. Physiol. Chem. 288: 90-103.

Kopp, M.S. and Nataraja, K. 1990. *In vitro* plant regeneration from shoot tip cultures of *Tamarindus indica* L. Indian J. For. 13: 30-33.

Kopp, M.S and Nataraja, K. 1992. Regeneration of plantlets from excised nodal segments of *Tamarindus indica* L. My Forest. 28: 231-234.

Kosakun, T. and Phothipathama, K. 1986. Tissue culture of *Acacia mangium* Willd. National Forest Seminar, Bangkok.

Kumar, A. 1989. Silver nitrate promoted proliferation of *Dalbergia sissoo* Roxb. Callus. Indian J. Plant physiol. 32: 387-388.

Kumar, A. 1992. Micropropagation of a mature leguminous trees *Bauhinia purpurea*. Plant Cell Tiss. Org. Cult. 31: 257-259.

Kumar, A., Tandon, P. and Sharma, A. 1991. Morphogenic response of cultured cell of cambial origin of mature tree, *Dalbergia sissoo* Roxb. Plant Cell Rep. 9: 703-706.

Kumar, K.P. and Nair, G.M. 1990. *In vitro* plantlet regeneration from different explants of *Sesbania grandiflora* L. var. *coccinia*. In: Biotechnological approaches for upgradation of Agricultural and Horticultural crops. BARC, Bombay.

Kumar, K.P. and Nair, G.M. 1991. *In vitro* organogenesis and micropropagation of *Sesbania aculeata* Poir. Multidisciplinary Research Review. 1(1): 20-24.

Kumar, S., Agrawal, V. and Gupta, S.C. 2002. Somatic embryogenesis in the woody legume *Calliandra tweedi*. Plant Cell Tiss. Org. Cult. 71: 77-80.

Kumar, S., Sarkar, A.K. and Kunchi Kanan, C. 1998. Regeneration of plants from leaflet explants of tissue culture raised safed siris (*Albizia procera*). Plant Cell. Tiss. Org. Cult. 54:137-143.

Kumar, S. and Singh, N. 2009. Micropropagation of *Prosopis cineraria* (L.) Druce – A multipurpose desert tree. Researcher 1(3): 28-32.

Lai, J., Zhou, C., Ye, C., Huan, S., Wei, P., Chen, X and Chen, F. 2003. *In vitro* micropropagation of *Acacia crassicarpa*. J. Sichuan Univ. 2003–05.

Lakshmanan, P. and Taji, A. 2000. Somatic embryogenesis in leguminous plants. Plant Biol. 2: 136-148.

Lakshmi Devi, U.N.. 1994. *In vitro* differentiation of plantlets from seedling explants of *Peltophorum pterocarpum* (DC.) Heyne. M.Phil. dissertation, S.K.University, Anantapur.

Lakshmi Devi, U.N., Anuradha, M., Sreedevi, E. and Pullaiah, T. 1996. *In vitro* shoot regeneration from seedling explants of *Peltophorum pterocarpum* (DC.) Heyne. In: Irfan A. Khan (ed.). Frontiers in plant science. pp. 333 – 337.

Lakshmi Sita, G. 1996. Direct somatic embrygenesis from immature embryos of rose wood (*Dalbergia latifolia* Roxb.). *In vitro* Cell. Dev. Biol. Plant. 32:91.

Lakshmi Sita, G., Chattopadhyay, L.S. and Tejavathi, D.H. 1986. Plant regeneration from shoot callus of rose wood (*Dalbergia latifolia* Roxb.). Plant Cell. Rep. 5: 266 – 268.

Lakshmi Sita, G. and Chattopadhyay, S. 1986. Regulation of plantlet formation in *Dalbergia latifolia* and *D.sissoo*. Int. Congr. Plant. Tiss. Cell. Cult. 6th Meet, 40.

Lakshmi Sita, G. and Raghava Swamy, B.V. 1992. Application of cell and tissue culture technology for mass propagation of elite tree with special reference to rose wood (*Dalbergia latifolia* Roxb.). Indian For. 118: 36-47.

Lakshmi Sita and Raghava Swamy, B.V. 1993. Regeneration of plantlets from leaf disc cultures of rose wood. Control of leaf abscission and shoot tip necrosis. Plant. Sci. 88: 107-112.

Lakshmi Sita and Raghava Swamy, B.V. 1998. Application of biotechnology in forest trees – clonal multiplication of Sandalwood, rosewood, eucalyptus, teak and bamboos by tissue culture in India. In: S. Puri (ed.). Tree Improvement: Applied research and technology transfer. Science Publishers, Enfield, USA pp 233-248.

Lakshmi Sita, G., Sreenatha, K.S. and Sujatha, S. 1992. Plantlet production from shoot tip cultures of red sandal wood (*Pterocarpus santalinus* L.). Curr. Sci. 62: 532-534.

Li, Z., Zhijian, Z., Yingchang, Z., Lhua, Z. and Zheng, G. 1996. The nursing of *Acacia cincinnata* by tissue culture. Forestry Science and Technology of Guandong Province

Lin, Y. and Wanger, M.R. 1995. A novel protocol in micropropagation of new Mexico locust *Robinia neomexicana*. *In vitro* Cell. Dev. Biol. Plant. 31: 222.

Liu, W.N., Torisky, R.S., Mc. Allister, K.R., Avdiushko, S., Hildebrand, D., Collins, G.B. and Liu, W.N. 1996. Somatic embryo cycling; evaluation of novel transformation and assay system for seed specific gene expression in soybean. Plant Cell Tiss. Org. Cult. 47: 33-42.

Luan, Q.-S. and Luo, F.-X. 2002. Shoot tissue culture of *Robinia pseudoacacia* f. *decaisneana*. J. For. Res. 13(1): 51-55.

Mamun, A.N., Matin, M.N., Bari, M.A., Siddique, N.A., Sultan, R.S., Rahman, M.H. and Musa, A.S. 2004. Micropropagation of woody legume *Albizia lebbeck* through tissue culture. Pakistan J. Biol. Sci. 7: 1099-1103.

Marguerite, Q., da Silva, M.S., Martins, K.G. and de Oliveira, D.E. 2001. Multiplication of black wattle by microcuttings. Plant Cell Tiss. Org. Cult. 66: 199-205.

Mascarenhas, A.F., Nair, S., Kulkarni, N.M., Agarwal, D.C., Khuspe S.S. and Mehta, V.J. 1987. Tamarind cell and tissue culture in forestry. Vol. III, In : J.M. Bonga and Durzan, D.J. (eds.) General principles of Biotechnology. Martinus Nijhoff. Dordreht. pp.316-325.

Mathews, H., Rao, P.S. and Bhatia, C.R. 1988. Increased frequency of normal plant regeneration from crown gall callus of *Sesbania rostrata*. Curr. Sci. 57: 1301-1306.

Mathur, J. and Chandra, N. 1983. Induced regeneration in plantlets of *Acacia nilotica*. Curr. Sci. 52: 882-883.

Mathur, J. and Mukuntha Kumar, S. 1992. Micropropagation of *Bauhinia variegata* and *Pakinsonia aculeata* from nodal explants of mature Trees. Plant Cell. Tiss. Org. Cult. 28: 119-121.

Mayer, H.J. and Staden, J.V. 1987. Regeneration of *Acacia melanoxylon* plantlets *in vitro*. South African J. Bot. 53(3): 206-209.

Mc Clean, G.D., Atherton, J.G. and Coaldrake, J.E. 1966. Tissue culture from Australian *Acacia, Acacia harphophylla* F.V.M. (brigalow). Phyton. 23: 101.

Mehta, U.J., Shasrabudhe, N. and Hazra, S. 2005. Thiadiazuron-induced morphogenesis in tamarind seedlings. *In vitro* Cell. Dev. Biol. Plant 41(3): 240-243. DOI: 10.1079/IVP2004611.

Merkle, D.B. 1995. Strategies for dealing with limitations of somatic embryogenesis in hard wood trees. *In Vitro* Cell. Dev. Bio.. 31:18.

Merckle, S.A. 1992. Somatic embryogenesis in black locust. In: Black Locust: Biology, Culture and Utilization. Proc. International conference on Black Locust, June 17-21, 1991. East Lansing, MI. USA, pp. 136-146 (eds. J.W.Hanover, K.Miller and S.Plesko). Michigan State University.

Mekcle, S.A. and Trigiano, R.N. 1992. Propagation of hard woods (In applications of vegetative propagation in forestry). Proc. SRIEG Biennial Symp. for Genets held on 8-10 July (USDA, New Orleans, LA). 23-37.

Merkle, S.A. and Wiecku, A.T. 1989. Regeneration of *Robinia pseudoacacia* via somatic embryogenesis. Can. J. For. Res. 19: 285-288.

Miah, S.A.R. and Rao, R. 1996. Regeneration of plantlets from excised roots of *Albizia lebbeck* Benth. Indian J. Exp. Biol. 34: 183-189.

Mithila, J. and Srivasuki, K.P. 1992. Growth responses of *Pterocarpus santalinus* Linn. In: K. Kesava Reddy (ed.), Vegetative propagation and Biotechnologies for tree improvement Natraj Publishers, Dehradun, pp.147-152.

Mittal, A., Agarwal, R. and Gupta, S.C. 1989. *In vitro* development of plantlets from axillary buds of *Acacia aurifuliformis*–a leguminous tree. Plant cell Tiss. Org. Cult. 19: 65-70. DOI: 10.1007/BF00037777.

Mohammed-Yassen, Y., Devwenport, T.L. and Splittstoesser, W.E. 1992. Tissue culture of some arboreal leguminous plants. Proc. Fla. State Hort. Soc. 105: 271-273.

Mohan Ram, H.Y., Mehta, U. and Ramanuja Rao, I.O. 1982. Tissue and protoplast culture and plantlet regeneration in legumes. In: Rao, A..N. (ed). Proc. costed symp. on Tissue culture of economically important plants Singapore. pp.66-69.

Monteuuis, O. 1988. Maturation concept and possible rejuvenation of arborescent species – Limits and promises of shoot apical meristems to ensure successful cloning. In Proc. IUFRO Conf. (Thailand) pp. 106-118.

Monteuuis, O.1995. *In vitro* grafting and *in vitro* micrografting of *Acacia mangium*: Impact of ortet age. Silvae Genet. 44: 190-193.

Monteuuis, O. 1996. *In vitro* shoot apex micrografting of mature *Acacia mangium*. Agrofor. Systems 34: 213-217.

Monteuuis, O. 2004. Micropropagation and rooting of *Acacia mangium* microshoots from juvenile and mature origin. *In vitro* Cell. Dev. Biol. –Plant. 40: 102-107.

Morel, G. 1948. Recherches sur la cultures associe de parasites obligatoires et de tissue vegetaux. Ann. Epiphyt. N.S. 14: 123-234.

Mukhopadhyay, A. and Bhojwani, S.S. 1986. Clonal multiplication and plant regeneration from stem callus of adult *Dalbergia sissoo* Int. Congr. Plant Tiss. Cell. Cult. 6th Meet. 112.

Mukhopadhyay, A. and Mohan Ram, H.Y. 1981. Regeneration of plantlets from excised roots of *Dalbergia sisoo*. Indian J. Exp. Biol. 19: 113-119.

Muralidhar Rao, M. and Lakshmi Sita, G. 1995. Direct somatic embryogenesis from immature embryos of rose wood (*Dalbergia latifolia* Roxb.). Plant Cell Reports. 15: 355-359.

Nagamani, R. and Venkateswaran, S. 1983a. *In vitro* culture of hypocotyl and cotyledon segments of *Leucaena*. *Leucaena* Res. Rep. 4: 88-89.

Nagamani, R. and Venkateswaran, S. 1983b. Morphogenic responses of cultured hypocotyl and cotyledonary segments of *Leucanea*. *Leucaena* Res. Rep. 19: 265.

Nagamani, R. and Venkateswaran, S. 1987. Plantlet regeneration in callus cultures of *Leucaena*. In: J.M. Bonga and Durzan, D.J. (ed.) General principles of biotechnology Cell. and Tissue culture in forestry, Martinus Nijhof. 3:285-291.

Nanda, R.M., Das, P. and Rout, G.R. 2004. *In vitro* clonal propagation of *Acacia mangium* Willd. and its evaluation of genetic stability through RAPD markers. Ann. For. Sci. 61: 381-396.

Nanda, R.M. and Rout, G.R. 2003. *In vitro* somatic embryogenesis and plant regeneration in *Acacia arabica*. Plant cell Tiss. Org. Cult. 73(2): 131-135. DOI: 10. 1023/A: 1022858118392.

Nandwani, D. 1995. *In vitro* micropropagation of a tree legume adapted to arid lands *Acacia tortilis* subsp. *raddiana*. Ann. For. Sci. 52(2): 183-189. DOI: 10.1051/forest:19950208.

Nandwani, D. and Ramawat, K.G. 1991. Callus culture and plantlets formation from nodal explants of *Prosopis juliflora* (Swartz) DC. Indian J. Exp. Biol. 29: 523-527.

Nandwani, D. and Ramawat, K.G. 1992a. High frequency plantlet regeneration from seedling explants of *Prosopis tamarugo*. Plant Cell Tiss. Org. Cult. 29: 173-178.

Nandwani, D. and Ramawat, K.G. 1992b. Regeneration potential of explants and callus morphogenesis in *Prosopis tamarugo*. Gartenbaiussenschaft. 57: 106-111.

Nandwani, D. and Ramavat, K.G. 1993. Plantlets formation through juvenile and mature explants in *Prosopis cineraria*. Indian J. Exp. Bio. 31: 156-160.

Nangia, S. and Singh, N. 1996. Micropropagation of *Leucaena leucocephala* (Lam.) de Wit. Annal. Biol. 12(1): 82-85.

Nataraja, K. and Sudhadevi, A.M. 1985. Induction of plantlets from seedling explants of *Dalbergia latifolia* Roxb. Bietr. Biol. Pflanz. 59:341-349.

Naujoks, G., Zaspel, I. and Behrendt.2001 Microorganisms acting in tissue culture of Black locust (*Robinia pseudoacacia* L.). Internation symposium on methods and markers for quality assurance in micropropagation.

Noguchi, H. and Sankawa, U. 1982a. Germichrysone formation in tissue culture of *Cassia torosa* induction of secondary metabolism in the lag phase. Plant Tissue Culture 5th meet. pp.377-378.

Noguchi, H. and Sankawa, U. 1982b. Formation of germichrysone by tissue cultures of *Cassia torosa*: Induction of secondary metabolism in the lag phase. Phytochemistry 21(2): 319-323.

Ortiz, B., Reyes, M. and Balch, E. 2000. Somatic embryogenesis and plant regeneration in *Acacia fernesiana* and *A. schaffneri*. *In vitro* Cell. Dev. Biol.-Plant. 36: 268-272.

Parmajit, K., Gharyal, G. and Maheswari, S.C. 1990. Differentiation in explants from mature leguminous trees. Plant Cell. Rep. 8: 550-553.

Patri, S., Bhatnagar, S.P. and Bhojwani, S.S. 1988. Preliminary investigations on micropropagation of a leguminous timber tree *Pterocarpus santalinus*. Phytomorphology. 38: 41-45.

Pellegrineschi, A. and Tepfer, D. 1993. Micropropagation and plant regeneration in *Sesbania rostrata*. Plant Sci. 88: 113-119.

Pereira-de-sa, S.P. and Caldas, L.S. 1991. Micropropagation of *Dalbergia miscolobium* (Leguminosae). *In vitro* Cell. Dev. Bio. Plant. 27: 106.

Philomina, N.S., Mithilam, J. and Srivasuki, K.P. 1992. *In vitro* multiplication of *Albizia lebbeck* Benth. In: K. Kesava Reddy (ed.) Vegetative propagation and biotechnologies for tree improvement, Nataraj Publishers, Dehradun. 110: 164-166.

Phukan, M.K. and Mitra, G.C. 1982. Regeneration of *Albizzia odoratissima* Benth., a shade tree for tea plantation of North-East India. Two and a Bud. 30:54-58.

Pradhan, C., Kar, S., Patnaik, S. and Chand, K. 1988. Propagation of *Dalbergia sissoo* Roxb. through *in vitro* shoot proliferation from cotyledonary nodes. Plant Cell. Rep. 18: 122-126.

Pradhan, C. Pattnaik, S., Diwari, M., Pattnaik, S.N. and Chand, P.K. 1998. Efficient plant regeneration from cell suspension derived callus of East Indian Rosewood (*Dalbergia latifolia*). Plant Cell Rep. 18: 138-142.

Prakash, E., Sha valli Khan, P.S., Sreenivasa Rao, J.V. and Singh, M.E. 2006. Micropropagation of red sanders (*Pterocarpus santalinus* L.) using mature nodal explants. J. Forest. Research. 11: 329-335.

Praveen, S. and Shahzad, A. 2010. A micropropagation protocol for *Cassia angustifolia* Vahl from root explants. Acta Physiologia Plantarum. DOI: 10.1007/S11738-101-0603-x.

Praveen, S., Shahzad, A. and Saena, S. 2010. *In vitro* plant regeneration system for *Cassia siamea* Lam., a leguminous tree of economic importance. Agroforestry systems 80(1): 109-116. DOI: 10.1007/s 10057-010-9301-3.

Preuss, P.W., Colevito, L. and Weinstein, L.H. 1970. The synthesis of monofluoroacetic acid by tissue culture of *Acacia georginae*. Experientia 26(10): 105-106. DOI: 10.1007/BF02112670.

Quoirin, M., Bittencourt, J.M., Zanette, F. and de Oliveira, D.E. 1998. Effect of growth regulators on indirect organogenesis of *Acacia mearnsii* tissues cultured *in vitro*. Revista Brasileira de Fisiologia Vegetal 10(2): 101-105.

Quoirin, M., Franche, O., Duhoux, E. and Galiana, A. 2002. Advances in tissue culture and genetic transformation of four tropical *Acacia* species: *A. crassicarpa, A. mearnsii, A. mangium* and *A. albida*. In: Recent Research development in Physiology. Trivandrum, Research Signpost, pp. 7-25.

Rady M. R. and N. M. Naif, 1997. Response of explant type of proliferation and anthraquinone accumulation in *Cassia acutifolia*. Fitoterapia. 68: 349-354.

Raghava Swamy, B.V., Himabindu, K. and Lakshmi Sita, G. 1992. Micropropagation of elite rose wood (*Dalbergia latifolia* Roxb.). Plant Cell Rep. 11(3): 126-131.

Rahman, S.M., Hossain. M., Biswas, B.K., Joarder, O.I. and Islam, R. 1993. Micropropagation of *Caesalpinia pulcherrima* through nodal bud culture of mature tree. Plant Cell Tiss. Org. Cult. 32: 363-365.

Rai, P.P. and Shok, M. 1982. Anthracene derivatives in tissue cultures of *Cassia* species indigenous to Nigeria. Plant tissue cell 5th meet. pp.277-278.

Rajadurai, D., Rao, A.N. and Loh C.S. 1989. *In vitro* culture studies on two leguminous species. Tissue Culture–Forest species. 104- 128.

Rajeswari, V. and Paliwal, K. 2004. *In vitro* adventitious shoot organogenesis and plant regeneration from seedling explants of *Albizia odoratissima* (L.f.) Benth. *In vitro* cell. Dev. Biol. Plant 44(2): 78-83. DOI: 10.1007/s11627-008-9120-7.

Rajeswari, V. and Paliwal, K. 2006. *In vitro* propagation of *Albizia odoratissima* (L.f.) Benth. from cotyledonary node and leaf nodal explants. *In vitro* cell Dev. Bio. Plant 42(5): 399-404.

Rajeswari, V. and Paliwal, K. 2007. *Albizia odoratissma* (L.f.) Benth. Micropropagation. In S.M.Jain and H. Haggman (eds.). Protocols for Micropropagation of Woody trees and fruits pp. 201-211.

Rama Subbu, R., Chandraprabha, A. and Sevugperumal, R. 2008. *In vitro* clonal propagation of vulnerable medicinal plant *Saraca aoca* (Roxb.) De Wilde. Nat. Prod. Radiance 7(4): 338-341.

Ramamurthy, N. and Savithramma, N. 2003. Shoot bud regeneration from leaf explants of *Albizia amara* Boiv. Indian J. Plant Physiol. 8(4): 372-376.

Ranga Rao, G.V. and Prasad, M.N.V. 1991. Plantlet regeneration from the hypocotyls callus of *Acacia auriculiformis* – multipurpose tree legume. J. Plant Physiol 37: 625-627.

Rao, K.S., Lakshmisita, G. and Vaidyanathan. 1985. Cloning of *Dalbergia latifolia* Roxb. (Indian rose wood). Basic Life Sci. 32:346.

Rao, P.V.L. and De, D.N. 1987. Haploid plants from anther culture of leguminous tree, *Peltophorum pterocarpum* (DC.) K. Heyne (Copper pod). Plant cell Tiss. Org. Cult. 11:167-177.

Rao, P.V.L. and De, D.N. 1987. Tissue culture propagation of tree legume *Albizia lebbeck* (L.) Benth. Plant Sci. 51(2-3): 263-267.

Rao, G.V.R. and Prasad, M.N.V. 1991. Plant regeneration from hypocotyl callus of *Acacia auriculiformis* multipurpose tree legume. J. Plant Physiol. 137: 625-627.

Rastogi, S., Rizvi, S.M.H., Singh, R.P. and Dwivedi, U.N. 2008. *In vitro* regeneration of *Leucaena leucocephala* by organogenesis and somatic embryogenesis. Biologia Plantarum 52(4): 743-748.

Rath, S.P. 1998. *In vitro* studies of tree legumes: *Pterocarpus marsupium* Roxb. Adv. Plant Sci. 11: 5-9.

Ravi Shankar, G.A., Walli, A. and Grewal, S. 1983. Plantlet formation through tissue culture of *Leucaena leucocephla*. *Leucaena* Res. Rep. 4:37.

Ravi Shankar Rai, V. and Jagadish Chandra, K.S. 1988. Regeneration of plantlets from shoot callus of mature tree of *Dalbergia latifolia*. Plant Cell. Tiss. Org. Cult. 13: 77-83.

Ravi Shankar Rai, V. and Jagadish Chandra, K.S. 1989. Micropropagation of Indian rose wood by tissue culture Ann. Bot. 64: 43-46.

Reddy, K.K., Bhaskar, P. and Subash, K. 1995. Propagation of *Acacia holosericea* – forest tree. *In vitro* Cell Dev. Biol. Plant. 31:83.

Reddy, P., Patil, V., Prasad, T.G., Padma, K. and Udaya Kumar, M. 1995. *In vitro* axillary bud break and multiple shoot production in *Acacia auriculiformis* by tissue culture technique. Curr. Sci. 69: 495-496.

Rout, G.R. and Nanda, R. 2005. Protocol for somatic embryogenesis in *Acacia arabica* (Lamk.) Willd.. In: Protocol for somatic embryogenesis in woody plants. Forestry Sciences. Vol. 77. Section C. pp. 401-411. DOI: 10.1007/1-4020-2985-3–32.

Rout, G.R., Samantaray, S. and Das, P., 1995. Somatic embryogenesis and plant regenerartion from callus cultures of *Acacia catechu* – a multipurpose leguminous tree. Plant Cell Tiss. Org. Cult. 42: 283-285.

Rout, G.R., Senapati, S.K. and Aparanjita, S. 2008. Micropropagation of *Acacia chundra* (Roxb.) DC. Hort. Sci. (Prague) 35(1): 22-26.

Rubluo, A., Arriaga, E. and Brunner, I. 2002. Shoot production from cotyledons of *Prosopis glandulosa* var. *torreyana* cultured *in vitro*. Anales del Instituto, Universidad Nacional Autonoma de Mexico Series Botanica 73(1): 83-87.

Ruffoni, B. Massabo, F., Costantino, C., Arena, V. and Damiano, C. 1995. Micropropagation of *Acacia* "Mimosa". ISHS Acta Horticulturae 300: *In vitro* Culture XXIII. IHC. Pp. 95-102.

Sahni, R. and Gupta, S.C. 2002. Plantlet regeneration from seedling nodal explants of *Acacia catechu*. Indian J. Exp. Biol. 40: 1050-1055.

Saito, Y., Kojima, K., Ide, Y. and Sasaki, S. 1993. *In vitro* propagation from axillary buds of *Acacia mangium*, a legume tree in the tropics. Plant tissue culture letters 10(2): 163-168.

Sankhla, D., Davis, T.D. and Sankhla, N. 1993. Effect of Gibberellin biosynthesis inhibitors on shoot regeneration from hypocotyl explants of *Albizia julibrissin*. Plant Cell Rep. 13: 115-118.

Sankhla, D., Davis, T.D. and Sankhla, N. 1994. Thidiazuron induced shoot formation from roots of intact seedlings of *Albizia julbrissin*. Plant Growth Regul. 14:267-272. DOI: 10.1007/BF00024802.

Sankhla, D. S., Davis, T.D. and Sankhla, N. 1996. *In vitro* regeneration of silk tree (*Albizia julibrissin*) from excised roots. Plant Cell. Tiss. Org. Cult. 44:83-86. DOI: 10.1007/BF00045917.

Sarita, S., Bhatnagar, S.P. and Bhojwani, S.S. 1988. Preliminary investigations on micropropagation of leguminous timber tree *Pterocarpus santalinus*. Phytomorphology. 38: 41-45.

Sascha, L.B., Dunlop, R. and Van Staden, J. 1998. Micropropagation of *Acacia mearnsii* from *ex vitro* material. Plant Growth Regulation. 26: 143-148.

Sasmitamihardja, D., Hadisutanto, J.S. and Widiyanto, S.N. 2001. *In vitro* regeneration of *Paraserianthes falcataria* (L.) Nielsen. ISHS Acta Horticulturae 692. II International symposium on Biotechnologyof Tropical and Subtropical species.

Saxena, P.K. and Gill, R. 1987. Plant regeneration from mesophyll protoplast of the tree legume *Pithecellobium dulce* Benth. Plant Sci. 53: 257-262.

Schenk, R.V. and Hildebrandt, A.C. 1972. Medium and technique for induction and growth of monocotyledon and dicotyledonous cell culture. Can. J. Bot. 50:199-204.

Seelinger, I. 1959. Uber die Bildung Wurzelbartiger Sprosse and das Wachstum isolierter Wurzet der Robinie (*Robinia pseudoacacia* L.). Flora. 148: 218-254.

Setiawan, I., Umboh, M.I. and Suprianto. 1991. Growth and rooting system of *Acacia mangium* obtained by tissue culture. Biotropica 4: 1-8.

Shahid, M., Shahzad., A., Mallik, A. and Anis, M. 2007. Antibacterial activity of aerial parts as well as *in vitro* raised calli of the medicinal plant *Saraca asoca* (Roxb.) de Wilde. I. Can. J. Microbiol. 53: 75-81.

Shankar, S. and Mohan Ram, H.Y. 1987. Tissue culture and plantlet regeneration in tree legume *Sesbania*. Eur. Congr. Biotechnol. 2:390.

Shankar, S. and Mohan Ram, H.Y. 1990a. *In vitro* plantlet regeneration in the leguminous tree *Sesbania sesban.* Phytomorphology. 40:43-52.

Shankar, S and Mohan Ram, H.Y. 1990b. Plantlet regeneration from tissue culture of *Sesbania grandiflora.* Curr. Sci. 59:1:39-43.

Shankara Rao, K. 1986. Plantlets from somatic callus tissue of the East Indian Rose wood (*Dalbergia latifolia* Roxb.). Plant Cell. Rep. 5:199-201.DOI: 10.1007/ BF00269118.

Shanti, P. 2008. *In vitro* propagation of *Ormocarpum sennoides* (Willd.) DC. From shoot tip explant. Indian J. Plant Physiol. 13: 1.

Sharma, D., Pareek, L.K. and Chandra, N. 1997. High frequency shoot proliferation, efficient rooting of shoots and histological basis of shoot proliferation in cultured cotyledonary node segments of *Prosopis cineraria* – A leguminous desert tree. J. Indian Bot. Soc. 76: 207-210.

Sharma, S. and Chandra, N. 1988. Organogenesis and plantlet formation *in vitro* in *Dalbergia sissoo* Roxb. J. Plant Physiol 132:145-147.

Shasthree, T., Savitha, R., Venkateswarlu, K. and B.Mallaiah. 2009. *In vitro* propagation of legume tree *Erythrina varietata* L., via multipurple shoot formation. Plant Cell Biotech. Mol. Biol. 10: 75-78.

Shekhawat, N.S., Rathore, T.S., Singh, R.P., Deora, N.S. and Rao, S.R. 1993. Factors affecting *in vitro* clonal propagation of *Prosopis cineraria*. Plant Growth regulation 13: 273-280.

Shrivastava, N. and Kant. T. 2010. Micropropagation of *Pongamia pinnata* (L.) Pierre – a native Indian biodiesel tree from cotyledonary node. Int. J. Biotech. Biochem. 6(4): 555-560.

Shrivastava, N., Patel, T. and Srivastava, A. 2006. Biosynthetic potential of *in vitro* grown callus cells of *Cassia senna* L. Curr. Sci. 90(11): 1472-1473.

Siddique, I. and Anis, M. 2007. *In vitro* shoot multiplication and plant regeneration from nodal explants of *Cassia angustifolia* Vahl – a medicinal plant. Acta Physiologia Plantarum 29: 233-238.

Singh, A.K. and Chand, S. 2003. Somatic embryogenesis and plant regeneration from cotyledon explants of a timber-yielding leguminous tree, *Dalbergia sissoo* Roxb. J. Plant Physiol 160: 415-421.

Singh, A.K. and Chand, S. 2010. Plant regeneration from alginate-encapsulated somatic embryos of *Dalbergia sissoo* Roxb. Indian J. Biotech. 9(3): 319-324.

Singh, A.K., Chand, S., Pattnaik, S. and Chand, P.K. 2001. Adventitious shoot organogenesis and plant regeneration from cotyledons of *Dalbergia sissoo* Roxb. a timber yielding tree legume. Plant Cell Tiss. Org. Cult. 68(2): 203-209. DOI: 10.1023/A: 1013870803937.

Singh, H.P., Singh, R.P.S., Saxena, K.P. and Singh, R.K. 1993. *In vitro* buds break in axillary nodal segments of mature trees of *Acacia nilotica*. Indian Plant Physiol. 36: 21-24.

Sinha, R.K., Majumdar, K. and Sinha, S. 2000. *In vitro* differentiation and plant regeneration of *Albizia chinensis* (Osb.) Merr. *In vitro* Cell Biol. Plant 36: 370-373.

Sinha, R.K. and Mallick, R. 1991. Plantlets from somatic callus tissue of the woody legume *Sesbania bispinosa* (Jacq.) W.F. Wight. Plant Cell Rep. 10: 247-250.

Sinha, R.K. and Mallck, R. 1993. Regeneration and multiplication of shoot in *Albizia falcataria*. Plant cell. Tiss. Org. Cult. 32:259-261. DOI: 10.1007/BF00029851.

Skoleman, R.G. 1986. *Acacia* (*Acacia koa* Gray). In: Y.P.S.Bajaj (ed.). Biotechnology in Agriculture and Forestry. Vol. 1 Trees I. Springer-Verlag, Berlin pp. 375-384.

Skoleman, R.G. and Mapes, M.O. 1976. *Acacia koa* Gray plantlets from somatic callus tissue. J. Heredity. 67 : 114-115.

Skoleman, R.G. and Mapes, M.O. 1978. After care procedures required for field survival of tissue culture propagated *Acacia koa*. Proc. Int. Plant Pro. Soc. 28:156-164.

Skoleman, R.G. and Mapes, M.O. 1986. *Acacia* (*Acacia koa* Gray). Bajaj, Y.P.S. (eds,). Biotechnology in Agriculture and Forestry, Trees Springer Verlag, Berlin Heideberg. 1: 375-383.

Smith, M.A.L. and Obeidy, A.A. 1991. *In vitro* rescue of a mature male kentucky tree (*Gymnocladus dioicus* L.) genotype. Hort. Science. 26:749.

Solanki, K.R. and Kackar, N.L. 1992. Studies of micropropagation and vegetative propagation in *Prosopis cineraria*. In: K. Kesava Reddy (ed.) Vegetative propagation and Biotechnologies for tree improvement. Natraj Publishers, Dehradun, pp.153-159.

Sowgandhika, M. 2008. *In vitro* propagation of *Erythrina variegata* L. (Fabaceae). M.Phil. dissertation, S.K.University, Anantapur.

Sreedevi, E 1994. *In vitro* plantlet regeneration from seedling explants of *Dalbergia paniculata* Roxb. M.Phil. dissertation, S.K.University, Anantapur.

Sreedevi, E and Pullaiah, T. 1999. *In vitro* plant regeneration from seedling explants of *Dalbergia paniculata* Rox. In: Kavi Kishor, P.B.(ed.). Emerging trends in plant Tissue Culture and Molecular Biology, Universities Press, Hyderabad. pp. 112-117.

Sreelakshmi, L and Janardhan Reddy, K. 2005. Production of karanjin, A Furanoflavonoid in *Pongamia glabra* through plant cells cultures. National symposium on Recent Trends in Plant Sciences souvenir and abstracts poster presentation, Department of Botany, Sri Krishnadevaraya University, Anantapur. 29-31, October.

Sreelatha, V.R. 2005. Rapid micropropagation studies of *Cassia siamea* Lam. M.Phil. dissertation, S.K.University, Anantapur.

Sreelatha, V.R., Prasad, P.J.N., Karuppusamy, S. and Pullaiah, T. 2008. Morphogenic response of callus derived from hypocotyls explant of *Cassia siamea* Lam. J. Indian Bot. Soc. 87: 92-98.

Srivastava, A. and Aggarwal, A.A. 2007. Induction of photoautotrophy in regenerants of *Sesbania aculeata*. Plant cell Rep. 12: 629-633.

Su, J. and Zhang, F. 2008. Study on propagation technology about root sprout and tissue culture of *Acacia melanoxylon*. Guangdong Forestry Sci. Technol.

Subhan, S., Sharmila, P. and Pardhasaradhi, P. 1998. *Glomus fasciculatum* alleviates transplantation shock of micropropagated *Sesbania sesban*. Plant Cell Rep. 17(4): 268-272.

Sudhadevi, A.M. and Nataraja, K. 1987. *In vitro* regeneration and establishment of plantlets in stem culture of *Dalbergia latifolia* Roxb. Indian For. 113:501-505.

Sugla, T., Purkayastha, J., Singh, S.K., Solleti, S.K. and Sahoo, L. 2007. Micropropagation of *Pongamia pinnata* through enhanced axillary branching. *In vitro* cell. Dev. Biol. Plant 43(5): 409-414. DOI: 10.1007/s11627-007-9086x.

Sujatha, K., Panda, B.M. and Hazras, S. 2008. *De novo* organogenesis and plant regeneration in *Pongamia pinnata*, oil producing tree legume. Trees Structure and Function. 22(5): 711-716. DOI: 10.1007/s00468-008-230-y.

Sujatha, K. and Hazra, S. 2007. Micropropagation of mature *Pongamia pinnata* Pierre. *In vitro* Cell Dev. Biol. Plant 43(6): 608-613. DOI: 10.1007/s11627-007-9049-2.

Sukarutiningsih, Saito, Y. and Ide, Y. 2002. *In vitro* plantlet regeneration of *Paraserianthus falcataria* (L.) Nielsen. Bulletin of the Tokyo Univ. Forests 107: 21-28.

Swamy, S.L., Ganguli, J.L. and Puri, S. 2004. Regulation and multiplication of *Albizia procera* Benth. through organogenesis. Agroforestry Systems 60: 113-121.

Tabene, T.J. P. Felker, R.L. Bingham, Reyes, I. And Loughrey, S. 1986. Techniques in the shoot multiplication of leguminous trees *Prosopis alba* and clones B_2V_{50}. In:

Fleker (ed.), Tree planting in semi-arid regions Elsevier Science Publishers, Amsterdam, pp. 191-200.

Thakur, M., Sharma, D.R., Kumar, K. and Kanth, A. 2002. *In vitro* regeneration of *Acacia catechu* Willd. from callus and mature nodal explants – An improved method. Indian J. Exp. Biol. 40: 850-853.

Tiwari, S., Shah, P. and Singh, K. 2004. *In vitro* propagation of *Pterocarpus marsupium* Roxb. an endangered medicinal tree. Indian J. Biotech. 3: 422-425.

Tomar, V.K. and Gupta, S.C. 1988a. Somatic embryogenesis and organogenesis in callus cultures of a tree legume – *Albizia richardiana* King. Plant Cell. Rep. 7:70-73.

Tomar, V.K. and Gupta, S.C. 1988b. *In vitro* plant regeneration of leguminous trees (*Albizia* sps.). Plant Cell Rep. 7: 385-388.

Toruan Mathius, N. 1992. Micropropagation of *Leucaena in vitro*. Biotechnol Forest Tree improvement, pp. 99-109.

Uddin, S.M., Narirujjaman, K., Zaman, S. and Reza, M.A. 2005. Regeneration of multiple shoots from different explants, *viz.* shoot tip, nodal segments and cotyledonary node of *in vitro* grown seedlings of *Peltophorum pterocarpum* (DC.) Backer ex K. Heyne. Biotechnology 4(1): 35-38.

Upadyaya, S. and Chandra, N., 1983. Shoot and plantlet formation in organ and callus cultures of *Albizia lebbeck* Benth. Ann. Bot. 52: 421 – 424.

Vengadesan, G., Ganapathi, A., Anand, R.P. and Anbazhagan, V.R. 2004. *In vitro* organogenesis and plant formation in *Acacia sinuata*. Plant Cell Tiss. Org. Cult. 61(1): 23-28. DOI: 10.1023/A:1006442818277.

Vengadesan, G., Ganapathi, A., Anbazhagan, V.R. and Anand, R.P. 2002. Somatic embryogenesis in cell suspension culture of *Acacia sinuata* (Lour.) Merr. In *vitro* Cell Dev. Biol. Plant 38: 52-59.

Varghese, T.M. and Kaur, A. 1988. *In vitro* propagation of *Albizia lebbeck* Benth. Curr. Sci. 57:1010-1012.

Vinolya Kumari, R. 2000. Morphogenic studies of *Terminalia chebula* Retz., *T. pallida* Brandis and *Samanea saman* (Jacq.) Merr. Ph.D. thesis, S.K. University, Anantapur.

Vinolya Kumari, R. and Pullaiah, T. 2003. *In vitro* propagation studies of *Samanea saman* (Jacq). Merr. Proc. of A.P Academic of Sciences. 4: 245-248.

Vlachova, M., Bright, J.S. and De Brujin, F.J. 1987. The tropical legume *Sesbania rostrata*. Tissue culture, plant regeneration and infection with *Agrobacterium tumifaciens* and *rhizogenes* strains. Plant. Sci. 50:213-223.

Wainright, H and England, N. 1987. The micropropagation of *Prosopis juliflora* (Swartz) DC. establishment *in vitro*. ISHS Acta Horticulturae. 212:49-52.

Wang, Q.Z. Zhao, J.B. and Zhao, B.H. 1985. Fast propagation of superior clones of *Robinia pseudoacacia* through tissue culture. For. Sci. Technol. Sci. 8-9.

Watanabe, Y., Ide, Y. and Ikeda, H. 1994. Plant regeneration from axillary bud culture of one-year old seedling of *Acacia auriculiformis* grown in green house. Bull. Tokyo Univ. For. 92: 29-35.

Weaver, L.A. and Trigiano, R.N. 1991. Regeneration of *Cladrastis lutea* (Fabaceae) via somatic embryogenesis. Plant Cell Rep. 10: 183-186.

Woo, J.H. 1994. Utilization and Tissue culture of *Robinia pseudoacacia* L. in Korea. Ph.D. thesis, Department of Forestry. Kyungpook National University, Daegu, Korea, pp. 113.

Woo, J.H., Choi, M.S. and Park, Y.G. 1995a. Plant regeneration from callus cultures of black locust (*Robinia pseudoacacia* L.). J. Kor. For. Sci. 84: 145-150.

Woo, J.H., Choi, M.S., Joung, E.Y., Chung, W.I., Jo, J.K. and Park, Y.G. 1995b. Improvement of black locust (*Robinia pseudoacacia* L.) through tissue culture I. Micropropagation and somatic embryogenesis. J. Kor. For. Sci. 84: 41-47.

Wood, A. 1985. The potential for the *in vitro* propagation of a number of economically important plants for arid areas. In: G.E.Wickens, J.R. Goodin and D.v.Field (eds.) Plants for Arid areas, Goerge Allen and Unwin, London pp. 333-341.

Xie, D. and Hong, Y. 2001. *In vitro* regeneration of *Acacia mangium* via organogenesis. Plant cell Tiss. Org. Cult. 66(3): 167-173. DOI: 10.1023/A:1010632619342.

Xie, D.Y. and Hong, Y. 2009 Regeneration of *Acacia mangium* through somatic embryogenesis. Plant Cell Rep. 20(1): 34-40. DOI: 10.1007/s002990000288.

Xie, X., Yang, M., Shang, Y., He, X. and Zhang, F. 2006. Advances in *in vitro* multiplication and genetic engineering in *Acacia*. J. Beijing Forestry Univ.

Xu, Z.H., Yank, L.J. Wei Z. M. and Gao, M.X. 1984. Plant regeneration tissue culture of four leguminous species. Act. Biol. Expl. Sin. 17:483-486.

Yamamota, HS, Ichimura, M and Inoue, K. 1995. Stimulation of Prenylated flavanone production by mannose and acidic polysaccharides in callus cultures of *Sophora flavescens*. Phytochemistry. 40:77-81.

Yang, J.C., Lu, M.C. and Tsay, J.Y. 1989. *In vitro* culture of *Acacia auriculiformis, A. mangium* and their inter specific hybrid. Proc. 6th Intl. Cong. Taiwan. 865-868.

Yang, Li Jia Liming. 2005. A review of the reearch and development in the clonal rapid propagation technologies of woody legumes. World Forestry Research.

Yang, M., Xie, X., He, X. and Zang, F. 2006. Plant regeneration from phyllode explants of *Acacia crassicarpa* via organogenesis. Plant Cell Tiss. Org. Cult. 85: 241-245.

Yaseen,Md. T.N., Davenport and Splitt Stoesser, W.E. 1992. Tissue culture of some arboreal leguminous plants. Proc Fla. State Hortic. Soc. 105:271-273.

Yasodha, R., Sumathi, R., Malliga, P. and Gurumurthi, K. 2002. *In vitro* expression of juvenility in acacia hybrid. I: Nat. Symp. On Emerging trends in Modern biology, held on 10-12, Jan, 2002, Loyola college, Chennai, pp. 24.

Ye, L., Lai, Z., Su, Q., Lin, S. and Lin, Q. 2009. Callus culture and histological observations on *Acacia* spp. Chinese Agriculture Science Bulletin

Yi, Y.D., Batchelor, C.A. Kochlev, M.J. and Haris, P.J.C. 1989. *In vitro* regeneration of *Prosopis* species (*P.chilensis* and *P.juliflora*) from nodal explants. Chinese J. Bot. 1:89-97.

Yousoff, A.M., Davi, M.K., Fadillah, Z., Halilah, A.K. and Haliza, I. 1999. Establishing a protocol for commercial micropropagation of *Acacia mangium* X *Acacia auriculiformis* hybrids. J. Trop. For. Sci. 11: 148-156.

Zeijilmaker, F.C.J. 1972. Rep. 1971-72 on Plant Physiology and Pathology. Inst. Pietermaritzburg. S. Afr. Univ. Natal. Pp. 49.

Zhang, H.W., Huang, X.L., Fu, J., Yang, M.Q. and Chen, C.Q. 1995. Axillary bud culture and plantletlet regeneration of *Acacia auriculiformis* and *A. mangium*. J. Trop. Subtrop. Bot. 3: 62-68.

Zhang, J., Liu, Y., and Wang, H. 2007. Micropropagation of Black Locust (*Robinia pseudoacacia* L.). In: Protocols for micropropagation of Woody trees and fruits. Part I pp. 193-199. DOI: 10.1007/978-1-4020-6352-7_18.

Zhang, Z. 2007a. Tissue culture and rapid propagation of *Acacia baileyana*. J. Southwest Univ.

Zhang, Z. 2007b. The study of techniques for rapid propagation and tissue culture of *Acacia floribunda* (Vent.) Willd. J. Fujian Forestry Sci. & Tech.

Zhang, Z. and Liu, X. 2004. Tissue culture and plantlet regeneration derived from *Acacia implexa* stem segments with axillary buds. J. Southwest Agric. Univ.

Zhao, D.L., Guo, G.Q., Wang, X.Y. and Zheng, G.C. 2003 *In vitro* micropropagation of a medicinal plant species *Sophora flavescens*. Biologia Plantarum 47(1): 117-120.

Zhao, Y.-X, Harris, P.J.C. and Yao, D.-Y. 1995. Plant regeneration from protoplasts isolated from cotyledons of *Sesbania bispinosa*. Plant Cell Tiss. Org. Cult. 40(2): 119-123. DOI: 10.1007/BF00037664.

Zhao, Y.X., Yao, D.Y. and Horris, J.C.P. 1993. Plant regeneration from callus and explants of *Sesbania* species. Plant Cell Tiss. Org. Cult. 34:253-260.

Zhou, L. and Zhang, H. 1996. Micropropagation of hybrid (*Acacia mangium* x *A. auriculiformis*. Forestry Sci. Tech. of Guandong Province.

Chapter 2

In vitro Propagation Studies of Sterculiaceae: A Review

S. Anitha[1] and T. Pullaiah[2]

[1]Department of Biotechnology
[2]Department of Botany,
Sri Krishnadevaraya University, Anantapur – 515 003

Biodiversity is playing major role in maintaining the ecological balance of the Nature. Diversity is so intertwined with the life on earth, that there is no chance of evolution and origin of the new species without it. Plants in nature are not only the primary source of the energy but also source of food, clothing, shelter and medicines. Conservation of such crucial plant diversity is now gaining momentum. Keeping in view the importance of conserving the valuable plant diversity, conservation of diversity rich areas, threatened, endangered and endemic plants are given priority, which was reported by a number of researchers (Mascarenhas and Muralidharan, 1989; Krogstrup *et al.*, 1992). According to the WHO observation 80 per cent of the people in developing country still depend on the wild plants for medicines. In developed countries also after using synthetic drugs indiscriminately and being aware of the health hazards, now are, looking forward towards the plant based drugs. This growing demand increased overexploitation and improper harvesting of the medicinal plants by the pharmaceutical companies. Biotic and abiotic conditions in combination, contribute to the extinction of the plant diversity (Courrier, 1992). *In vitro* propagation is one of the best methods used for the conservation of this depleting diversity. In the present article we focus on the *in vitro* strategies adopted for the conservation of Sterculiaceae family till now. In spite of the number of medicinal and horticultural plants existing in this family, only few plants were opted for *in vitro* studies, which might be due to the recalcitrant nature. In Sterculiaceae major work was done only in *Theobroma cacao* while in all the other members reports are very few.

In vitro technique as known can be applied for developing disease free plants from diseased ones, retention of the regenerative potential for longer periods using the germplasm conservation methods, apart from rapid propagation using little parent material for developing homozygous and uniform lines (Litz and Gray, 1992). Artificial seed technology, one of the application, have many potential advantages in forestry, like rapid propagation of desirable lines, maintaining genetic uniformity of plants and easy handling of cultured materials, suitability to transport from one place to another, reduced space of storage etc. When we observe the importance and *in vitro* work done in this family, *Cola* genus has small evergreen plants cultivated in West Africa for alkaloids, caffeine and threobromine. Large quantities of nuts are exported to Europe and N.America as they are used for flavouring the cola drinks. Bark, leaves, twigs, flowers and fruits are used in traditional medicine preparation (Ayensu, 1978; Fereday *et al.*, 1997). *Cola nitida* is one of the species of this genus cultivated largely because of its commercial importance. Bark extracts of *Cola nitida* when tested against few pathogenic bacteria have showed inhibitory activity (Ebana *et al.*, 1991). Vegetative propagation using cuttings was reported in this species apart from seeds as conventional means of propagation. Multiple shoots were regenerated from axillary bud culturing by Doss *et al.* (1994).

Guazuma crinita is the medium sized fast growing tree species endemic to Amazon forests (Freytag, 1951). Soft wood of this plant is used for light construction, paneling, interior joinery, moulding, packaging and has various other miscellaneous uses. It is considered as one of the potential species for reforestation in low lands of Amazon. This species is propagated normally through seeds which don't provide adequate desirable selected seedlings for reforestation. Vegetative propagation methods available are also insufficient, so Maruyama *et al.* (1996) opted for tissue culture technique. They used shoot tip explants derived from aseptic seedlings as explant source and cultured on WPM supplemented with 4.44 µM of BAP. After 45 days shoots were transferred to WPM containing 10 µM of Z. For elongation and rooting these shoots were transferred to WPM containing 1 µM of Kn. Thus obtained plantlets were acclimatized in pots containing vermiculite inside the growth cabinet. Maruyama *et al.* (1997a) used shoot tips and axillary buds to produce artificial seeds for germplasm conservation. These scientists used 3-4mm long aseptically excised explants, immersed them in 4 per cent of sodium alginate and dropped into 1.4 per cent (w/v) calcium chloride and allowed to stay for 20 min. After decanting the calcium chloride solution rinsed in the standard WPM containing medium with 1 µM BAP + 1 µM Kn. To improve the rate of plant regeneration from artificial seeds, beads with single or double layer were tested, and always double layered beads containing medium at a concentration of 1000 (w/v) along with 0.5 per cent (w/v) activated charcoal in the inner layer and at concentration of 100 per cent (w/v) in outer layer performed better. After 6 weeks of incubation on water substratum solidified with 1 per cent (w/v) agar, the rate of bud emergence and shoot growth was 100- 80 per cent for *Gauzuma crinita*. Maruyama *et al.* (1997b) reported germplasm conservation using shoot tip encapsulation in calcium alginate. They stored at different temperatures *i.e.*, 12, 20 and 25°C studied the percentage of viability of these encapsulated shoot. After 12 months when shoot tips were stored on water solidified with 1 per cent (w/v) agar, 90 per cent of the encapsulated explants responded, when stored at 25°C. Maruyama *et al.* (1997c) reported micropropagation using root and petiole explants of seedlings when cultured on BAP containing medium and later transferred to 10 µM of Z. Complete plantlets were formed after 60 days from the petiole explants.

Helicteres isora is a shrub growing in deciduous forests. Fruit is used for treating intestinal complaints, colic pains and flatulence (Chopra *et al.*,1956), while root and bark are used to treat scabies, diabetes, dysentery and diarrhorea (Kirtikar and Basu, 1935). The antidiabetic property of this plant was experimentally proved by Kumar *et al.* (2009) with their experiments on rats. The antibiotic property of *Helicteres isora* was tested against 9 microbial strains and was found that strains like *Micrococcus luteus*, *Aspergillus niger* and *Candida albicans* are sensitive to aqueous ethanol extracts of this plants root by Venkatesh *et al.*(2007). Shriram *et al.* (2008) reported an efficient method of plant regeneration *via* shoot organogenesis from callus. They reported callus induction using nodal explants when placed on MS medium (Murashige and Skoog, 1962) containing individually supplemented auxins or cytokinins or in combination of these. They reported compact, hard, green and white callus on medium containing 13.32 µM BAP and 2.32 µM Kn. Shoot induction was observed from this callus when placed on medium containing 2.22 µM BAP and 2.32 µM Kn. These shoots rooted after 35 days of culture on ½ MS medium containing 4.90 µM IBA. They also studied about antioxidant enzymes, and biochemical parameters in regenerating and non regenerating calluses, to establish a correlation between these parameters and shoot morphogenesis. Finally reported that all the enzyme activities like catalase, peroxidase, polyphenol oxidase and biochemical parameters like hydrogen peroxide, reducing and non reducing sugars, proteins and proline contents were found more in regenerating callus than in non regenerating except phenols.

Heritiera fomes is a threatened mangrove plant in Indian Sundarbans (Naskar *et al.*, 1997). Both biotic and abiotic factors are responsible for the change of conditions that lead to the disappearance of this species. Preliminary study was done to know about *in vitro* behavior of leaf explants of *Heritiera fomes*. Different phytohormones and vitamins at various concentrations were tested for callus induction, using leaf explants on MS medium. Callus formation was reported by Mukhopadhyay *et al.* (1991) on MS basal medium supplemented with 0.5 mg/l 2,4-D, 1.0 mg/l BAP and 0.5 mg/l 2-iP. Das *et al.* (1997) reported significant increase in root number in the air layers and pregirdled stem cuttings of *H.fomes* treated with auxins combination of IBA (5000ppm) and NAA (2500 ppm). This combination increased the starch hydrolysis and reduced C/N ratio during root development. Auxin therefore influenced mobilization of nitrogen to the rooting zone and promoted rooting.

Hildegardia populifolia is a medium sized tree confirmed to tropical deciduous forest. Fiber is extracted from bark while leaf extract is said to have healing properties. According to Nayar and Sastry (1990), this is an endemic threatened species due to non apparent factors. Anuradha and Pullaiah (2001), reported shoot formation from nodal explants, when placed on MS medium containing 2 mg/l BAP. Their significant achievement was the production of somatic embryos from immature zygotic embryo explants, when placed on medium supplemented with 3mg/l NAA and 1000 mg/l CH, after 60 days of culture. These somatic embryos developed into plantlets when transferred to hormone free MS medium. Shoots regenerated from nodal explants developed roots when placed on ½ MS medium containing 2 mg/l NAA. These plantlets were later acclimatized to field conditions.

Table 2.1: *In vitro* Work Carried Over in the Family Sterculiaceae

Plant Name	Source of Explant	Result	Reference
Cola nitida	Axillary bud	Multiple shoots	Doss et al. (1994)
Guazuma crinita	Seedling shoot tip	Shoots, Plantlets	Maruyama et al. (1996)
Guazuma crinita	Seedling petiole, root	Adventitious buds, Plantlets	Maruyama et al. (1997c)
Guazuma crinita	Shoot tips	Encapsulation and cryopreservation	Maruyama et al. (1997a,b)
Helicteres isora	Nodal explants	Shoots	Shriram et al. (2008)
Heritiera fomes	Leaf explants	Callus	Mukhopadhyay et al. (1991)
Hildegardia populifolia	Seedling nodes, immature embryos	Shoots, somatic embryos, plantlets	Anuradha and Pullaiah (2001)
Sterculia alata	Immature embryos	Somatic embryos	Pandey (1998)
Sterculia foetida	Cotyledonary node	Shoots, plantlets	Anitha and Pullaiah (2001)
Sterculia foetida	Hypocotyl, shoot tips	Shoots, plantlets	Anitha and Pullaiah (2002)
Sterculia urens	Cotyledons	Shoots	Purohit and Dave (1996)
Sterculia urens	Seedling shoot apex	Multiple shoots	Bahadur and Shailaja (1997)
Sterculia urens	Node	Multiple shoots	Shailaja et al. (1997)
Sterculia urens	Mature tree nodes	Shoots, somatic embryos	Sunnichan et al. (1998)
Sterculia urens	Cotyledonary node	Shoots, plantlets, callus	Anitha and Pullaiah(1999)
Sterculia urens	Cotyledonary node	Adventitious shoots, plantlets	Hussain et al. (2007)
Theobroma cacao	Cambium	Callus	Archibald (1954)
Theobroma cacao	Mature embryo axes	Leaf, root development	Ibanez (1964)
Theobroma cacao	Stem, petiole	Callus, suspension culture	Townsley (1974)
Theobroma cacao	Seedling root, cotyledon, hypocotyl, stem, embryo	Callus, suspension culture	Hall and Collin (1975)
Theobroma cacao	Leaf, immature fruit	Callus	Searles et al. (1976)
Theobroma cacao	Immature embryos	Shoots and plantlets	Esan (1977)
Theobroma cacao	Embryo, anther	Callus	Prior (1977)
Theobroma cacao	Seedling shoot tips	Shoot	Orchard et al. (1979)
Theobroma cacao	Leaf, immature and mature embryos	Callus, roots, somatic embryos	Pence et al. (1979,1980)

Contd...

Table 2.1–Contd...

Plant Name	Source of Explant	Result	Reference
Theobroma cacao	Immature cotyledons	Callus and cell suspension	Tsai and Kinsella (1981, 2006)
Theobroma cacao	Embryos	Shoots from plumule callus	Esan (1982)
Theobroma cacao	Seedling shoot tips, axillary buds	Shoot growth proliferation	Passey and Jones (1983)
Theobroma cacao	Axillary buds	Multiple shoots	Blake and Mazwell (1984)
Theobroma cacao	Pericarp	Callus	Dublin (1984)
Theobroma cacao	Zygotic embryos, cotyledon callus	Somatic embryo-genesis germination	Kononwicz and Janick (1984a, b,c)
Theobroma cacao	Cotyledonary node	Bud out growth	Legrand *et al.* (1984)
Theobroma cacao	Zygotic embryos	Germination	Wang and Janick (1984)
Theobroma cacao	Asexual embryos	Fatty acid composition	Wright *et al.* (1984)
Theobroma cacao	Cotyledonary node	Shoots	Janick and Whipkey (1985)
Theobroma cacao	Shoot tips, axillary buds from mature tree	Bud proliferation	Litz (1986)
Theobroma cacao	Immature cotyledons	Somatic embryo-genesis, plantlets	Novak *et al.* (1986)
Theobroma cacao	Zygotic tissue	Callus, somatic embryos	Elhag *et al.* (1987)
Theobroma cacao	Zygotic tissue	Callus	Elhag *et al.*(1988)
Theobroma cacao	Meristem	Whole plant	Adu–Ampomah *et al.* (1988, 1992)
Theobroma cacao	Immature zygotic embryo	Somatic embryos, cryopreservation	Villalobos and Aguilar (1989)
Theobroma cacao		Suspension culture	Leathers and Scragg(1989)
Theobroma cacao	Axillary buds	Bud expansion and elongation	Flynn *et al.* (1990)
Theobroma cacao	Axillary buds	Shoots, plantlets	William *et al.* (1990)
Theobroma cacao	Shoot tip	Multiple shoots	Figueira *et al.* (1991a,b)
Theobroma cacao	Zygotic embryos	Growth suspension, cryopreservation	Pence (1991)
Theobroma cacao	Zygotic embryos	Somatic embryogenesis	Wen and Kinsella (1991)
Theobroma cacao	Cotyledons, somatic embryos	Micrografted somatic embryos on root stocks	Aguilar *et al.* (1992)
Theobroma cacao	Leaf hypocptyl, root	Callus, adventitious roots	Azahari – Othman *et al.* (1992)
Theobroma cacao	Seeds	Embryogenic cell differentiation	Chatelet *et al.* (1992)
Theobroma cacao	Seeds	Maturation of embryos	Pence (1992)
Theobroma cacao	Nucellus	Somatic embryos	Figueira and Janick (1993)

Contd...

Plant Tissue Culture: Emerging Trends

Table 2.1–Contd...

Plant Name	Source of Explant	Result	Reference
Theobroma cacao	Flower buds	Somatic embryogenesis	Lopez – Baez *et al.* (1993)
Theobroma cacao	Petals, nucellus	Somatic embryogenesis	Sondahl *et al.* (1993)
Theobroma cacao	Seedling explants	Shoots and somatic embryos	Figueira *et al.* (1994)
Theobroma cacao	Cotyledon	Embryoids, plantlets	Chantrapradist and Kancha-napoom (1995)
Theobroma cacao	Seedling explants	Somatic embryos	Figueira and Janick (1995)
Theobroma cacao	Petals of immature flower buds	Somatic embryogenesis	Tahardi and Mardiana (1995)
Theobroma cacao	Floral explants	Zygotic embryogenesis and Somatic embryogenesis	Alemanno *et al.* (1997, 2003)
Theobroma cacao	Floral explants	Plantlets via somatic embryogenesis	Li *et al.* (1998)
Theobroma cacao	Embryogenic callus	Cryopreservation	Florin *et al.* (2000)
Theobroma cacao	Nodes, shoot tips	Somatic embryogenesis, micropropagation and factors affecting genetic transformation	Traore (2000)
Theobroma cacao	Staminodes and petals	Somatic embryogenesis	Maximova *et al.* (2002, 2005, 2008)
Theobroma cacao	Staminodes	Repetitive embryogenesis	Santos *et al.* (2002)
Theobroma cacao	Floral explants	Somatic embryogenesis	Alemanno *et al.* (1997, 2003)
Theobroma cacao	Mature tissues	Somatic embryogenesis	Tan and Furtek (2003)
Theobroma cacao	Somatic embryos derived from nodal and apical explants	Shoots and plantlets	Traore *et al.* (2003)
Theobroma cacao	Somatic embryos	Cryopreservation	Fang *et al.* (2004)
Theobroma cacao	Staminodes, petals	Somatic embryogenesis	Guiltinan (2006, 2007)
Theobroma cacao		Cell suspension culture	Wen and Kinsella (2006)
Theobroma cacao		Zygoyic embryogenesis	Minyaka *et al.* (2007)
Theobroma cacao	Staminodes, petals	Somatic embryogenesis	Minyaka *et al.* (2008a,b)
Theobroma cacao		Somatic embryogenesis	Niemenak *et al.* (2008)
Theobroma cacao	Asexual embryos	Study of fatty acid composition in embryos	Pence *et al.* (2008)
Theobroma cacao	Virus infected explant tissue	Virus free somatic embryos	Quainoo *et al.* (2008)
Theobroma grandiflorum	Somatic embryos	Embryogenic callus	Janick and Whipkey (1988)

In *Sterculia* genus three species were tried for *in vitro* propagation, trees in this genus are moderate sized and evergreen, several species are grown in gardens for their handsome foliage and attractive fruits. *Sterculia alata* is moist deciduous forest tree whose seeds are edible and plant has medicinal value. In this plant somatic embryos were raised from immature embryos by Pandey (1998). *Sterculia foetida* is also ornamental tree but the foetid odour of flowers is a disadvantage. Wood of this plant is used in interior work and rough packing cases. Seed oil is used in surface coating and soap making industries. Coming to the medicinal importance of seed oil, it is used for treating skin diseases and rheumatism, leaves and bark are also having minor medicinal importance (Kirtikar and Basu, 1935). Anitha and Pullaiah (2001) used seedling explants to raise plantlets. Cotyledonary node was chosen as best explant source, when placed on the MS medium containing different concentrations of cytokinins alone or cytokinins in combination with different auxins. 4mg/l BAP was the best concentration to raise the shoots from cotyledonary nodes, while hypocotyls and shoot tip explants showed better response on the 2mg/l BAP supplemented MS medium (Anitha and Pullaiah, 2002). Shoots thus formed were placed on different auxins containing half strength medium for rooting. Among the tested concentrations of auxins 2mg/l IAA proved to be the best. Rooted plantlets were acclimatized to field conditions. Callus was produced using different auxins at various concentrations using seedling explants. Epicotyl and hypocotyl explants responded with white, compact callus at 4 mg/l 2, 4-D (Anitha and Pullaiah, 2001).

Sterculia urens is a moderate sized gum yielding tree. The gum of this tree is having high commercial value as it is used in textile, leather, cosmetic, food, dairy and pharmaceutical industries. Major part of gum is exported to USA. Timber is used for making toys and musical instruments. Bark and leaves are having medicinal importance, while seeds are edible. Over exploitation by blazing too deeply and exposing wood to borer attacks are the main reasons for the destruction of these trees. Purohit and Dave (1996) used cotyledonary node of seedling explant and reported shoot induction. They placed the explants on MS medium supplemented with 2mg/l BAP, for rooting they transferred these shoots to ¼ MS medium, after pulse treatment with 500 mg/l IBA for 10 minutes. Bahadur and Shailaja (1997) and Shailaja *et al.* (1997) reported shoot formation using seedling shoot tip and nodal explants. Sunnichan *et al.* (1998) used mature nodal explants as the source material and reported 6 shoots on average when placed on MS medium containing 6.62 µM BAP. These shoots rooted when transferred to ¼ MS medium containing 9.82 µM IBA. They reported nodular callus formation from hypocotyl explants cultured on MS medium supplemented with 4.52 µM 2,4-D and 8.90 µM BAP. Somatic embryos were formed from the nodular callus when transferred to medium supplemented with 0.45 µM TDZ for two days. 30 per cent of somatic embryos formed into plantlets when placed on ¼ MS basal medium without any growth regulators. Shoots and somatic embryos were acclimatized later. Anitha and Pullaiah (1999) reported that 2 mg/l BAP was suitable concentration for shoot regeneration. They studied the effect of coconut milk as additive to the shoot improvement, but the result was not encouraging. They rooted the shoots on ½ MS medium supplemented with 1 mg/l IBA. In callus studies carried by them profuse callus was observed from cotyledonary explants placed on 2 mg/l 2, 4-D supplemented medium. Hussain *et al.* (2007) reported highest shoot multiplication using cotyledonary node only, when placed on 2.27 µM TDZ$_+$0.1 per cent ascorbic acid supplemented MS medium. They rooted shoots on ¼ MS medium containing 9.80 µM IBA after 8 weeks and acclimatized the plants using autoclaved vermiculite. Shoot culture was established by them by repeatedly sub-culturing the original cotyledonary node on medium containing 0.45 µM TDZ, even after second harvest.

Theobroma cacao is tree grown in humid tropical regions, which is a cash crop, as it is an important raw material of cocoa used in chocolate industry and for extraction of cocoa butter. Seeds and clonal propagation with root cuttings and grafting are conventional means of propagation (Eskes, 2005). First report of *in vitro* culture was by Archibald (1954). He cultured cambium and raised callus when grown on Gautheret (1952) and White (1943) media supplemented with coconut milk. Ibanez (1964) cultured matured embryos on Rudolph and Cox medium (1943), he used glucose replacing sucrose and observed leaf and root development. Townsley (1974) cultured stem and petiole explants and was able to develop callus and suspension cultures on Gamborg and Eveleigh medium (1968). Searles *et al.* (1976) raised callus by culturing leaf and immature fruits on MS + Cysteine(10 mg/l) + Kn (2-6 mg/l) or Z (0.25 – 1.0 mg/l)+ IAA (0.5mg/l). Hall and Collin (1975) used seedling explants root, cotyledon, hypocotyl, stem and embryo. They reported callus and suspension cultures by placing the explants on MS medium supplemented with Vitamins (2X) + CM (10 per cent) + IAA (2 mg/l + IBA (2mg/l) + Kn (0.1 mg/l). Prior (1977) used embryo and anther explants, he observed callus induction from these explants on medium formulated by Hall and Collin (1975). Esan (1977) cultured nucellus, ovule and anther explants on MS medium supplemented with Sucrose (5 per cent) + CH (500 mg/l). He observed callus development only from anthers and nucellus while necrosis was observed in ovule culture. Pence *et al.* (1979 and 1980) cultured leaf on MS supplemented with CH (2000 mg/l)+ CM (10 per cent) + IAA, NAA or 2, 4-D (0.1-2.0 mg/l) + BAP (1 mg/l) or Z (0.1- 2.0 mg/l) and reported callus and root formation. By culturing embryos which were mature, immature and near mature on MS medium containing CH (2000mg/l) + CM (10 per cent) + IAA (13.6 mg/l) or NAA (1.5- 6 mg/l) or 2,4-D (1.6 mg/l), and raised somatic embryos. Orchard *et al.* (1979) cultured seedling shoot tips and reported shoot growth on LS medium (Linsmaier and Skoog, 1965), supplemented with GA$_3$ (5-10 mg/l). Tsai and Kinsella (1981) reported callus and suspension culture placing immature cotyledons on MS medium supplemented with CM (10 per cent) + 2, 4-D (1mg/l) + Kn (0.2 mg/l) or on (Gamborg *et al.*, 1968) B$_5$ medium containing 2,4-D (1 mg/l) + Kn (0.2 mg/l). Pence *et al.* (1981) reported the effect of sucrose in regulation of fatty acid composition. Esan (1982) reported shoot formation from plumule callus using embryos as explant source on MS medium augmented with calcium nitrate (300 mg/l) + Peptone (10-100 mg/l) + Sucrose (2-5 per cent) + NAA (0.01- 2 mg/l) + Kn (0.01 + 1 mg/l). Passey and Jones (1983) used seedling shoot tips and axillary buds and reported shoot growth and callus proliferation when they used MS medium supplemented with BAP (0.2 mg/l) or Z (2 mg/l) or Zeatin riboside (3.5 mg/l). Dublin (1984) placed pericarp explants on MS medium containing 2, 4- D (0.5- 1 mg/l) + BAP (1 mg/l) or Z (1 mg/l) and reported callus formation. Kononowicz and Janick (1984a, c) used asexual embryos as explants and studied the influence of carbon source on the growth and development. Kononowicz *et al.* (1984b), reported somatic embryogenesis using the somatic embryo cotyledon callus, when placed on MS medium containing CH (2000 mg/l) + CM (10 per cent) + 2, 4- D (0.001- 0.01 mg/l).

Blake and Mazwell (1984) reported propagation of cocoa using axillary buds. Wang and Janick (1984) reported somatic embryo germination placing on ½ MS medium. Wright *et al.* (1984) investigated the effect of different factors affecting fatty acid content and composition in asexual embryos. Legrand *et al.* (1984) used cotyledonary node buds and reported bud outgrowth when placed on Heller's medium (1953) supplemented with Iron (3X) + CM (10 per cent) + NAA (0.1 mg/l). Janick and Whipkey (1985) used cotyledonary node as explant and reported axillary bud proliferation. Novak *et al.* (1986) used immature cotyledons and reported somatic embryogenesis using MS medium supplemented with CH (2000 mg/l) + NAA (1.8 mg/l). Plantlets were reported by them when placed on MS medium containing Z (0.22 mg/l) + NAA (0.0018 mg/l). Litz (1986) reported bud proliferation using shoot tips and axillary buds as explant source obtained from mature trees, when placed on MS + Z (2.2 mg/l) or BAP (0.2- 2 mg/l). Elhag *et al.* (1987) made comparative study on effect of sucrose, glucose and fructose, on cacao somatic embryogenesis using calli induced from zygotic tissues. They observed that there was increasing embryogenesis when sucrose was replaced by glucose or fructose. Adu-Ampomah *et al.* (1988) reported initiation and growth of somatic embryos in cacao. Leathers and Scragg (1989) studied the effect of different temperatures to determine pattern, extent of change in culture growth, lipid content and fatty acid composition. They found 30°C as optimum temperature. Temperature between 15-20°C increased the levels of polyunsaturated fatty acids particularly linoleic acid of total lipid extract in suspension cultures when compared to mature leaf tissues. Flynn *et al.* (1990) and William *et al.* (1990) used axillary bud for inducing shoots. Wen and Kinsella (1991) reported somatic embryos using zygotic embryos. Figueria *et al.* (1991 a,b and 1995) reported effect of CO_2 and light on shoot and somatic embryo development.

Aguilar *et al.* (1992) micrografted somatic embryos without cotyledons to seedling root stocks and observed complete plantlet formation, these plantlets were later transferred to soil. Li *et al.* (1992) and Chatelet *et al.* (1992) reported somatic embryogenesis using nucellus as explant source, while Chatelet *et al.* also used inner integument as explant source. Adu- Ampomah *et al.* (1992) used meristem tips excised from terminal buds of young cocoa plants (2-4 week old) cultured with 1 µM Z and various concentration of NAA + GA_3. Gamma ray irradiation at low doses (1-4 Gy) helped to reduced callus formation. Roots were induced on ½ MS medium supplemented with 10 µM IBA + 0.1 µM IAA and reduced sucrose of 15000 mg/l. Azahari-Othman *et al.* (1992) demonstrated the role of plant growth regulators in the regeneration process. Pence (1992) investigated the role of abscisic acid in cacao embryo maturation.

Janick (1993) stressed the applications of somatic embryos in agriculture. Sondahl *et al.* (1993) demonstrated the advantage of nonsexual explants like petal and nucellus in producing somatic embryos. Thereby *in vitro* method can play important role to establish superior germplasm of *Theobroma cacao*. Chantrapradist and Kanchanapoom

(1995) reported maximum callus growth on MS medium supplemented with 2, 4 –D (0.01 mg/l) which was white and compact. When cotyledon explants were placed on medium containing CM+ NAA, white nodular embryoids were observed. When these embryoids were transferred to medium devoid of hormones developed normal plantlets. Figueira and Janick (1993) also reported somatic embryos form nucellar tissue, while Lopez- Baez *et al.* (1993), reported somatic embryogenesis from immature floral explants like staminode and petals. They placed these explants on DKW medium (Driver and Kuniyuki, 1984) supplemented with glucose and sucrose. Chandel *et al.* (1995) reported desiccation and freezing sensitivity in cocao seeds. To improve the late phases of embryogenesis, Alemanno *et al.* (1997) compared zygotic embryogenesis and somatic embryogenesis. They developed somatic embryos using immature floral explants and observed that morphological abnormalities of somatic embryos lacked starch and protein reserves but increased water content. Addition of sucrose and abscisic acid to the maturation medium was effective in increasing reserve synthesis and resulted in higher germination conversion and acclimatization rates.

Li *et al.* (1998) reported somatic embryogenesis, using staminode explants cultured on DKW salt supplemented medium with glucose (20 g/l) + 2,4-D (9 µM) + TDZ (22.7 nM) for callus and subcultured this callus to WPM + glucose (20 g/l) + 2,4-D (9 µM) + Kn (1.4 nM) to raise somatic embryos, later these were transferred to hormones free DWK medium. TDZ concentration in the medium influenced the callus growth and the frequency of embryogenesis. Florin *et al.* (2000) reported cryopreservation using embryogenic callus in *Theobroma cacao*. Guiltinan *et al.* (2000) reported the result of their experiments on acclimatization and field evaluation of cacao plants produced via somatic embryogenesis and other *in vitro* propagated plants. Maximova *et al.* (2002) reported the efficiency, genotypic variablility and cellular origin of primary and secondary embryogenesis of *T.cacao*. Traore (2000) and Traore *et al.* (2003) reported mass propagation of *Theobroma cacao* using somatic embryos developed from nodal and apical stems. They used glucose and sucrose for plantlet formation from somatic embryos. Nodal explants were cultured on TDZ containing medium while for root induction short treatment of IBA was required. Tan and Furtek (2003) investigated that the induction of somatic embryogenesis from tissues in unopened flower buds with respect to physiological age, type of floral explant, genotype, medium composition and hormones. Finally 2-3 old staminodes were found to be the suitable explant source for somatic embryogenesis. Alemanno *et al.* (2003) reported localization and identification of phenolic compounds in cacao somatic embryogenesis.

Fang *et al.* (2004) demonstrated the importance of sucrose and ABA in desiccation and cryopreservation of cocoa somatic embryos. In this study different concentrations of sucrose (0.5, 0.75 and 1M) were used for pre culture, regardless of level of ABA combination, with drying duration of 4h allowed highest embryo survival after cryopreservation. Pre culture for minimum of 7 days and early cotyledonary stage has improved survival rate of embryos after desiccation. According to them better understanding of cryoprotectant action, embryo developmental stage and recovery pathway may further contribute to the development of cryopreservation of other tropical species with recalcitrant seeds.

Maximova *et al.* (2005) reported an integrated system for propagation of *Theobroma cacao*. They presented a protocol after studying various experiments of cocoa somatic embryogenesis. Their final conclusion is that somatic embryogenesis depends on genotypic variations. Protocol developed and optimized for one genotype can yield more number of somatic embryos in that genotype only, while other genotypes produce fewer embryos during primary response, but increases in number during secondary embryogenesis. Therefore, protocol of one needs further optimization for the other genotype. Salazar *et al.* (2005) cultured anthers on MS+ Sucrose (45 g/l) + CM (10 per cent) + 2,4-D (3mg/l) containing medium with varying concentrations of CH (1, 3 and 5g/l). They finally reported that the growth of callus decreased with increasing concentrations of tested CH. Addition of CM proved to be beneficial, while increasing CH accelerated browning of tissues. Guiltinan (2006) investigated the effect of five different carbon sources (Glucose, fructose, maltose, sorbitol and sucrose) and 2 explant sources (petals and staminodes) on cacao somatic embryogenesis and reported staminodes as best explant source. Among four of the carbon sources, except maltose, responded similarly in weight of the embryos, total shoot and root production, but shoots produced on glucose had normal leaves, while cotyledon like leaves were produced on other carbon sources incorporated medium. Wen and Kinsella (2006) studied that fatty acid composition of suspension cell cultures of *T.cacao* altered by culture conditions. Tsai and Kinsella (2006) investigated changes of lipids in growth and maturation of cells and calli in culture. Minyaka *et al.* (2007) reported an implication of cysteine, glutathione and cysteine synthase in cacao zygotic embryogenesis. Fang *et al.* (2007) has conducted screening experiments for knowing about the genetic fidelity of cocoa somatic embryos after cryopreservation. Guiltinan (2007) presented the applications of the cocoa in forestry.

Maximova *et al.* (2008) conducted field test for plants derived from somatic embryogenesis and reported that these plant growth parameters are similar to plants propagated by traditional methods except that they are shorter in height. They have investigated the effect of different concentration of $MgSO_4$ and K_2SO_4 on somatic embryo differentiation in cocoa plants. They used DKW basal salts supplemented with glucose (20 g/l) + 2, 4 -D (18 µM) + TDZ (45.4 nM) for primary callus development with varying K_2SO_4 and $MgSO_4$. For secondary callus growth they used K_2SO_4 (8.946 mM) and $MgSO_4$ (3.0 mM) added along with DKW salts and 0.5 ml/l DKW vitamins + glucose (20 g/l) + 2,4-D (9 µM)+ Kn (250 µM)+ phyta gel (0.2 per cent w/v). In screening the effect of K_2SO_4 and $MgSO_4$ in 3 different steps *i.e.*, i) primary callus, ii) secondary callus and iii) embryo development, it was found that sulphate influence was most effective. According to them sulphate concentration in the medium is responsible for direct and indirect somatic embryogenesis. Indirect somatic embryogenesis was reported when K_2SO_4 and $MgSO_4$ concentrations are below 71.568 mM and 24.0 mM, while direct somatic embryogenesis occurred from cells when $MgSO_4$ concentration was 24.0mM. Expression of direct somatic embryogenesis increases with sulphate and may help in the inhibition of callogenesis, which may also lead to directly compete and determined the change of their differentiation pathway to become embryogenic. This result also supports the observation that stress related compounds are required for somatic cells to become embryogenic. Repetitive primary somatic embryo clusters were observed at high sulphate content. Their results also showed that to overcome the recalcitrance of many cacao genotypes a screening for suitable $MgSO_4$ and K_2SO_4 concentration was required.

Minyaka *et al.* (2008a,b) when investigated the effect of sulphate as K_2SO_4 found to improve somatic embryogenesis of the tested 5 genotypes using DKW salts. Niemenak *et al.* (2008) used a temporary immersion bioreactor for multiplication of somatic embryos of Cacao. The number of embryos formed after 3 months of culture was significantly higher than in solid medium and it also improved embryo development. Mature embryos derived from temporary immersion system pretreated with 6 per cent sucrose converted into plantlets after direct sowing. Quainoo *et al.* (2008) reported that somatic embryos when developed from infected callus tissue contained no detectable virus and therefore suggested that *in vitro* method is an effective one for the elimination of virus from infected trees.

Somatic embryos were used for producing embryogenic callus by Janick and Whipkey (1988) in *Theobroma grandiflorum.*

References

Adu–Ampomah, Y., F.J.Novak., R.Afza., M.Van – Duren. 1992.Meristem tip culture of cacao (*Theobroma cacao* L.). Tropical Agriculture. 69: 268-272.

Adu – Ampomah, Y., F.J.Novak., R.Afza., M.Van – Duren and M.Perea – Dallos. 1988. Initiation and growth of somatic embryos of cacao (*Theobroma cacao* L.). Cafe Cacao The. 32: 187-200

Aguilar, M.E., V.M.Villalobs and N.Vasquez. 1992. Production of cocoa plants (*Theobroma cacao* L.) via micrografting of somatic embryos. *In vitro* cellular and Dev. Biol.- Plant. 28: 15-19

Alemannno L., M.Berthouly, N.Michauz-Ferriere. 1997. A comparison between *Theobroma cacao* L. zygotic embryogenesis and somatic embryogenesis from floral explants. *In vitro* Cell Dev Biol- Plant. 33: 163-172.

Alemanno.L, T.Ramos, A.Gargadenec, C.Andry and N.Ferriere. 2003. Localization and Identification of phenolic compounds in *Theobroma cacao*. L. somatic embryogenesis. Ann Bot. 92: 613-623.

Anita, S. and T.Pullaiah. 1999. *In vitro* propagation studies of *Sterculia urens* Roxb. In: P.B.Kavi Kishor (ed).Plant Tissue Culture and Biotechnology, Emerging Trends. Universities Pres Ltd., Hyderabad, India. pp. 146-150.

Anitha, S. and T.Pullaiah. 2001. *In vitro* propagation of *Sterculia foetida* Linn. (Sterculiaceae). Plant Cell Biotechnology and Molecular Biology. 2: 139-144.

Anitha, S. and T.Pullaiah. 2002. Shoot regeneration from hypocotyl and shoot tip explants of *Sterculia foetida* L. derived from seedlings. Taiwania. 47: 62-69.

Anuradha, T and T.Pullaiah. 2001. Effect of hormones on the organogenesis and the somatic embryogenesis of an endangered tropical forest tree- *Hildegardia populifolia* (Roxb.) Schott. & Endl. Taiwania. 46: 62-74.

Archibald, J.F.1954. Culture *in vitro* of cambial tissue of cocoa. Nature. 173: 351-352.

Ayensu, E. S.1978. Medicinal plants of West Africa. Reference Publication International, Michigan, Michigan.

Azahari- Othman, Radzali- Muse and Marziah- Mahmood. 1992. *In vitro* regeneration of cocoa, *Theobroma cacao* L. treated with plant growth regulators. Proceedings of the Seventeenth Malaysian Biochemical Society Conference, Bamgi, Selangor, Malaysia.

Bahadur, B and A.Shailaja. 1997. Multiple shoot formation from shoot apex of karaya gum, *Sterculia urens* Roxb. J. Swamy Bot. Cl. 14: 114- 116.

Blake, J and P.Mazwell. 1984. Tissue culture propagation of cacao by the use of axillary buds. International Conference on Cocoa and Cocounts. pp.1-12.

Chandel, K.P.S., R.Chaudhury., J.Radhamani and S.K.Malik. 1995. Desiccation and freezing sensitivity in recalcitrant seeds of tea, cacao and jack fruit. Ann. Bot. 76: 443-450.

Chantrapradist, C. and K.Kanchanaoom.1995. Somatic embryo formation from cotyledonary culture of *Theobroma cacao* L. J.Sci.Soc.Thailand. 21: 125-130.

Chatelet, P., N.Michaux- Ferriere and P.Dublin. 1992. Embryogenic potential in nucellus and inner integument tissue cultures of immature cacao seeds. Comptes Rendus De l. Academie Des Sciences Serie III Science De La Cie 315: 55-62.

Chopra, R.N., S.L.Nayar and I.C.Chopra.1956. Glossary of Indian medicinal plants, 1st edition. CSIR, New Delhi, India.pp.131.

Courrier, K.1992. Losses of biodiversity and their causes. In: K.Courrier (ed). Global Biodiversity Strategy- Guidelines for action to save, study and use earths biotic wealth sustainably and equitably. pp.7. WRI, IUCN& UNEP Publication, U.K.

Das, P., U.C.Basak and A.B.Das.1997. Metabolic changes during rooting in pregirdled stem cuttings and air-layers of *Heritiera*. Bot.Bull.Acad.Sin. 38: 91-95.

Doss, E.L., B.Bertrand and A.Sidam.1994. Micropropagation *in vitro* of *Cola nitida* Schott & Endl. Cafe Cacao The 38:57-60.

Driver, J.A. and A.H.Kuniyuki. 1984. *In vitro* propagation of Paradox walnut root stock. Hortscience.19: 507-509.

Dublin, P. 1984. Cacao. In: P.V.Ammirato, D.A.Evans, W.R.Sharp and Y.Yamada (ed.) Handbook of plant cell culture, Vol. 3. Macmillan, New York, pp. 541-563.

Ebana, R. U. B., Madunagu, B. E., Ekpe, E. D., and Otung, I. N. 1991. Microbiological exploitation of cardiac glycosides and alkaloids from *Garcinia kola, Borreria ocymoides, Kola nitida* and *Citrus aurantifolia*. J.Applied Bacteriol. 71:398-401.

Elhag, H.M., A.Whipkey and J.Janick.1987. Induction of somatic embryogenesis from callus in response to carbon source and concentration. Rev.Theobroma. 17: 153-162.

Elhag, H.M., A.Whipkey and J.Janick.1988. Factors affecting asexual embryogenesis via callus in *Theobroma cacao*. Arab Gulf J. Scientific Research Agr. Biol.Sci., B6: 31- 43

Esan, E.B. 1977. Tissue culture studies on cacao (*Theobroma cacao* L.). A supplementation of current research. Proceedings, Fifth International Cacao Research Conference, Cacao Res. Inst., Ibadan, Nigeria. pp. 116-125.

Esan, E.B. 1982. Shoot regeneration from callus derived from embryo axis cultures of *Theobroma cacao in vitro*. Turrialba. 32: 359-364.

Eskes, A. 2005.Proceedings of the International Workshop on Cocoa Breeding for Improved Production Systems: INGENIC International Workshop on Cocoa Breeding, F.Bekele, M.End, & A.Eskes, eds (INGENIC and Ghana Cocoa BSoard, Accra, Ghana), pp.1-10.S

Fang, J.Y., A.Wetten, P.Hadley.2004. Cryopreservation of cocoa (*Theobroma cacao* L.) somatic embryos for long-term germplasm storage. Plant Sci. 166: 669-675.

Fang, J.Y., C.R.Lopes and A.Wetten.2007. Screening for genetic fidelity of cocoa (*Theobroma cacao* L.) somatic embryos following cryopreservation. Cryobiology. 55: 368.

Fereday, N., Gordon, A and G. Oji.1997. Domestic market potential for tree products from farms and rural communities: experience from Cameroon: NRI Socio-Economic Serices Report No. 13. Natural Resources Institute (NRI), Chatham.

Figueira, A and J.Janick. 1993. Development of nucellar somatic embryos of *Theobroma cacao* L. Acta Horticulturae. 336: 231 – 236.

Figueira, A and J.Janick. 1995. Somatic embryogenesis in cacao (*Theobroma cacao* L.). In: S.Jain, P, Gupta and R.Newton (Eds.), Somatic Embryogenesis in Woody Plants, vol.2. Netherlands, Klumer Academic Publishers, pp. 291-310.

Figueria, A., A.Whipkey and J.Janick. 1991a. Increased CO_2 and light promote *in vitro* shoot growth and development of *Theobroma cacao*. J.Am. Soc. Hort. Sci. 116: 585-589.

Figueria, A., A.Whipkey and J.Janick.1991b. Effect of CO_2 levels on *in vitro* growth and development of shoots and somatic embryos of *Theobroma cacao* L. Hort.Science. 26:757.Abstr.535.

Figueira, A., A.Whipkey and J.Janick.1994. Elevated CO_2 facilitates micropropagation of *Theobroma cacao* L. Proc. Int.Cacao Conference, Challenges in the 90s. Kuala Lumpur 1991.

Florin, B., E.Brulard and V.Petiard. 2000. *In vitro* cryopreservation of cacao genetic resources. In: F.Engellmann, H.Takagi (Eds.), Cryopreservation of Tropical Plant Germplasm, IPGRI, pp. 344-347.

Flynn, W.P., L.J.Glicenstein and P.J.Fritz. 1990. *Theobroma cacao* L. An axillary bud *in vitro* propagation procedure. Plant Cell Tiss. Org. Cult. 20: 111-117.

Freytag.G.F. 1951. A revision of the genus *Guazuma, Ceiba* (Hond.). 1: 193-225.

Gamborg O.L and D.E.Eveleigh.1968. Culture methods and detection of gluconases in suspension cultures of wheat and barley. Can. J. Biochem. 46: 417-421.

Gamborg, O.L., RA.Miller and K.Ojima. 1968. Nutrient requirements of suspension cultures of soybean root cells. Exp. Cell. Res. 50: 151-158.

Gautheret, R.J.1952. Remarques sur l' emploi du lait de coco pour realization des cultures de tissues vegetaux. C.R.Acad.Sci.Paris. 235:1321-1324.

Guiltinan, M.J. 2006. Effects of Carbon source and explant type on somatic embryogenesis of four Cacao Genotypes. Hort. Science. 41: 753-758.

Guiltinan, M.J.2007. Cacao. In E.C.Pua and M.R.Davey. (Eds), Biotechnology in Agriculture and Forestry, Transgenic crops. Springer Verlag, Berlin Heidelberg, pp. 497-518.

Guiltinan, M.J., C.Miller, A.Traore and S.Maximova. 2000. Greenhouses and field evaluation of orthrotropic cacao plants produced via somatic embryogenesis, micro and macro propagation: 13th International Cocoa Research Conference.

Hall, T.R.H and H.A.Collin. 1975. Initiation and growth of tissue culture of *Theobroma cacao*. Ann. Bot. 39: 555-570.

Heller, R. 1953. Recherches sur la nutrition minerale des tissues vegetaux cultivies *in vitro*. Ann. Sci. Nat. Bot. Bio. Veg Ser 2, 14: 1-223.

Hussain, T.M., T.Chandrasekhar and G.R.Gopal.2007. High frequency shoot regeneration of *Sterculia urens* Roxb. an endangered tree species through cotyledonary node cultures. African J.Biotech. 6: 1643-1649.

Ibanez, M.L.1964.The cultivation of cacao embryos in sterile cultures. Trop.Agri.(Trinidad).41:325-328.

Janick, J.1993. Agricultural uses of somatic embryos. Acta Horticulturae.336: 207-215.

Janick, J and A. Whipkey.1985. Axillary proliferation from cotyledonary nodal tissue of Cacao. Rev. Theobroma.15: 125-131.

Janick, J and A.Whipkey.1988. Somatic embryogenesis in *Theobroma grandiflorum* Hort.Sci. 23: 807.

Kirtikar, K.R and B.D. Basu. 1935. Indian Medicinal Plants. Vol.1, Lalit Mohan Basu, Allahabad., India, pp389.

Kononwicz, A.K and J. Janick. 1984a. *In vitro* development of zygotic embryos of *Theobroma cacao*. J. Am. Soc. Hortic.Sci. 190: 266-269.

Kononwicz, A.K and J. Janick. 1984b. Asexual embryogenesis via callus of *Theobroma cacao* L. Z.Pflanzen Physiol. 113: 347-358.

Kononwicz, A.K and J. Janick. 1984c. The influence of carbon sources on growth and development of asexual embryos of *Theobroma cacao*. Physiol. Plant. 61: 155-162.

Krogstrup, P., S.Baldursson and J.V.Norgard.1992. *Ex situ* genetic conservation by use of tissue culture. Opera Bot. 113:49-53.

Kumar, G., G.S.Banu and A.G.Murugesan. 2009. Anti- diabetic activity of *Helicteres isora* L. bark extracts on streptozotocin-induced diabetic rats. International Journal of Pharmaceutical Sciences and Nanotechnology.1: 379-382.

Leathers, R.R. and A.H.Scragg.1989. The effect of different temperatures on the growth, lipid content and fatty acid composition of *Theobroma cacao* cell suspension cultures. Plant science. 62: 217-227.

Legrand, B.C.Cilas and E.Mississo. 1984. Comportement des tissu de *Theobroma cacao* L. var. *amelonada* cultives *in vitro*. Cafe Cacao The. 28: 245-250.

Li, Z., A.Traore, S.Maximova and M.J.Guiltinan. 1998. Somatic embryognesis and plant regeneration from floral explants of cacao (*Theobroma cacao* L.) using thidiazuron. *In vitro* Cell.Dev.Biol.Plant. 34: 293-299.

Linsmaier, E.M and F.Skoog. 1965. Organic growth factor requirements of tobacco tissue cultures. Physiol Plant. 18: 100-127.

Litz, R.E. 1986. Tissue culture studies with *Theobroma cacao*. In: P.S.Dimick (ed). Proceedings Cacao Biotechnology Symposium, Pennsylvania State Univ., Park. pp.111-120.

Litz. R.E and D.J.Gray. 1992. Organogenesis and somatic embryogenesis. In: F.A.Hammer Schlag and R.E.Litz (eds.). Biotechnology of perennial fruit crops. CAB International, Wallingford, UK.pp.3-34.

Lopez- Baez.O., H.Bollon., A.Eskers and V.Petriard. 1993. Somatic embryogenesis and plant regeneration from flower parts of cocao (*Theobroma cacao* L). Comptes Rendus de 1 Academis des Sciences. Series 3 Sciences de la vie, 316: 519-584.

Maruyama, E., K.Ishii, I.Kinosuita, K.Ohba and A.Saito.1996. Micropropagation of Bolaina blanca (*Guazuma crinita* Mart.), a fast- growing tree in the Amazon region. J.For.Res.1:211-217.

Maruyama, E., I.Kinosuita, K.Ishii, H.Shigenaga, K.Ohba and A.Saito.1997a. Alginate-encapsulated Technology for the Propagation of the Tropical Forest Trees: *Cedrela odorata* L., *Guazuma crinita* Mart., and *Jacaranda mimosaefolia* D.Don. Silvae Genetica 46: I7-23.

Maruyama, E., I.Kinosuita, K.Ishii, H.Shigenaga, K.Ohba and A.Saito.1997b. Germplasm conservation of the tropical forest trees, *Cedrela odorata* L., *Guazuma crinita* Mart., and *Jacaranda mimosaefolia* D.Don.,by shoot tip encapsulation in calcium – alginate and storage at 12-25°C. Plant Cell Reports.16:393-396.

Maruyama, E., K.Ishii, I.Kinosuita,, H.Shigenaga, K.Ohba and A.Saito.1997c. Micropropagation of *Guazuma crinita* Mart. by root and petiole Culture. *In Vitro* Cell Dev.Biol-Plant. 33: 131-135.

Mascarenhas, A.F and E.M.Muralidharan. 1989. Tissue culture of forest trees in India. Curr. Sci. 58: 606-610.

Maximova, S.N., L.Alemanno, A.Young, N.Ferriere, A.Traore and M.J.Guiltinan.2002. Efficiency, genotypic variability, and cellular origin of primary and secondary somatic embryognesis of *Theobroma cacao* L. *In Vitro* Cell Dev Biol. Plant. 38: 252-259.

Maximova, S.N., A.Young, S.Pishak, C.Miler, A.Traore and M.J.Guiltinan. 2005. Integrated system for propagation of *Theobroma cacao* L. In: S.M.Jain and P.K. Gupta. (Eds). Protocol for Somatic Embryogenesis in Woody Plants, Series: Vol.77, Springer, Dordrecht, The Netherlands. Pp 209-229.

Maximova, S.N., A.Young, S.Pishak and M.J.Guiltinan. 2008. Field performance of *Theobroma cacao* L. plants propagated via somatic embryogenesis. *In Vitro* Cell.Dev.Biol.Plant. 44: 487-493.

Minyaka, E., N.Niemenak, Fotso, A.Sangare and D.N.Omokolo. 2008a. Effect of $MgSO_4$ and K_2SO_4 on somatic embryo differentiation in *Theobroma cacao* L. Plant Cell Tissue Organ Culture. 94: 149-160.

Minyaka, E., N.Niemenak, E.K.Koffi,A.E.Issali A.Sangare and D.N.Omokolo. 2008b. Sulphate supply promotes somatic embryogenesis in *Theobroma cacao* L. Journal of Biological Sciences. 8: 306-313.

Minyaka, E., N.Niemenak, N.M.S.Soupi, A.Sangare and D.N.Omokolo. 2007. Implication of cysteine, glutathione and cysteine synthase in *Theobroma cacao* L. zygotic embryogenesis. Biotechnology. 6: 129-137.

Mukhopadhyay, K., J.Choudhury and S.Maity.1991. National seminar on Conservation and management of mangrove ecosystem with special reference to Sunderbans.(Dec,6-8). pp35.

Murashrige.T and F.Skoog. 1962. A revised medium for rapid growth and bioassays with tobacco tissue culture. Physiol. Plant. 15: 473-497.

Naskar, K.R., D.Ghosh, N.Sen, R.N.Mandal and A.K.Sarkar.1997. Mangrove ecology of the Indian Sundarbans: its impact on the rural economy and coastal environment. J. Interacademicia. 1: 49-60.

Nayar, M.P.and A.R.K.Sastrry.1990. Red Data Book of Indian Plants.Vol.3. Botanical Survey of India, Calcutta.

Niemenak, N., K.Saaresurminski, C.Rhosius, D.Omokolo Noloumou and R.Lieberei. 2008. Regeneration of somatic embryos in *Theobroma cacao* L. in temporary immersion bioreactor and analyses of free amino acids in different tissues. Plant cell Reports. 27: 667-676.

Novak, F.J., B.Donini, G.Owusu. 1986. Somatic embryogenesis and *in vitro* plant development of cocoa (*Theobroma cacao*.) In: Proc Int Sym Nucelar techniques and *in vitro* culture for plant improvement. IAEA, Vienna, pp. 443-449.

Orchard, J.E., H.A.Collin and K.Hardwick. 1979. Culture of shoot apices of *Theobroma cacao*. Physiol. Plant. 47: 207-210.

Pandey, S. 1998. Plant regeneration through somatic embryogenesis in two tropical trees, *Mangifera indica* and *Sterculia alata* Roxb. Ph.D.Thesis, Banaras Hindu University, Varanasi.

Passey, A.J and O.P.Jones. 1983. Shoot proliferation and rooting *in vitro* of *Theobroma cacao* L. Amelonado J.Hort. Sci. 58: 589-592.

Pence, V.C.1991. Cryopreservation of immature embryos of *Theobroma cacao*. Plant Cell Rep. 10: 144-147.

Pence, V.C. 1992. Abscisic acid and the maturation of cacao embryos *in vitro*. Plant Physiol. 98: 1391-1395.

Pence, V.C., P.M.Hasegawa and J.Janick. 1979. Asexual embryogenesis in *Theobroma cacao* L. J. Am. Soc. Hort. Sci.104: 145- 148.

Pence, V.C., P.M.Hasegawa and J.Janick. 1980. Initiation and development of sexual embryos of *Theobroma cacao* L. *In vitro.* Z. Pflanzen Physiol. 98:1-14.

Pence, V.C., P.M.Hasegawa and J.Janick. 1981. Sucrose- mediated regulation of fatty acid composition in asexual embryos of *Theobroma cacao.* Physiol Plant. 53: 378-384.

Prior, C.1977. Growth of *Oncobasidium theobromae* Talbot & Keane in dual culture with callus tissue of *Theobroma cacao* L. J.Gen. Microbiol. 99: 219-222.

Purohit, S.D and A.Dave. 1996. Micropropagation of *Sterculia urens* Roxb. An endangered tree species. Plant Cell Rep. 15: 704-706.

Quainoo, A.K., A.C.Wetten and J.Allainguillaume. 2008. The effectiveness of somatic embryogenesis in eliminating the cacoa swollen shoots virus from infected cocoa trees. Journal of virological methods. 149: 91-96.

Rudolph, L.F and L.C.Cox.1943. Factors influencing the germination of iris seed and the relation of inhibiting substances to embryo dormancy. Proc.Am.Soc. Hortic. Sci. 43: 284-300.

Salazar, E., D.Torrealba, Luis and M.C.Torrealba. 2005. Hydrolyzed casein inhibits the development of callus from anthers of cocoa cultivated *in vitro.* Agronomia rrop. 56: 497-505.

Santos, M., E. Albuquerque de Barros, M.Tinoco, A.Brasileiro and F.Aragao.2002. Repetitive somatic embryogenesis in cacao and optimization of gene expression by particle bombardment. J. Plant Biotechnol. 4: 71-76.

Searles, B.R., P.de T.Alvim and W.R.Sharp.1976. Hormonal control of cellular proliferation in cultured callus derived from *Theobroma cacao* L. Rev. Theobr. 6: 77-81.

Shailaja, A., D. Laxmi, E.Chamundeswri and B.Bahadur.1997. Multiple shoot induction from nodal explants of *Sterculia urens* Roxb. an important gum yielding tree. J.Swamy Bot.Cl.14:98-100.

Shriram, V., V.Kumar and M.G.Shitole. 2008. *In vitro* propagation through nodal explants in *Helicteres isora* L. a medicinally important plant. J. Plant Biotech. 34: 1-7.

Sondahl, M.R., S.Liu., C.Ballato and A.Bragin. 1993. Cacao somatic embryogenesis. Acta Horticulture, 336: Second International Symposium on *in vitro* culture and horticultural breeding, Baltimore. M.D.USA. pp.245-248.

Sunnichan, V.G., K.R.Shivanna and H.Y.Mohan Ram. 1998. Micropropagation of gum karaya (*Sterculia urens*) by adventitious shoot formation and somatic embryogenesis. Plant cell Rep. 17: 951-956.

Tahardi, J.S and N.Mardiana. 1995. Cacao regeneration via somatic embryogenesis Menara-Perkebunan.63:3-7.

Tan, C.L and D.B.Furtek. 2003. Development of an *in vitro* regeneration system for *Theobroma cacao*, from mature tissues, Plant Sci. 164: 407-412.

Townsley, P.M.1974. Chocolate aroma from plant cells. J.Inst.Can.Sci.Technol. Aliment. 7:76-78.

Traore, A. 2000. Somatic embryogenesis, embryo conversion, micropropagation and factors affecting genetic transformation of *Theobroma cacao* L. Ph.D thesis. Pa. State Univ., University Park.

Traore, A., S.N.Maximova and M.J.Guiltinan. 2003. Micropropagation of *Theobroma cacao* using somatic embryo-derived plants. *In Vitro* Cell Dev.Biol.Plant. 39: 332-337.

Tsai, C.Hand J.E.Kinsella. 1981. Initiation and growth of callus and cell suspensions of *Theobroma cacao* L. Ann.Bot. 48: 549-558.

Tsai, C.Hand J.E.Kinsella. 2006. Tissue culture of cocoa bean (*Theobroma cacao* L. changes in lipids during maturation of beans and growth of cells and calli in culture. Lipids. 16: 577-582.

Venkatesh, S., K.Sailaxmi, B.M.Reddy and M.Ramesh.2007. Antimicrobial activity of *Helicteres isora* root. Indian J. Pharmaceutical Sci. 69: 687-689.

Villalobos, V.M.and M.E.Aguilar. 1989. Tissue culture, Micropropagation and *In vitro* conservation of Cacao germplasm. Seminario Manejo de Germplasma de Cacao Turrialba (Coasta Rica). 19-20.

Wang, Y.C and J.Janick 1984. Inducing precocious germination in asexual embryos of cacao. Hort. Sci. 19: 839-841.

Wen, M.C and J.E.Kinsella. 1991. Somatic embryogenesis and plantlet regeneration of *Theobroma cacao*. Food Biotechnology. 5:119 – 138.

Wen, M.C and J.E.Kinsella. 2006. Fatty acid composition of suspension cell cultures of *Theobroma cacao* are altered by culture conditions. J.Food Science. 57: 1452-1453.

White, P.R. 1943. A hand book of plant tissue culture. Cattel, Lancaster, Penn.

William, P.F., J.G.Leon and J.F.Paul 1990. *Theobroma cacao* L. an axillary bud *in vitro* propagation procedure. Plant Cell Tiss. Org. Cult. 20: 111-117.W

Wright, D.C., A.K.Kononowicz and J.Janick. 1984. Factors affecting *in vitro* fatty acid content and composition in asexual embryos of *Theobroma cacao*. J. Am. Soc.Hort.Sci. 109:77-81.

Chapter 3

Studies on *De Novo* Shoot Buds Regeneration, Organogenesis and Somatic Embryogenesis in *Curculigo orchioides* Gaertn.: An Endangered Medicinal Herb

K.S. Nagesh and C. Shanthamma

Department of Studies in Botany, University of Mysore, Manasagangothri, Mysore – 570 006, Karnataka

ABSTRACT

Curculigo orchioides Gaertn. is an endangered medicinal plant with antioxidant and anticancer properties. The rhizome and tuberous roots of the plant have been used extensively in India in indigenous medicine. Due to its multiple uses, the demand for *Curculigo orchioides* is constantly on the rise; however, the supply is rather erratic and inadequate. Destructive harvesting, combined with habitat destruction in the form of deforestation has aggravated the problem. The plant is now considered 'endangered' in its natural habitat. Therefore, the need for conservation of this plant is crucial. Present investigation highlighted a successful protocol for *in vitro* propagation of *C. orchioides*. Proximal rhizome discs were potential for induction of *de novo* shoot bud formation, maximum number of shoot buds were regenerated from proximal rhizome disc due to synergic effects of 6–Benzylaminopurine (BAP) and Kinetin (Kn) at 1 mg/l each. Morphogenic callus was induced on medium with 0.5 – 3 mg/l of 2,4-Dichlorophenoxy acetic acid (2,4-D) containing BAP (0.5 mg/l). Maximum number of somatic embryos were induced on Murashige and Skoog's (MS) medium + BAP (2 mg/l) and maximum number of shoots were regenerated on MS medium supplemented with BAP (3 mg/l) and NAA(0.3 mg/l). Root induction was optimized on half-strength MS liquid medium with 1 mg/l of indole-3-butyric acid (IBA). Rooted shoot-lets were acclimatized on sand and soil. Present study therefore provided a successful *in vitro* propagation of *C. orchioides* for conservation as well as extraction of medicinal compounds for medicinal uses.

Keywords*: Proximal rhizome disc, Shoot regeneration, Somatic embryogenesis, Curculigo orchioides.*

Introduction

Curculigo orchioides Gaertn. (Hypoxidiaceae) is popularly known as black musali in India. The rhizome as well as tuberous roots of the plant have been extensively used in indigenous system of medicine in India, Pakistan and China, for the treatment of various diseases including jaundice, asthma and diarthrosis (Dhar *et al.*,1968). The juice extracted from the rhizome has also been used as a tonic to overcome impotency (Chopra *et al.*, 1956), to prevent bone loss (Cao *et al.*, 2008) for antidiabetic activity (Madhavan *et al.*, 2007), antitumor activity (Singh *et al.*, 2008) and antibacterial activity (Nagesh and Shanthamma, 2009).

C. orchioides is a small geophilous, perennial herb with long cylindrical rhizome. The plant is found from near sea level to 2300 m especially on moist laterite soil. The active principles that have been reported are flavones, glycosides, steroids, saponins, triterpenoids (Misra *et al.*, 1984; Misra *et al.*, 1990; Xu *et al.*, 1992). Conventionally the plant is propagated through seeds and grows only during rainy season. Poor seed setting and poor seed germination restricts the natural multiplication. Associated with these, over-exploitation has lead to the present endangered status of this plant (Ansari, 1993; Anonymous, 2000). Plant tissue culture has been useful as a tool for the conservation and rapid micropropagation of rare and endangered medicinal plants (Sanyal *et al.*, 1998).

Direct somatic embryogenesis and direct shoot formation from rhizome and leaf culture has been reported in *C. orchioides* (Augustine and Souza, 1997; Suri, *et al.*, 1999; Thomas and Jacob, 2004; Nagesh, 2008; Nagesh *et al.*, 2008). However mass propagation of this plant for commercial purpose requires a simple, economical, reproducible and faster multiplication protocol to overcome the constraints of natural multiplication (Anonymous, 2000). Thus the present investigation was undertaken with objectives to improve and standardize rapid and efficient *in vitro* techniques for *de novo* shoot bud regeneration and somatic embryogenesis from proximal rhizome disc culture of *C. orchioides*.

Materials and Methods

Explants Source

Mature plants approximately 12 cm lengths were collected from Biligiri Rangana Hills (altitude about 600-1300 m) in Karnataka. Shoot tip of 0.4 cm length and remaining rhizome portion (shoot axis) of approximately 10 cm length were collected and washed with neutral detergent, teepol (10 per cent v/v) for 10 min. Explants were surface disinfected by a mixture of disinfectants of cetrimide (0.25 per cent w/v), ampicillin (0.15 per cent w/v) for 10 min followed by surface sterilization with mercuric chloride (0.1 per cent w/v). The disinfected explants were washed with sterile distilled water to remove the traces of sterilants after each treatment.

Culture Media

Murashige and Skoog's (1962) basal medium, supplemented with 3 per cent w/v sucrose and 0.7 per cent agar (Bacteriological, Qualigens, India) was used for all experiments. The medium was supplemented with different concentrations of plant growth regulators either individually or in combination. 6-benzylaminopurine (BAP), kinetin (Kn), 1-naphthalene acetic acid (NAA) for shoot regeneration and somatic embryogenesis; indole-3-butryic acid (IBA), 1-naphthalene acetic acid (NAA) for root induction; and 2,4-dichlorophenoxyacetic acid (2-4,D), indole acetic acid (IAA) for callus induction and root induction were employed. pH of the medium was adjusted to 5.7 prior to addition of the gelling agent.

Aliquots of the medium were distributed in 100 ml Erlenmeyer flasks and autoclaved at 121°C for 20 min. Cultures were incubated at 26 ± 2°C under cool white fluorescent light for 16 h of daily photoperiod.

Determination of Potentiality of Rhizome Disc of Proximal to Distal End for De Novo Shoot Bud Regeneration

In the preliminary experiment, rhizome disc, 0.5 cm thickness each, from proximal to distal end of shoot axis were tested for *de novo* shoot bud regeneration on MS + BAP (1mg/l). Since the proximal rhizome disc (PRD) produced more number of shoot buds, further experiments were carried out with PRD only.

Callus Induction for Shoot Regeneration and Somatic Embryogenesis

Rhizome discs were cultured on MS basal medium with different concentrations of 2,4-D (0.5 – 3 mg/l) or IAA (0.5 – 3 mg/l) individually and in combination with BAP (0.5 – mg/l) or Kn (0.5 mg/l) for induction of callus.

Morphogenic cultures obtained from MS medium containing 2,4 –D combined with BAP or Kn. After 4 weeks, they were sub cultured on MS medium with BAP (0.5 – 4 mg/l) and also Kn (0.5 – 4 mg/l) individually for induction of somatic embryogenesis. For maturation and germination of somatic embryos, cultures were transferred to MS basal medium or supplemented with BAP (1 mg/l). Only cultures that had produced both a root and a shoot were recorded after 2-3 weeks subculture.

The calli obtained from media containing 2,4-D (0.5-4mg/l) and BAP (0.5-2.5 mg/l) or Kn (0.5-2.5 mg/l) was tested on medium fortified with NAA (0.1-0.5 mg/l) and high concentrations of BAP (0.5-4 mg/l) or Kn (0.5-4 mg/l) for shoot bud regeneration.

Effect Auxins on Root Induction

In vitro derived shoots were transferred to half-strength MS liquid medium with different concentrations of IBA (0.5 – 2 mg/l) and NAA (0.5 – 2 mg/l) individually for root induction.

Acclimatization

The plantlets were transferred to pots containing sand and soil (1:1) and irrigated with quarter strength of MS medium salt solution. The potted plants were maintained at culture room condition for one week and transferred to green house and to ambient condition.

Results

Determination of Potentiality of Rhizome Disc from Proximal to Distal End for *De Novo* Shoot Bud Regeneration

In the present investigation we have successfully developed a protocol for shoot multiplication and somatic embryogenesis of *C. orchioides*. Responses of rhizome discs from proximal to distal end revealed that shoot bud regeneration *de novo* on MS+BAP (1 mg/ml), maximum of 11 at the proximal end than distal end of the shoot axis (Figures 3.1A-B and Figure 3.2). Further experiments were therefore carried out using only proximal rhizome discs.

Table 3.1: Effect of Different Concentrations and Combinations on Shoot Multiplication from Rhizome Discs Explants of *C. orchioides*

MS + Growth Regulators (mg/l)	No. Shoots/Rhizome Disc Explant (M ± S.E)
Control	
BAP	00.00 ± 0.00
0.5	08.00 ± 0.12
1	11.35 ± 0.13
2	09.50 ± 0.12
3	07.50 ± 0.11
4	05.55 ± 0.11
Kn	
0.5	1.90 ± 0.23
1	2.30 ± 0.11
2	3.10 ± 0.25
3	1.50 ± 0.42
4	1.50 ± 0.16
BAP+ Kn	
0.5 + 0.5	08.50 ± 0.45
1 + 1	13.50 ± 0.48
2 + 2	10.40 ± 0.56
3 + 3	09.20 ± 0.26
4 + 4	07.40 ± 0.24

Data given are Mean of triplicates ± standard error.

Effect of Cytokinins on Shoot Bud Multiplication

Proximal rhizome discs on MS medium with different concentrations and combinations of BAP and Kn were tested for shoot multiplication (Table 3.1). Maximum number of (11) shoot buds regenerated from the proximal rhizome disc on MS (BAP 1 mg/l). On the other hand, medium with different concentrations of Kn (0.5 – 3 mg/l) induced, less significant number (1-2) of shoot buds.

The combination of both cytokinins BAP and Kn at different levels induced multiple shoots buds. Maximum of 13 shoot buds were produced from rhizome disc on MS + BAP (1 mg/l) and Kn (1 mg/l), where as shoot tip produced maximum of 8 shoot buds under similar conditions.

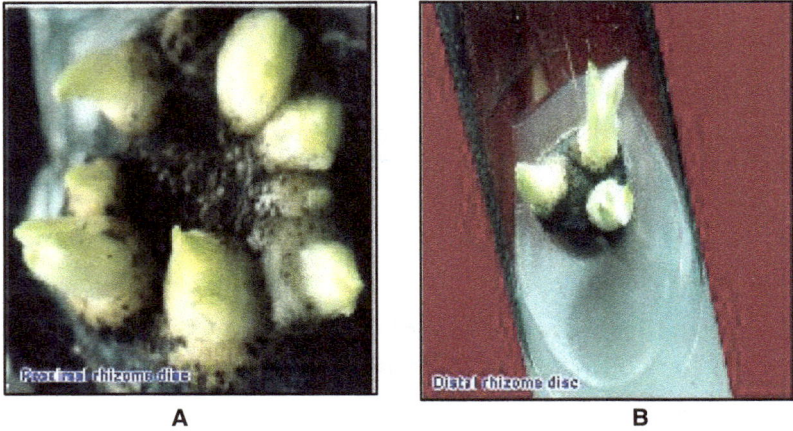

A   B

Figures 3.1A–B: *De novo* Shoot Buds Regeneration from Rhizome Disc from
Proximal to Distal End of the Shoot Axis.

A: Maximum numbers of shoot buds formation on MS medium with BAP (1 mg/l)
from proximal rhizome disc; B: Few number of shoot buds regeneration on MS +
BAP (1 mg/l) from distal rhizome disc.

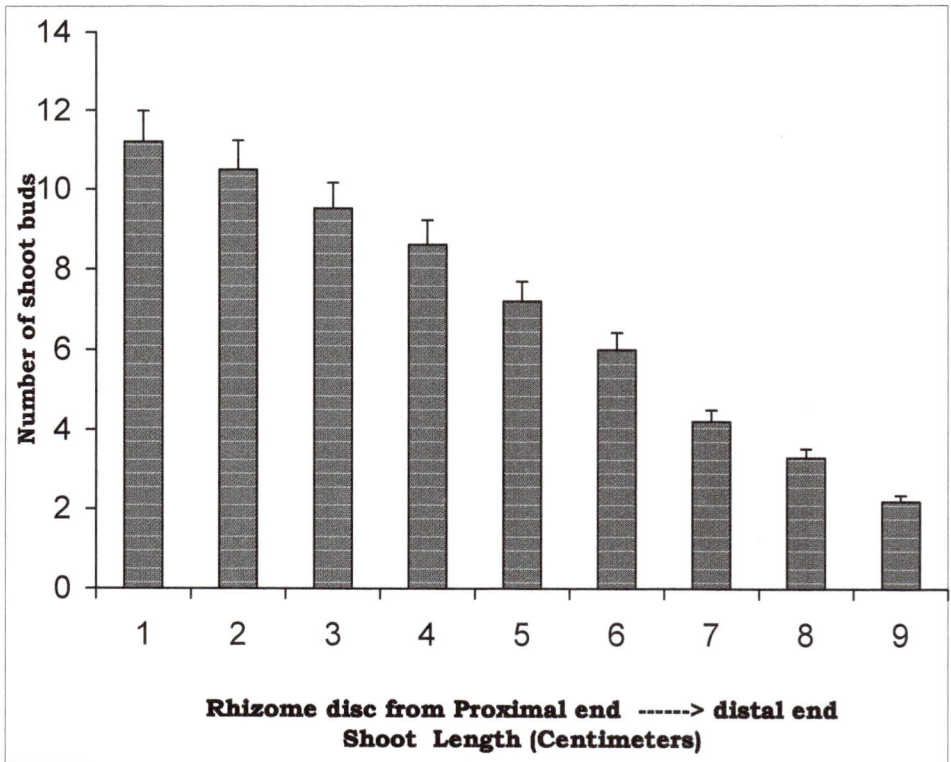

Figure 3.2: Effect of Polarity (from proximal to distal end of the shoot axis)
on *de novo* Shoot Bud Regeneration from Rhizome Disc on
MS Medium Supplemented with BAP(1 mg/l)

Callus Induction for Organogenesis and Somatic Embryogenesis

Observations of callusing response of the rhizome disc explants incubated with varying concentrations of 2,4-D (0.5 – 3 mg/l) and IAA (0.5 – 3 mg/l) individually or combined with BAP (0.5 – 3 mg/l) or Kn (0.5 – 3 mg/l) are summarized in Table 3.2. Lower levels of 2,4-D (0.5 – 3 mg/l) induced meager amount of callus, while higher levels of 2,4-D (2-3 mg/l) induced high amount of whitish callus. However, minimum or maximum amount of callus induced on medium with 2,4-D was non embryogenic and not suitable for induction of somatic embryos. Higher concentrations of 2,4-D (3 mg/l) found to be toxic and cultures turned brown after 2-3 weeks of incubation.

Table 3.2: Effect of Different Concentration and Combination of Growth Regulators on Induction of Morphogenic Callus from Rhizome Explants of *Curculigo orchioides*

MS+Growth Regulators (mg/l)		Fresh Weight (mg ± SE)	Type of Response
Control	–	00.0 ± 0.00	
2,4-D	0.5	600.01 ± 0.16	Unorganized callus
	1	1058.00 ± 0.13	
	2	2400.00 ± 0.14	
	3	2100.00 ± 0.51	
2,4-D+BAP	0.5+0.5	1102.70 ± 0.17	Morphogenic callus
	1+1	2188.00 ± 0.00	
	1.5+2	3504.10 ± 0.78	
	2+3	4103.20 ± 0.12	
2,4-D+Kn	0.5+0.5	800.85 ± 0.19	Morphogenic callus
	1+1	1502.25 ± 0.20	
	1.5+2	2109.80 ± 0.78	
	2+3	3001.00 ± 0.01	
IAA	0.5	300.00 ± 0.21	Unorganized callus
	1	500.08 ± 0.11	
	2	655.50 ± .0.12	
	3	550.00 ± 0.08	
	0.5+0.5	501.35 ± 0.48	
IAA+BAP	1+1	802.05 ± 0.13	Hyperhydric callus
	1.5+2	887.00 ± 0.29	
	2+3	721.40 ± 0.68	
IAA+Kn	0.5+0.5	401.40 ± 0.15	Hyperhydric callus
	1+1	600.85 ± 0.20	
	1.5+2	622.80 ± 0.20	
	2+3	713.20 ± 0.015	

Data given are Mean of triplicates ± standard error.

However, addition of BAP (0.5 – 3 mg/l) along with different levels of 2,4-D, not only increased the amount of callus but also embryogenic in nature, whereas on medium 2,4-D with other cytokinins such as Kinetin (0.5 – 3 mg/l) also produced whitish nodular embryogenic callus, though quantitatively lesser than with those of BAP. MS medium with IAA (0.5 mg/l) induced low amount of callus, while on higher concentrations of IAA (1 – 3 mg/l) increased callus along with roots initiation was observed. Addition of IAA along with BAP (0.5 – 3 mg/l) or Kn (0.5 – 3 mg/l) induced whitish hyperhydric callus, which was an unorganized and not suitable for further experiments.

Three weeks old mophogenic callus from medium containing 2,4-D (0.5-4 mg/l) and BAP (0.5 – 2.5 mg/l), or Kn (0.5 – 2.5 mg/l), when subcultured on MS medium with low levels NAA (0.1-0.5 mg/l) and high levels of BAP (1- 5 mg/l) or Kn (1-5 mg/l) the regeneration of shoot bud was observed. The data recorded after 30 days and presented in Table 3.3. The whitish nodular calli, on medium supplemented with NAA (0.1-0.5 mg/l) and BAP (1-5 mg/l), turned into greenish opaque structure within 4 weeks of incubation (Figure 3.3A). These greenish structures further grown and developed into shoot buds within 6 weeks of incubation. Maximum number of shoot buds induction were noticed in the presence of BAP (3 mg/l) and combined with 0.3 mg/l of NAA and 90 per cent response within 4 weeks of incubation (Figure 3.3B). Shoot buds were stout and well developed with 2-3 leaves within 8 weeks of incubation. As the levels of BAP and NAA increased, the number of shoot buds decreased. On medium with NAA (0.1-0.5 mg/l) and Kn (1-5 mg/l), also regeneration of shoot bud was noticed, however the number of shoot buds was lesser than that of medium containing BAP and also shoot buds were weak. Maximum numbers of shoot buds were regenerated on medium with Kn (2 mg/l) and NAA (0.2 mg/l) (Table 3.3). However, decrease or increase beyond the optimal levels of Kn (2 mg/l) and NAA (0.2 mg/l) decreased the regeneration of number of shoot buds (Table 3.3). The results of the present investigation indicated that low concentration of NAA and higher concentration of cytokinin (BAP or Kn), influenced positively on regeneration of multiple shoot buds, however on higher concentrations of NAA roots were induced along with low number of shoot buds.

Table 3.3: Effect of Different Concentrations of NAA and Cytokinins on Shoot Regeneration from Rhizome Derived Callus of Curculigo orchioides

Growth Regulators (mg/l)	NAA+BAP	NAA+Kn
	Number of Shoot Regeneration (M±S.E)	Number of Shoot Regeneration (M±S.E)
0.1+1	12.0±0.22[e]	9.0±0.21[e]
0.2+2	29.65±0.22[b]	20.0±0.35[b]
0.3+3	43.30±0.37[a]	18.15±0.18[a]
0.4+4	22.95±0.45[c]	14.0±0.11[c]
0.5+5	20.0±0.11[d]	20.0±0.23[d]

Data given are Mean of triplicates ± standard error

Values followed by superscript letters through columns differs significantly at (p< 0.001) 1 per cent level when subjected DMRT

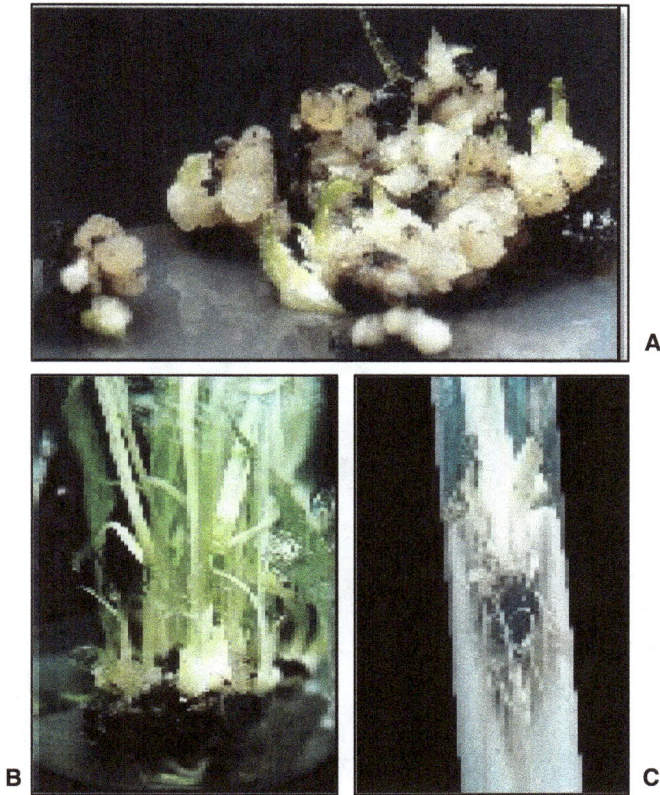

Figure 3.3

A: Regeneration of shoots from rhizome derived callus on MS medium with NAA(0.3 mg/l) and BAP(3 mg/l) within 3 weeks of incubation;
B: Maximum number of well developed shoots on MS medium with NAA(0.3 mg/l) and BAP (3 mg/l) within 6 weeks of incubation;
C: Maximum number of sturdy roots developed on MS half-strength liquid medium with IBA (1 mg/l).

Root Induction

In vitro derived shoots were rooted on half-strength MS liquid medium with IBA (1 mg/l) produced maximum number of healthy and sturdy roots than those with NAA (Figures 3.3C and 3.4).

The morphogenic cultures obtained from the medium containing 2,4-D and different levels of BAP or Kn were used for further experiment to induce somatic embryogenesis on medium with different concentrations of BAP (0.5 – 4 mg/l) or Kn (0.5 – 4 mg/l). The morphogenic response was presented in Table 3.4. MS medium containing higher level of BAP (2 mg/l) induced maximum number of embryoids and were mostly of globular and rarely heart shaped during 6 weeks of culture. The

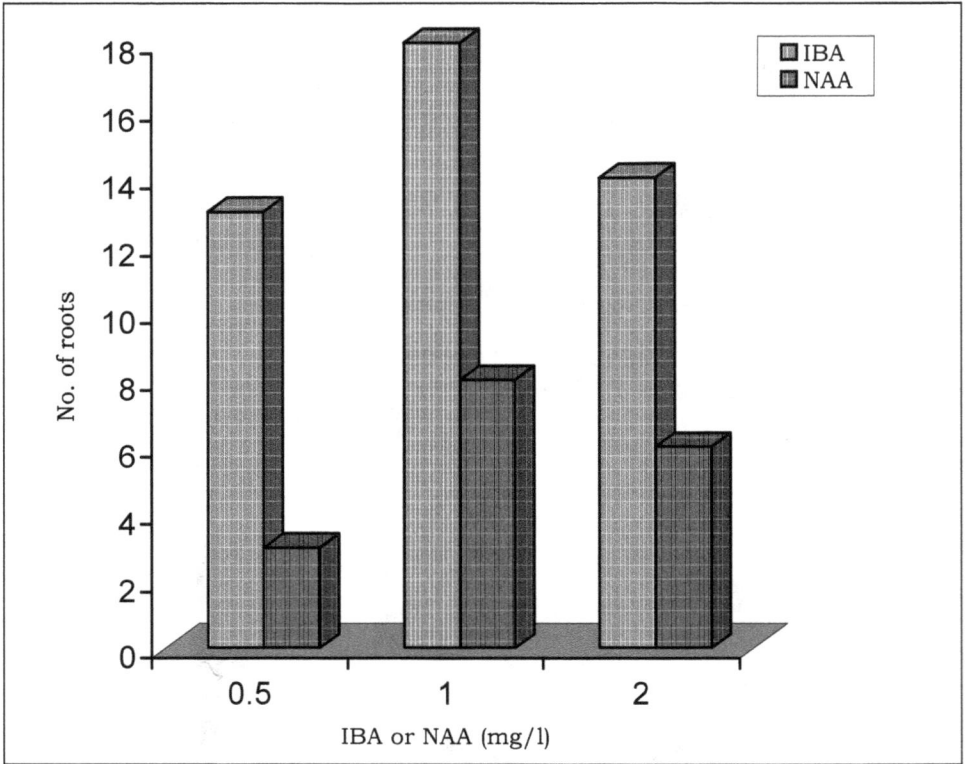

Figure 3.4: Effect of IBA or NAA on Induction of Roots

Table 3.4: Effect of Cytokinins on Somatic Embryogenesis

Growth Regulators (mg/l)	BAP	Kn
	Number of Somatic Embryoids/Callus (M±S.E)	Number of Somatic Embryoids/Callus (M±S.E)
0.5	10.40 ± 0.11[d]	04.30 ± 0.12[e]
1	18.40 ± 0.11[b]	06.23 ± 0.23[b]
2	24.10 ± 0.26[a]	10.13 ± 015[a]
3	14.40 ± 0.12[c]	8.02 ± 0.19[c]
4	8.85 ± 0.23[e]	7.00 ± 0.17[d]

Data given are Mean of triplicates ± standard error

Values followed by superscript letters through columns differs significantly at (p< 0.001) 1 per cent level when subjected DMRT

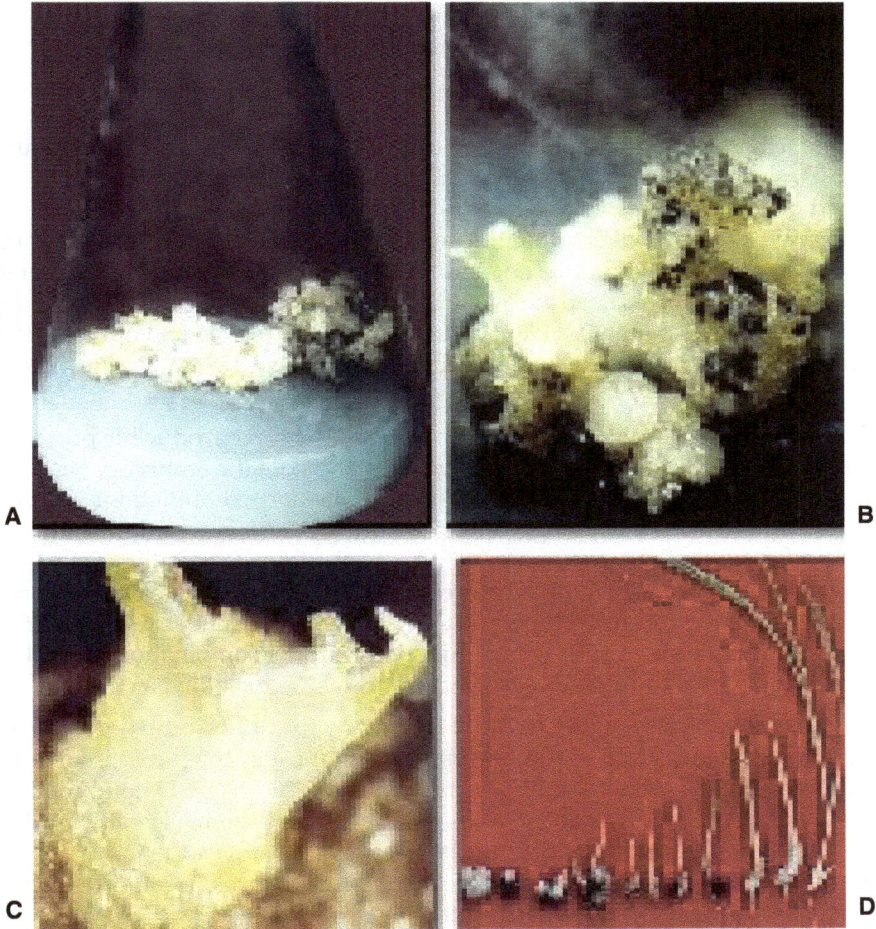

Figure 3.5

A & B: Maximum number of somatic embryos developed on MS medium supplemented with BAP (2 mg/l);
C: Somatic embryo with well developed cotyledon;
D: Plantlets with well established root and shoot system

maturation and germination of globular shaped embryos occurred on the same medium or on fresh MS medium with 2 mg/l of BAP within 6 weeks of subculture. Globular shaped embryos developed (Figure 3.5A) into heart and torpedo shaped structure with differentiation of cotyledons, hypocotyls and radicle portion (Figure 3.5B). Well-developed somatic embryos (Figure 3.5C) were transferred to pots containing sand and soil (1:1) and nourished with quarter strength of MS basal salt solution and was found to be beneficial to early growth and vigor of the plant. Further growth occurred when they were transferred to pots containing sand and soil (1:1), where they grew luxuriantly and showed good survival rate of over 90 per cent and 60 per cent in micropropagated shoots and somatic embryos respectively.

Discussion

In India the Ayurvedic system of medicine has been in use for over three thousand years. Charaka and Susruta, developed samhitas based on herbal sources and is still esteemed even in this day as treasure of indigenous medicines. These indigenous medicines are preferred over allopathic medicines since the latter causes lot of side effects due to its synthetic nature. The choice of today's therapy is therefore exploration of plant drugs. However, due to the over exploitation of the medicinal plants, many of them have become endangered. Plant biotechnology has played an important role in the mass multiplication for conservation and also to procure innumerable content of drugs.

In the present investigation we were able to develop a reproducible protocol for *de novo* shoot buds regeneration, organogenesis and somatic embryogenesis using proximal rhizome disc on different concentrations and combinations of growth regulators. Rhizome disc explants were more suitable for mass multiplication than shoot tip culture, because continuous excision of meristematic region may threaten the existence of mother plant, as has been reported in *Dendrobium moschatum* (Kanjilal *et al.*, 1999). In contrast to earlier report on *C. orchioides* (Augustine *et al.*, 1997), efficient numbers of multiple shoot buds were achieved by adding combination of cytokinins (BAP and Kn), which could be due to synergistic interaction between them.

A strong auxin such as 2,4-D is used frequently for callus induction, its further proliferation and induction of somatic embryos in most of the plant tissue culture (George *et al.*, 1993). Interestingly, in the present study, 2,4-D combined with cytokinins, produced embryogenic callus as observed in Brahmi (Tiwari *et al.*, 1998). Further cytokinins particularly BAP played a crucial role for induction, maturation and germination of maximum number of somatic embryos as reported in *Abies nordmanniana* (Norgard and Krogstrup 1991).

The effects of different combinations of NAA and BA concentrations on shoot organogenesis, indicated that maximum of shoots on MS medium containing high concentrations of BAP (3 mg/l) and low concentrations of NAA (0.3 mg/l), shoot multiplication rate generally increased with increased BA and NAA concentrations. It appeared that BA and NAA have synergistic effects on shoot multiplication. The synergistic effect of BA in combination with an auxin has been reported for many medicinal plants such as *Holostemma annulare* (Sudha *et al.*, 1998), *Hemidesmus indicus* (Sreekumar *et al.*, 2000), and *Ceropegia candelabrum* (Beena *et al.*, 2003).

This *in vitro* protocol could be useful for not only preventing further depletion of their population in nature and also large-scale production of plant source throughout the season for the extraction of drugs.

The study also showed that NAA was more effective than IBA for root induction as reported in *Ixora singaporenesis* (Lalit and Tewari, 1998).

A crucial aspect of *in vitro* propagation is to acquire regenerated plants that are capable of surviving outside the sterile and protected *in vitro* environment. A substantial number of micropropagated plants do not survive transfer from *in vitro* conditions to greenhouse or field environment. The greenhouse and field have substantially lower relative humidity, higher light septic environment that are stressful to micropropagated plants compared to *in vitro* conditions. The benefit of any micropropagation system can, however, only be fully realized by the successful transfer of plantlets from tissue-culture vessels to the ambient conditions found *ex vitro* (Hazarika, 2003). In the present study, the plantlets were transferred to plastic cups containing growing medium for hardening. The data generated here is of great use in establishing improved protocols for mass propagation of this endangered plant.

Thus the present communication clearly describes *in vitro* protocol that can be used for large-scale propagation, provide source for extraction of medicine and also for its *ex-situ* conservation of this over exploited endangered medicinal plant.

References

Anonymous, 2000. In Plant tissue from Research to commercialization (Decade of support), Publisher DBT Ministry of Science and Technology India, pp. 33-34.

Ansari, A.1993. Threatened medicinal plants from Madhaulia forest of Gorakhpur. J. Econ. Taxon. Bot. 17 : 23-24.

Augustine, A C and Souza, L D. 1997. Regeneration of an anticancer herb *Curculigo orchioides* Gaertn. *In vitro* Cell. Dev. Bio.- Planta. 33: 111-113.

Beena, M. R., K.P. Martin, P.B. Kirti and M. Hariharan, 2003. Rapid *in vitro* propagation of medicinally important *Ceropegia candelabrum*. Plant Cell Tiss. organ Cult. 72:285–289.

Cao, D.P., Y.N. Zheng, L.P. Quin, T. Han, H. Zhang, K. Rahman and Q.Y. Zhang 2008. *Curculigo orchioides*, a traditional Chinese medicinal plant, prevents bone loss in ovariectomized rats. Maturitas 59 (4): 373-80.

Chopra, R N., S. L. Nayar and I. C. Chopra. 1956. Glossary of Indian medicinal plants, CSIR, New Delhi, pp 84.

Dhar, M. L., M. N. Dhar, B. N. Dhawan, D. N. Mehrota, and C. Ray. 1968. Screening of Indian plants for biological activity part-I. Indian J. Exp. Biol. 6: 232-249.

George E F, 1993. Plant propagation by tissue culture part II. The Technology, England Exergetics Ltd (Ed), pp 433-434.

Hazarika, B. N. 2003. Acclimatization of tissue-cultured plants. Curr. Sci. 85:1704-1712.

Kanjilal, B. D. De Sarker, J. Mitra and K. B. Datta. 1999. Stem disc culture; Development of rapid mass propagation method for *Dendrobium moschatum* (Buch. – Ham.) Swartz-An endangered orchid. Curr. Sci. 77 : 497-500.

Lalit, M and Tewari, 1998. Micropropagation of *Ixora singaporensis* Linn. – An ornamental shrub. *Curr. Sci.* 75 : 545-547.

Madhavan, V., R. Joshi, A. Murali, and S.N. Yoganarasimhan 2007. Antidiabetic Activity of *Curculigo orchioides* root tuber. Pharm. Biol. 45: 18 – 21.

Misra, T.N., R S. Singh, J. Upadhya and D. N. M. Tripathi. 1984. Aliphatic compounds from *Curculigo orchioides*. Phytochem. 23 : 1643-1645.

Misra, T.N., R. S. Singh, Tripathi and S. C. Sharma. 1990. Curculigo a cycloartane, triterpene alcohol from *Curculigo orchioides*. Phytochem. 29 : 929-931.

Murashige, T. and F. Skoog, 1962. A revised medium for rapid growth and bioassay with tobacco tissue culture. Phy. Plant. 15 : 473-497.

Nagesh, K. S. and C. Shanthamma. 2009. Antibacterial activity of *Curculigo orchioides* rhizome extract on pathogenic bacteria. African J. Microbl. Res. 3(1): 5-9

Nagesh, K.S., H. Nayaka., S. M. Dharmesh., C. Shanthamma and T. Pullaih 2008. *In vitro* propagation and antioxidant activity of *Curculigo orchioides* Gaertn. J. Trop. Med. Plants. 9(2): 404-410.

Nagesh, K.S. 2008. High frequency of multiple shoot induction of *Curculigo orchioides* Gaertn. Shoot tip V/S Rhizome Disc. Taiwania. 53(3):242-247.

Norgard and Krogstrup, 1991. Cytokinin induced somatic embryogenesis from immature embryos of *Abies nordmanniana*. Plant cell Report. 9 : 509-513.

Singh, R and A.K. Gupta. 2008. Antimicrobial and antitumor activity of the fractionated extracts of Kali musli (*Curculigo orchioides*) Int. J. Green Pharmacy. 2 (1) : 34-36.

Sreekumar, S., S. Seeni and P. Pushpangadan. 2000. Micropropagation of *Hemidesmus indicus* for cultivation and production of 2-hydroxy 4-methoxy benzaldehyde. Plant Cell Tiss. Org. Cult. 62:211–218.

Sudha C.G., P.N. Krishnan and P. Pushpangadan 1998. *In vitro* propagation of *Holostemma annulare* (Roxb.) K. Schum., a rare medicinal plant. *In Vitro* Cell Dev. Biol. Plant. 33:57–63.

Suri, S. S., Sunithajain and K.G. Ramawat. 1999. Plant regeneration and bulbils formation *in vitro* from leaf and stem explants of *Curculigo orchioides* Gaertn.–an endangered medicinal plant. Sci. Hort. 79 :127-134.

Thomas, T.D and A. Jacob. 2004. Direct somatic embryogenesis of *Curculigo orchioides* Gaertn., an endangered medicinal herb. J. Plant Biotechnol. 6 (3): 193-198.

Tiwari, Deosingh and B. Nath. 1998. Shoot regeneration somatic embryogenesis from different explants of Brahmi (*Bacopa monnieri* (L.) Pennell). Plant Cell Rep. 17: 538-543.

Xu Jun-ping, Ren Sheg Xu, and X. Yuli, 1992. Glycosides of cycloartane sapogenine from *Curculigo orchioides*. Phytochem. 31: 233-236.

Chapter 4

Micropropagation of *Talinum cuneifolium* (Vahl) Willd. through Axillary Bud Culture

N. Savithramma, Beena Prabha, A. Sasikala, A. Kedarnath Reddy and J. Saradvathi

Department of Botany, S.V.University, Tirupati – 517502, A.P

ABSTRACT

The present work describes a reproducible and efficient protocol for mass propagation of *Talinum cuneifolium*. This multipurpose tuberous shrub has been successfully *in vitro* propagated through axillary bud as explant on MS (Murashige and Skoog) medium. More number of multiple shoots were obtained on MS medium containing 3.0 mg l^{-1} BA and in combination of 0.2 mg l^{-1} IAA (Indole acetic acid) 3.0 mg l^{-1}. BA (Benzyladenine) and 0.2 mg l^{-1} NAA (Naphthalene acetic acid) showed less number of shoots when compared with the IAA combination with BA. Maximum elongation of isolated shoots on MS medium containing 1 to 2 mg l^{-1} Kn followed by IAA, BA and NAA within 6 weeks time. 83 per cent shoot regeneration occurred in MS medium supplemented with 3.0 mg l^{-1} and 0.2 mg l^{-1} BA and IAA respectively. The isolated plantlets were induced more number of roots with maximum length at high frequency after 4 weeks in half strength MS medium supplemented with 1.0 mg l^{-1} IBA (Indole butyric acid). The rooted plantlets were gradually acclimatized to lab and greenhouse environmental conditions. Hardened plantlets were transplanted in 1:1 (sand : soil) with ¼ MS medium for highest survival rate. An attempt was also made to determine the extent of clonal purity of the *in vitro* regenerated plants at the biochemical level by employing peroxidase isozyme as marker, in order to get an insight into the impact of somaclonal variations in the course of their regeneration.

Keywords: Micropropagation, Talinum cuneifolium, Axillary buds, Isozymes.

Introduction

Talinum cuneifolium (Vahl) Willd. (= *Talinum portulacifolium* (Forsk.) Asch. ex Schw.) – an erect shrub with subterranean tuber belongs to the family Portulacaceae. The leaves of *Talinum cuneifolium*, commonly known as Ceylone bachalli, are eaten as a cooked vegetable or raw as a salad, alone or with young stem parts. It is cultivated in Africa (like spinach), and is used as a green leafy vegetable due to its rich vitamin A and mineral content (Anon, 2004). The leaves can also be stored dry for later use. The plant is a palatable fodder for cattle and goats. It is also an important medicinal plant in the local system of medicine. Indian System of Medicine (ISM) refers that the leaves and roots of this plant are medicinally important. The supplementation of the leaves of this plant is reported to be a better diet for strengthening the body. The powdered leaves are used in treatment of diabetic, mouth ulcer, and aphrodisiac; roots are used for cough, gastritis, diarrhoea and pulmonary tuberculosis (Janapati *et al.*, 2008). In Ethiopia the leaves are applied medicinally against eye diseases and the root against cough and gonorrhoea. This valuable plant has markedly depleted to satisfy the local food needs and no attempts were made for its replenishment. The growth was very slow and takes long time. One of the constraints associated with the conventional propagation was very short span of seed viability. A low survival rate by stem cuttings in *Talinum cuneifolium* restricts its mass propagation via conventional methods. No alternative mode of multiplication was available to propagate and to conserve genetic stock of this plant. *In vitro* multiple shoot regeneration may give higher rate of propagation within very short time and space.

Isoenzymes can be considered to be the direct expression of the gene function of cells during differentiation and their variations were often associated with somaclonal variations. A detailed analysis of their changing patterns during development may lead to some understanding of the basic mechanism of cellular differentiation to obtain efficient plant regeneration *in vitro*. Peroxidases and esterases were widely distributed among higher plants. The application of isozymes as markers in morphological and regeneration studies has been reported by several workers (Bhatt *et al.*, 1992; Feuser *et al.*, 2003) to detect clonal fidelity. The current problem facing the regenerating system in plant tissue culture was the occurrence of uncontrollable somaclonal variations that were undesirable in any clonal propagation and conservation programmes. Propagation by all methods of indirect organogenesis carries a risk that the regenerated plants will differ genetically from each other and from the mother plant (George, 1993). It has been over emphasized that the regenerants raised from resident meristems like nodes, shoot tips would be true–to–type (Hu and Wang, 1983; Bajaj *et al.*, 1988). However, in some reports, variations were detected even when shoot meristems were used (Pramanik and Datta, 1986). There have been no reports of a regeneration system for *T. cuneifolium*. Therefore, the present study was undertaken with an aim of establishing an efficient protocol for *in vitro* plant regeneration from axillary bud explants.

Materials and Methods

Axillary buds of *Talinum cuneifolium* (Vahl) Willd. were collected from S.V.University Botanical gardens, Tirupati, A.P. Explants were initially washed under running tap water and with Teepol solution (5 per cent v/v) for 15 min. followed by 4-5 washings with distilled water. Disinfestation of these explants was then made under laminar air flow chamber by keeping them in 70 per cent alcohol for 60 sec followed by rinsing for 3 times in sterile distilled water. Finally the explants were immersed in 0.1 per cent $HgCl_2$ (Mercuric chloride) for 3 min. The surface sterilization was followed by 5-6 rinses in sterile distilled water. The surface sterilized explants were cultured on MS basal medium (Murashige and Skoog, 1962) containing 3 per cent (w/v) sucrose and 0.8 per cent (w/v) agar. Explants were implanted in different combinations and concentrations of growth regulators (BA, Kn, IAA and NAA) singly as well as in combinations for shoot proliferation. The pH of the medium was adjusted to 5.8 using 0.1 N HCl or 0.1 N NaOH (Sodium hydroxide) solutions before autoclaving. All cultures were incubated in a culture room at $25 \pm 2°C$ with a relative humidity of 50-60 per cent and 16 h photoperiod at a photon flux density of 15-20 mE m^2/s^{-1} from white cool fluorescent tubes. For each treatment 12 replicates were used and each experiment was repeated at least thrice. The cultures were examined periodically.

These *in vitro* elongated multiple shoots were excised and transferred on MS medium supplemented with IBA, NAA and IAA in different concentrations and combinations separately for root induction. These *in vitro* raised plantlets with well developed roots were taken from the culture tubes and washed thoroughly to remove the traces of agar. They were then planted in plastic cups containing mixture of soil rite and soil (1:1) and maintained in the hardening chamber with controlled temperature, light and relative humidity. The acclimatized complete plantlets were then transferred to the field and its survival frequency was recorded. SDS-PAGE (Poly Acrylamide Gel Electrophoresis) of Peroxidase isozyme was carried out to test the clonal purity as per the method of Van Eldic *et al.* (1980).

Results and Discussion

In the present study, among the different types of sterilents used [H_2O_2 (Hydrogen peroxide), NaOCl (Sodium hypo chlorite) and $HgCl_2$], 0.1 per cent $HgCl_2$ for 10 min was identified as an appropriate surface sterilizing agent for producing aseptic shoots from mature nodal explants. Murashige and Skoog (1962) medium has been designated for tissue culture of tobacco and a wide variety of shrubs. Shoot regeneration efficiency from these axillary bud explants was analysed by supplementing different cytokinins (BA, Kn) at various concentrations either used alone or in combination with auxins (IAA and NAA) (Table 4.1) (Plate 4.1). Irrespective of the concentrations of BA used, the axillary buds sprouted within 10–15 days after culture initiation. Earlier studies found BA to be the most effective cytokinin for inducing shoot development (Heloir *et al.*, 1997). The frequency of shoot regeneration and the number of shoots per explant increased with increase in the concentration of BA (Plate 4.1C-F). Maximum frequency of shoot regeneration and maximum number of shoots per explant was found at 3.0 mg l^{-1} BA (Plate 4.1F). Further increase in the concentration of BA decreased the number of shoots (Plate 4.1G). Kinetin alone on MS medium did not evoke good shoot multiplication and had low frequency of regeneration. MS medium supplemented with 2 mg l^{-1} Kn resulted in the formation of 3.13 shoots per explant with a maximum mean shoot length of 1.65 cm (Plate 4.1H).

Table 4.1: Effect of Different PGR in MS Medium on Shoot Multiplication from Axillary Bud Explants of *Talinum cuneifolium*

PGR (mg^{-1})				Mature Bodal Explants		
BA	KN	IAA	NAA	%of Shoot Regeneration	Mean Number of Shoots/Explant	Mean Length of Shoots (cm)
–	–	–	–	41.66 ± 0.88[a]	1.20 ± 0.07[a]	0.93 ± 0.04[d]
0.5	–	–	–	66.00 ± 0.57[d]	2.06 ± 0.06[b]	1.11 ± 0.02[e]
1.0	–	–	–	70.64 ± 0.38[f]	3.38 ± 0.18[c]	0.63 ± 0.09[ab]
2.0	–	–	–	72.39 ± 0.32[g]	8.75 ± 0.1[i]	0.97 ± 0.01[d]
3.0	–	–	–	78.26 ± 0.13[i]	10.04 ± 0.05[j]	0.67 ± 0.01[bc]
5.0	–	–	–	65.14 ± 0.15[d]	5.08 ± 0.06[e]	1.21 ± 0.02[efg]
–	1	–	–	44.35 ± 0.22[b]	1.35 ± 0.20[a]	1.75 ± 0.02[i]
–	2	–	–	58.04 ± 0.11[c]	3.13 ± 0.08[c]	1.65 ± 0.04[h]
1	2			68.33 ± 0.45[e]	4.08 ± 0.06[d]	0.54 ± 0.01[a]
3	2			71.56 ± 0.12[fg]	6.31 ± 0.27[f]	1.16 ± 0.01[ef]
3.0	–	0.1	–	77.27 ± 0.14[i]	10.5 ± 0.03[k]	1.27 ± 0.01[g]
3.0	–	0.2	–	82.59 ± 0.35[j]	12.54 ± 0.21[l]	1.23 ± 0.02[fg]
3.0	–	–	0.1	69.12 ± 0.28[e]	7.27 ± 0.06[g]	0.95 ± 0.01[d]
3.0	–	–	0.2	74.02 ± 0.10[h]	8.04 ± 0.03[h]	0.76 ± 0.03[c]

Values represented above are the means of 12 replicates. ('±' indicates the standard error).

Observations after 6 weeks of culture. Mean values having the same letter in each column don't differ significantly at P ≤ 0.05 (Duncans Test)

In most of the plants, micropropagation could be performed readily by using axillary buds as explants (Villarreal and Rojas, 1996; Ramamurhy and Savithramma, 2002). Direct shoot regeneration of *T. cuneifolium* was achieved by proliferation of axillary buds. The production of multiple shoots from axillary buds through *in vitro* propagation was caused by stimulating precocious axillary shoots to overcome the dominance of shoot apical meristems. Phytohormones were the crucial factors affecting regeneration of shoots from axillary buds. Cellular differentiation and organogenesis in tissue and organ culture have been found to be controlled by an interaction between different phytohormone concentrations. In the present study, it was found that BA was more effective for shoot multiplication from axillary bud and

Plate 4.1

A) Aseptic seedling germination of *Talinum cuneifolium* on MS medium containing BA 1.0 mg l⁻¹ (1 cm Bar = 6.09 mm); B) Shoot indication on MS basal medium (1 cm Bar = 6,25 mm) Multiple shoot induction from axillary bud on MS medium supplemented with C) 0.5 mg l⁻¹ BA (1 cm Bar = 5.31 mm); D) 1.0 mg l⁻¹ BA (1 cm Bar = 5.68 mm); E) 2 mg l⁻¹ BA (1 cm Bar = 5.68 mm); F) 3 mg l⁻¹ BA (1 cm Bar = 4.80 mm); G) 5 mg l⁻¹ BA(1 cm Bar = 6.25 mm); H) 2 mg l⁻¹ Kn (1 cm Bar = 5.95 mm); I) 1 mg l⁻¹ BA + 2 mg l⁻¹ KN (1 cm Bar = 5.68 mm); J) 3 mg l⁻¹ BA 2 mg l⁻¹ KN (1 cm Bar = 3.84 mm); K) 3 mg l⁻¹ BA + 0.2 mg l⁻¹ IAA (1 cm Bar = 3.90 mm); L) 3 mg l⁻¹ BA + 2 mg l⁻¹ NAA (1 cm Bar = 5.55 mm)

Plate 4.2

A) Induction of tubers from the root of *Talinum cuneifolium* on half strength medium supplemented with 0.5 mg l⁻¹ IBA and 100 g/l sucrose (1 cm Bar = 5.10 mm); B) Magnified view of transparent spindle shaped mini tuber in the roots of *Talinum cuneifolium* after 30 days (1 cm Bar = 2.80 mm); C, D) Elongation and rooting of plantlets on half strength MS medium supplemented with 0.5 mg l⁻¹ IBA (1 cm Bar = 5.10 mm, 5.10 mm); E, F) Transplanted plantlet E) in a paper cup after 2 weeks F) in a plastic pot after 1 month.

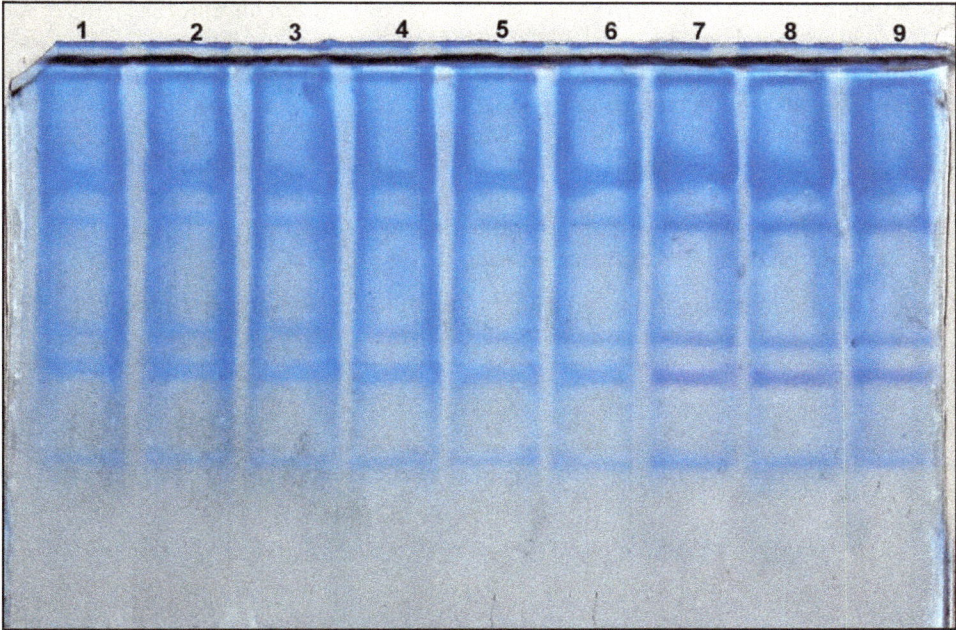

Plate 4.3

Banding pattern of peroxidase isozyme in leaf tissue of *Talinum cuneifolium*
1,2,3 Field grown; 4,5,6 Direct regenerated shoots; 7,8,9 Indirect regenerated shoots

the effectiveness of BA in promoting *in vitro* axillary shoot production was well documented in different plants by Nobre *et al.* (2000), Mandal *et al.* (2001) and Gisele and Thomas (2005). Amongst all the combinations of cytokinins used the multiplication of axillary buds was highest at 3 mg l⁻¹ BA. Similar results were obtained by Avani *et al.* (2006). In *T. cuneifolium* for effective shoot regeneration BA was found to be superior cytokinin than Kn. Increasing concentrations of BA (> 3.0 mg l⁻¹) reduced the number of shoots. Higher concentrations of growth regulators in *T. cuneifolium* had a toxic effect on shoot regeneration.

Efficient rooting of *in vitro* regenerated plants and subsequent field establishment was the last and crucial stage of rapid clonal propagation. An *in vitro* rooting experiment reiterates the importance of auxins on root induction. About 3-4 cms long microshoots were isolated from proliferating bud cultures growing on MS medium and used for *in vitro* rooting. Among the three auxins used, IBA alone was found to be most effective when compared with IAA and NAA for root induction (Table 4.2). The IBA improved rooting efficiency. Microshoots when subjected to rooting on ½ strength MS medium supplemented with 0.5 mg l⁻¹ IBA also exhibited elongation at the earlier stages of inoculation while at the later stages produced a bunch of small thin roots with lateral roots (8.64) without basal callus within two weeks (Plate 4.2C,D). High frequency (85.7 per cent) of root regeneration and maximum mean number (12.46) of thin slender lengthy roots were also induced from the base of the shoots with out interference of callus on ½ strength MS medium supplemented with 1.0 mg l⁻¹ IBA.

The superiority of IBA in rhizogenesis was seen in *Lonicera tatarica* (Palacios *et al.*, 2001), *Artemisia judaica* (Liu *et al.*, 2002), apricot (Tornero and Burgos, 2000), *Piper longum* (Soniya and Das, 2002), mung bean (Tivarekar and Eapen, 2001) and *Cunila galioides* (Fracaro and Echeverrigaray, 2001). Efficiency of IBA at lower concentration in *in vitro* rooting has been reported by Das and Rout (2002), Soniya and Das (2002), Martin *et al.* (2003) and Prasad (2004). Heloir *et al.* (1997) reported that IBA provided a suitable auxin for *in vitro* rooting of *Vitis vinifera*. The observations coincide with the present study as the higher number of roots obtained in the MS medium supplemented with higher concentration of IBA than NAA.

Table 4.2: Effect of Different Auxins on the Root Induction of *In vitro* Raised Shoots of *Talinum cuneifolium* on Half Strength MS Medium

Plant Growth Regulators (mg l⁻¹)			Frequency of Regeneration	Mean Number of Roots	Mean Length of the Root
NAA	IBA	IAA			
	0.5		78.81 ± 0.09^k	8.64 ± 0.01^j	3.50 ± 0.04^h
	1.0		85.70 ± 0.07^l	12.46 ± 0.02^k	4.60 ± 0.01^i
		0.5	46.70 ± 0.02^b	2.18 ± 0.01^a	1.28 ± 0.03^b
		1.0	54.36 ± 0.04^e	3.36 ± 0.02^c	1.50 ± 0.03^c
0.5			58.21 ± 0.05^g	4.62 ± 0.08^f	2.15 ± 0.02^d
1.0			67.41 ± 0.03^i	7.90 ± 0.05^i	3.26 ± 0.02^g
0.5	0.1		74.64 ± 0.02^j	5.31 ± 0.01^g	3.03 ± 0.10^f
1.0	0.1		63.92 ± 0.01^h	7.60 ± 0.03^h	3.30 ± 0.03^g
	0.5	0.1	52.63 ± 0.02^d	4.40 ± 0.01^e	1.60 ± 0.04^c
	1.0	0.1	55.05 ± 0.03^f	5.32 ± 0.06^g	2.55 ± 0.06^e
0.1		0.5	44.71 ± 0.01^a	2.83 ± 0.09^b	1.02 ± 0.02^a
0.1		1.0	52.23 ± 0.02^c	3.60 ± 0.04^d	1.18 ± 0.01^b

Values represented above are the means of 12 replicates. ('±' indicates the standard error.)

Observations after 4 weeks of culture. Mean values having the same letter in each column don't differ significantly at P ≤ 0.05 (Duncans Test)

The primary target of a micropropagation system was the best acclimatization and field establishment of regenerated plants. Rooted plantlets were gradually acclimatized with an increase in temperature from 25-28°C and decrease in relative humidity from 80-50 per cent for a period of 15-20 days. Therefore, in the present study, the paper cups containing *in vitro* derived plantlets were kept in the culture room and cups were covered with polythene bags to maintain high humidity and kept in mist chamber covered with coir mat. Survival of *in vitro* plants after planting was largely dependent on the components of the potting media. In the present study, garden soil and sand (1:1) was found suitable for highest survival of plants. These plants were irrigated with 1/4 strength MS salts and exposed gradually to external environment.

The observations obtained with peroxidase isoenzyme pattern in the current study substantiated the uniformity of the multiple shoots derived from axillary bud explant (Plate 4.3) which was most desirable in any micropropagation system. These findings were also supporting the view that biochemical traits such as isozyme provides an evidence to study the extent of somaclonal variations in a manner analogous to their use in elucidating genetic variation in natural population (Bhaskaran *et al.*, 1987; Ramamurthy and Savithramma, 2003). Plants derived from tissue culture as having superior field performance to those derived from stem cuttings in terms of survival rate, fruit yield, rhizome production and total plant weight. After two weeks, by which time a fresh leaf appeared from the potted plants, were transferred to green house and then transferred to field condition. 70-75 per cent of regenerated plants were successfully acclimatized to natural environment (Plate 4.2E, F). The direct regeneration of plants from axillary buds results in the maintenance of genotypic stability without the risk of somaclonal variations normally associated with adventitious regeneration via callus (Levieille and Wilson, 2002). *In vitro* derived plantlets were morphologically similar to *in vivo* plants. The present study has successfully established a high frequency, mass propagation system for *Talinum cuneifolium* a valuable leafy vegetable having known medicinal benefits. This protocol provides a successful and rapid technique that can be used for *ex situ* conservation.

Acknowledgements

The authors are thankful to APCOST, Hyderabad for financial support.

References

Anon 2004. The Wealth of India, A dictionary of Indian raw materials and industrial products, first supplement series, Niscair. vol 5, pp. 184.

Avani, K., P. Harish, S. Neeta and B.V. Patel. 2006. *Ex situ* conservation method for *Clerodendrum inerme*: a medicinal plant of India. African J. Biotechnol. 5:415-418.

Bajaj, Y.P.S., M. Furmanowa and O. Olszowskao. 1988. Biotechnology of the micropropagation of medicinal and aromatic plants. In: Bajaj, Y.P.S. (ed.), Biotechnology in Agriculture and forestry, Springer and Verlag, New York, vol. 4, pp. 60-103.

Bhaskaran, S., R.H. Smith, S. Paliwal and K.F. Schert. 1987. Somaclonal variation from *Sorghum bicolor* (L.) Moench cell culture.Plant Cell Tiss. Org. Cult. 9:189-196.

Bhatt, S.R., A. Kackar and K.P.S. Chandel. 1992. Plant regeneration from callus cultures of *Piper longum* L. by organogenesis. Plant Cell Rep. 11: 5252-5254.

Das, G. and G.R. Rout. 2002. Direct plant regeneration from leaf explants of *Plumbago* species. Plant Cell Tiss. Org. Cult. 68: 311-314.

Donnelly, D.J. and L. Tindall. 1993. Acclimatization strategies for micropropagated plants, In: Ahuja, M.R. (ed.), Micropropagation of Woody Plants, Kluwer Academic Publishers, Dordrecht, pp. 153-166.

Feuser, S., K. Meler, M. Daquinta, M.P.Guerra and R.O.Nodari. 2003. Genotypic fidelity of micropropagated pineapple (*Ananas comosus*) plantlets assessed by isozyme and RAPD markers. Plant Cell Tiss. Org. Cult. 73: 221-227.

Fracaro, F. and S. Echeverrigaray. 2001. Micropropagation of *Cunila galioides*, a popular medicinal plant of south Brazil. Plant Cell Tiss. Org. Cult. 64:1-4.

George, E.F. 1993. Plant Propagation by Tissue Culture Part I, 2nd ed., Exegetics Ltd, Edington, England.

Gisele Schoene. and Thomas Yeager. 2005. Micropropagation of sweet viburnum (*Viburnum odoratissimum*). Plant Cell Tiss. Org. Cult. 83:271-277.

Heloir, M.C., J.C. Fournioux, L.Oziol and R.Bessis. 1997. An improved procedure for the propagation *in vitro* of grape vine (*Vitis vinifera* cv. Pinot noir) using axillary bud microcuttings. Plant Cell Tiss. Org. Cult. 49: 223-225.

Hu, C.Y. and P.J. Wang. 1983. Meristem, shoot tip and bud culture. In:, Evans, D.A., Sharp W.R., Ammirato, P.V. and Yamata, Y. (ed.), Handbook of Plant Cell Culture. McMillan Publ., Co., New York, vol. 1, pp. 117-227.

Janapati, Y.K., A.Rasheed, K.N.Jayaveera, K.Ravindra Reddy, A.Srikar and Manohar Siddaiah. 2008. Anti-diabetic activity of alcoholic extract of *Talinum cuneifolium* in rats. Pharmacology. 2: 63-73.

Levieille, G. and G. Wilson. 2002. *In vitro* propagation and iridoid analysis of the medicinal species *Harpagophytum procumbens* and *H. zeyheri.* Plant Cell Rep. 21:220-225.

Liu, C.Z., S.J.Murch, M.Demerdash and P.K. Saxena. 2002. Regeneration of the Egyptian medicinal plant *Artemisia judaica* L. Plant Cell Rep. 21:525-530.

Mandal, A.K.A., S.D. Gupta and A.K.Chatterji. 2001. *In Vitro* plant regeneration in safflower. In: Irfan A Khan and Khanum A (ed.), Role of Biotechnolgy in Medicinal and Aromatic Plants, Ukaaz, Hyderabad. vol. 4, pp. 159-167.

Martin, K.P., M.R.Beena, and Domini Joseph. 2003. High frequency axillary bud multiplication and *ex vitro* rooting of *Wedelia chinensis* (Osbeck) Merr. A medicinal plant. Indian J. Exp. Biol. 41:262-266.

Murashige, T. and F. Skoog. 1962. A revised medium for rapid growth and bioassay with tobacco tissue culture; Physiol. Plant. 15:473-497.

Nobre, J., C.Santos and A.Romano. 2000. Micropropagation of the Mediterranean species *Viburnum tinus.* Plant Cell. Tiss. Org. Cult. 60:75-78.

Palacios, N., P. Christou and M.J. Leech. 2001. Regeneration of *Lonicera tatarica* plants via adventitious organogenesis from cultured stem explants. Plant Cell Rep. 20:808-813.

Pramanik, T.K. and S.K. Datta. 1986. Plant regeneration and ploidy variation in culture derived plants of *Asclepias curassavica* L. Plant Cell Rep. 5:219-222.

Prasad, P.J.N. 2004. *In vitro* studies of *Cryptolepis buchanani* Roem. & Schult. and *Sarcostemma intermedium* Dcne. (Asclepiadaceae). Ph.D. Thesis submitted to S.K. University, Anantapur, Andhra Pradesh, India.

Ramamurthy, N. and N.Savithramma. 2002. *In vitro* regeneration of a medicinal plant *Cassia alata* L. through axillary bud. J.Plant Biol. 29(2): 215-218.

Ramamurthy, N. and N.Savithramma. 2003. Shoot bud regeneration from leaf explants of *Albizzia amara* Boiv. Indian J.Plant Physiol., 8(4): 372-376.

Soniya, E.V. and M.R. Das. 2002. *In vitro* micropropagation of *Piper longum* – an important medicinal plant. Plant Cell Tiss. Org. Cult. 70:325-327.

Thakur, R., P.S. Rao and V.A. Bapat. 1998. *In vitro* plant regeneration in *Melia azedarach* L. Plant Cell Rep. 18:127-131.

Tivarekar, S. and S. Eapen. 2001. High frequency plant regeneration from immature cotyledons of mungbean. Plant Cell Tiss. Org. Cult. 66:227-230.

Tornero, O.P. and L.Burgos. 2000. Different media requirements for micropropagation of apricot cultivars. Plant Cell Tiss. Org. Cult. 63:133-141.

Van Eldic, L. J., A.R. Grossman, D.B.Iversion. and Watterson D.M. 1980. Isolation and characterisation of calmodulin from spinach leaves and *in vitro* translation mixtures. Proc.Natl.Acad.Sci. USA. 77: 1912-1916.

Villarreal, M.L. and G. Rojas. 1996. *In vitro* propagation of *Mimosa tenuiflora* (Willd.) Poiret, a Mexican medicinal tree. Plant Cell Rep. 16:80-82.

Chapter 5

Hormones and Plant Growth Regulators on *in vitro* Somatic Embryogenesis: A Review

K. Jaganmohan Reddy* and M. Venkateshwarlu

Department of Botany, Kakatiya University,
Warangal – 506 009, Andhra Pradesh, India

Somatic embryogenesis is the process of a single cell or a group of cells initiating the developmental pathway that leads to reproducible regeneration of non-zygotic embryos capable of germinating to form complete plants. According to Sharp *at al.* (1982) somatic embryogenesis is intiated either by 'pre-embryogenic determined cells' (PEDCs) or by 'induced embryogenic determined cells' (IEDCs). In PEDCs, the embryogenic pathway is predetermined and the cells appear to only wait for the synthesis of an inducer (or removal of an inhibitor) to resume independent mitotic divisions in order to express their potential. Such cells are found in embryonic tissues (including scutellum of cereals), certain tissues of young *in vitro* grown plantlets, the nucellus and the embryo sac (within ovules of mature plants). IEDCs, on the other hand, require redetermination to the embryogenic state by exposure to specific growth regulators such as 2, 4-D. These cells are differentiated generally in microspore (anther) cultures and callus cultures. Once the embryogenic state has been reached both cell types proliferate in a similar manner as *embryogenic determined cells* (EDCs). Plantlets are then produced directly by following the full embryogenic pathway as a co-ordinated group of EDCs.

* Corresponding Author: E-mail: reddykankanala@yahoo.com

Embryos formed in cultures have been variously designated as accessory embryos, adventive embryos, embryoids and supernumerary embryos. Kohlenbach (1978) has proposed the following classification of embryos.

1. Zygotic embryos–those formed by fertilised egg or the zygote.
2. Non-zygotic embryos–those formed by cells other than the zygote.

   (*a*) Somatic embryos–those formed by the sporophytic cells (except zygote) either *in vitro* or *in vivo.*

   (*b*) Parthenogenetic embryos–those formed by unfertilised egg.

   (*c*) Androgenic embryos–those formed by the male gametophyte (microspore pollen grains).

Somatic Embryogenesis in Dicotyledonous Cultures

Totipotent embryogenic cells have been most commonly obtained from explants of embryonic or young seedling tissues. Excised small tissues from young inflorescences (before maturation of floral primordia) are equally effective for the induction of somatic embryogenesis in cultures. Other explants used are the scutellum, young roots, petioles, immature leaf, and immature hypocotyls. In *Ranunculus scleratus* various floral (including anthers) and vegetative tissues proliferate to form a callus on a medium containing coconut milk (10 per cent) with or without IAA. Within 3 weeks, numerous embryos appear from the peripheral and deep-seated cells of the callus. A high yield of embryogenic calli can also be obtained from isolated fully differentiated mesophyll cells or protoplasts in a defined culture medium. *Citrus* nucellar cells have a natural potential for somatic embryogenesis, which is also manifested in their cultures.

Somatic embryos germinate *in situ* or when they are excised and cultured individually on a fresh semi-solid medium. A special and noteworthy feature may be the development of a fresh crop of adventive embryo (numbering 5-50) which originates from single epidermal cell on the stem surface of the plantlets obtained from germinating embryos. Age, physiological state, genotype and orientations of the explant, while in contact with the medium, influence the induction of somatic embryogenesis. These aspects govern the disruption of explant tissue integrity, callus friability, isolation of cells and other requirements in order to enhance somatic embryogenesis in various species.

Somatic Embryogenesis in Monocotyledonous Cultures

Many monocotyledonous plants are of agricultural and medicinal importance. Unlike dicots, the vegetative parts of a monocot plant do not readily proliferate in cultures. Therefore, explants are best taken from embryogenic or meristematic tissues (young inflorescences and leaves). The procedures have been developed to induce somatic embryogenesis in suspension cultures of other monocot species, such as *Dioscorea (D.floribunda, D. bulbifera).*

Young caryopses (10-15 days after pollination) or seeds are sterilised by a 30 s rinse in 70 per cent ethanol, followed by 10-20 min soak in 20 per cent commercial bleach to which a few drops of detergent have been added as a wetting agent. Some species may require a further 30-60 s rinse in mercuric chloride (0.01-0.1 per cent) to eliminate the contamination problem. Explants are then washed at least three times in sterile distilled water and zygotic embryos removed aseptically with the unaided eye (maize embryo) or using a dissecting microscope (for young and immature embryos). Excised embryos are now transferred to a culture vial containing MS medium supplemented with 2, 4-D (0.5-2.5 mg l^{-1} in the case of cereals or 18 µM for *Dioscorea*) and sucrose (2-6 per cent). Cultures are incubated in diffused light or complete darkness. A small and slow-growing callus will appear in 4-6 weeks. In gymnosperm (conifer) species, immature zygotic embryos have embryogenic tissue arising from cells in suspensor region which is initiated into embryogenic cultures.

Premeiotic inflorescences, with the primordia of the individual florets just beginning to protrude, have been observed to be the most suitable material in some systems. The inflorescences, generally 1-2 cm in length, are sterilised according to the procedure described for zygotic embryos. Following sterilisation, each inflorescence is exposed by a vertical incision through the surrounding leaves and then cut into 1-2 mm thick segments. Individual segments are then cultured on a medium containing 2, 4-D for proliferation and initiation of an embryogenic callus.

Factors Affecting Somatic Embryogenesis

Conventionally, somatic embryogenesis is regarded as a two step process: 'induction of embryogenesis' and 'embryo development', both requiring different culture conditions. Recently, the third step of 'embryo maturation' has been identified, during which the embryo is prepared for germination.

Explant

The success in obtaining regenerating cultures of several plant species which were once regarded recalcitrant, such as cereals, grain legumes and forest tree species, has been possible largely due to a shift in emphasis from media manipulation to explant selection. Immature zygotic embryos have proved to be the best explant to raise embryogenic cultures of these plants. In the cultures of embryonic explants SEs may arise directly or after slight callusing. Cotyledons from SEs of soybean gave considerably higher embryogenic response than those from zygotic embryos. In cereals, zygotic embryos exhibit the potential to form SEs only during shortly after histogenesis and prior to embryo maturation, which corresponds to a period from 11-14 days post-anthesis (DPA) in *Triticum aestivum*. During this period embryogenic callus is readily induced from the tissue of the scutellum. The loss of competence to form somatic embryos 14 days after anthesis is correlated with rapid accumulation of storage proteins within the scutellum.

Genotype

Genotypic effect on somatic embryogenesis occurs as for regeneration via shoot bud differentiation. Genotypic variations could be due to endogenous levels of hormones. Carman and Campbell cultured the spike-bearing culms of a recalcitrant cultivar of *Triticum aestivum* in media containing high levels of zeatin. Immature embryos (12 DPA) from such spikes were less prone to produce embryogenic cultures and more prone to germination. ABA and IAA had a similar effect. Elimination of hormones caused a tenfold increase in the embryogenic response. The ovules of highly embryogenic lines of *Zea mays* contained 16-20 times less auxin and 10- 15 per cent less cytokinin than those of non-embryogenic and poorly embryogenic lines. The optimum concentration of 2, 4- D required for the formation of embryogenic callus in rice varied with the cultivar.

Somatic embryogenesis/caulogenesis in alfalfa is a genetically controlled process. Most cultivars of alfalfa contain genotypes capable of regenerating in cultures; on an average the regeneration frequency is 10 per cent but some cultivars, such as Rangelander, exhibit a much higher frequency. Inheritance studies have shown that the capacity to regenerate plants in alfalfa is controlled by two dominant genes. Somatic embryogenesis in orchardgrass is also shown to be a heritable dominant trait. Highly regenerating genotypes of alfalfa have been produced using conventional breeding approach, suggesting that it is possible to genetically combine regeneration capacity with agronomic performance. A highly regenerating tetraploid line 'Regan-s' (67 per cent regeneration) was produced by crossing two poorly responding parents 'Du Puits' (10 per cent regeneration) and 'Sarnac' (14 per cent regeneration) followed by recurrent selection. McCoy and Bingham selected a diploid line of alfalfa 'HG2' by chromosome manipulation and breeding which showed even greater regeneration ability (96 per cent) than 'Regan'.

Growth Regulators

Auxin

All the well studied somatic embryogenic systems, such as alfalfa, carrot, celery, coffee, orchardgrass, and most of the cereals, require a synthetic auxin for the induction of somatic embryogenesis followed by transfer to an auxin-free medium for embryo differentiation. 2, 4-D has been the most commonly used auxin for the induction of somatic embryogenesis. However, Kamada *et al*. (1979) and Smith *et al*. (1974) succeeded in establishing embryogenic cultures of carrot without a growth regulator. Whereas Kamada *et al*. (1979) achieved it with high concentration of sucrose, Smith *et al*. (1974) managed it by manipulating the pH of the medium. At pH 4 embryogenic clumps continued to proliferate without the appearance of embryos. Embryos developed when the pH was increased to 5.6.

Generally, the embryogenic cultures of carrot are initiated and multiplied in a medium containing 2,4-D in the range of 0.5-1 mg l^{-1}. On such a medium ('proliferation medium') callus differentiates localized groups of meristematic cells called 'proembryogenic masses' (PEMs). In repeated subcultures on the proliferation medium the ECs continue to multiply without the appearance of embryos. However, if the PEMs are transferred to a medium with a very low level of auxin (0.01- 0.1 mg l^{-1}) or no auxin at all ('embryo development medium'; ED- medium), they develop into embryos. The presence of an auxin in the proliferation medium seems essential for the tissue to develop embryos in the ED medium. The tissue maintained continuously in auxin- free medium would not form embryos. Therefore, the proliferation medium is regarded as the 'induction medium' for somatic embryogenesis and each PEM an unorganized embryo.

All the major species of cereals and grasses have been reported to regenerate plants *in vitro* via somatic embryogenesis. Among the large number of growth regulators tested, 2, 4-D is by far the most effective for producing embryogenic cultures. The cultures are initiated on a medium containing 1-2.5 mg l^{-1} 2, 4-D and the embryo development generally occurs when the concentration of 2, 4-D is reduced to 5-10 per cent of the initial concentration. Orchardgrass (*Dactylis glomerata*) is an exception, where 30 μ M dicamba (3, 6-dichloro-o-anisic acid) has been reported to give best response. 2-(2, 4-Dichlorophenoxy) propionic acid was found to be the most effective auxin for the induction of somatic embryogenesis in alfalfa.

The importance of auxin in embryogenesis is also suggested by the detailed work on habituated callus tissues of *Citrus sinensis*. Originally, the nucellar callus of *C. sinensis* required IAA and kinetin for its growth and embryo differentiation. In repeated subcultures the callus showed a gradual decline of embryogenic potential, and after about 2 years some tissue lines appeared which was phytohormone autonomous. In these habituated tissues the presence of as low as 0.001 mg l^{-1} of IAA in the medium inhibited embryogenesis. On the other hand, any treatment which checked auxin concentration in the cells, such as auxin synthesis inhibitors (2-hydroxy-5-nitro-benzyl bromide or 7-aza-indole) and irradiation significantly improved embryo differentiation. Irradiation is known to break down auxin in the tissues exposed to irradiation levels higher than 16 kR the auxin, which otherwise inhibited embryo formation, turned out to be promotive. All these observations suggest that a high level of endogenous auxin was the embryogenic potential of the habituated iration of subculture (ranging from 6 to 14 weeks) and sucrose starvation in the preceding passage also considerably promoted embryo formation in the habituated *Citrus* callus.

Cytokinin

Halperin (1970) reported that the presence of BAP in the proliferation medium may promote cell division, but it inhibits the embryogenic potential of carrot cultures. This may be due to selective stimulation of multiplication of non-embryogenic cell-components of the cultures. However, some workers have recorded a promotive effect of cytokinins on embryogenesis. In carrot itself, Fujimura and Komamine (1975) observed that whereas BAP and kinetin were inhibitory for embryogenesis, zeatin at a concentration of 0.1 μ M promoted the process. Zeatin was especially promotive when supplied to the embryogenic callus during days 3 and 4 after their transfer from the proliferation medium to the ED medium. Embryogenesis could be induced in coffee in the presence of a single cytokinin. The induction medium for alfalfa contains kinetin (5 μ m) in addition to 2, 4-D (22.6-45.2 μm). The relative concentrations of the two growth regulators in the induction medium determines the type of morphogenic differentiation after transfer to hormone-free medium. Whereas high 2, 4-D to kinetin favours embryo/shoot differentiation the reverse ratio favours rooting.

Others

Gibberellin inhibits somatic embryogenesis. IAA, ABA and GA_3 have been reported to suppress embryogenesis in carrot and *Citrus*. In this context it is interesting to note that the ovules of monoembryonate *Citrus medica* contain significantly higher levels of IAA, ABA and GA_3 than those of polyembryonate *C. reticulata*.

Nitrogen Source

The form of nitrogen in the medium significantly affects *in vitro* embryogenesis. Halperin (1970) reported that in the cultures of wild carrot raised from petiolar segments, embryo development occurred only if the medium contained some amount of reduced nitrogen. The calli initiated on a medium with KNO_3 as the sole source of nitrogen failed to form embryos upon removal of auxin. However, the addition of a small amount (5 µM) of nitrogen in the form of NH_4Cl in the presence of 55 µM KNO_3 allowed embryo development. Halperin also demonstrated that the presence of reduced nitrogen was critical only in the induction medium. Meijer and Brown (1985) found an absolute requirement for ammonium during induction and differentiation of SEs in alfalfa, with 5 µM being optimum for induction and 10-20 µM optimum for differentiation of embryos. White's (lacking NH_4^+) and SH (with 2.6 µM NH_4^+) media are non-inductive for somatic embryogenesis in carrot and orchard grass, respectively. Addition of 2.5-3 µM NH_4^+ makes these media inductive. In orchard grass the number of embryos formed in the presence of optimum concentration of NH_4^+ (12.5 µM) was substantially higher than that with any other form of reduced nitrogen but the embryos were of poor quality.

Polyamines

There is some evidence to suggest that polyamines are required for embryo development *in vivo* and *in vitro*. Nucellus of polyembryonate cultivars of mango contain significantly higher levels of the three polyamines, putrescine, spermidine and spermine, than those of monoembryonate cultivars. Increase in the endogenous level of polyamines and the enzymes for their biosynthesis concomitant with the induction of somatic embryogenesis in carrot and the suppression of somatic embryogenesis by the inhibitors of polyamine biosynthesis suggest the involvement of polyamines in somatic embryogenesis. Globular embryos of cerels showed 37-fold higher polyamine content than the plantlets; of the three polyamines studied (putrescine, spermidine and spermine) putrescine showed the most dramatic increase (6-fold). Similarly, embryogenic cultures of mango showed substantial increase in the level of putrescine. However, the increase in the cultures of monoembryonate variety was almost double that in the cultures of polyembryonate variety. It was even higher in the non-embryogenic cultures. The role of the observed changes in polyamine content and biosynthesis and their causal relationship to somatic embryogenesis remains to be established.

Oxygen Concentration

Oxygen tension has been shown to promote embryogenic development in cultures. Kessel *et al.* (1977) reported that the amount of dissolved oxygen (DO_2) in the medium should be below the critical level of 1.5 mg 1^{-1} to allow embryo development in carrot. Higher levels of DO_2 favored rooting. The need for reduced DO_2 could be substituted by the addition of ATP to the medium, suggesting that, probably, oxygen tension enhanced the level of cellular ATP. Similarly, immature embryos of *Triticum aestivum*, on callus induction medium for 28 days, in 8 per cent O_2 produced about 3600 SE g^{-1} of scutellar tissue which was 6 times higher than when cultures were incubated at atmospheric level (21 per cent) of oxygen. Incubation in a low O_2

environment also reduced the amount of 2, 4-D required to initiate embryogenic callus. With low 2, 4-D concentration (1 μM), under atmospheric O_2 level a large number of SEs were converted into uni polar root structures. This abnormality is considerably reduced at low O_2 levels. Low O_2 level also significantly decreases precocious germination of SEs and the frequency of abnormal scutellar enlargement. In contrast to these observations, in alfalfa somatic embryogenesis was better in DO_2 higher than 70 per cent ; no embryos were formed in 21 per cent DO_2. Kessel *at al.* (1977) reported that at DO_2 concentration of 88 per cent or more 394 SE m l^{-1} were formed as against 19 embryos m l^{-1} at a DO_2 concentration of 18 per cent.

Electrical Stimulation

Very young (spherical) SEs of carrot exhibit electric gradient along the future longitudinal axis. Rathore and Goldsworthy reported stimulation of shoot bud differentiation in tobacco and wheat callus cultures by exposure to mild electric field. Exposure of freshly isolated mesophyll protoplasts to an electric field (0.02 V DC current for 20 h) considerably promoted the embryogenic response in alfalfa. The pattern of embryogenesis in protoplast cultures of alfalfa varies with the genotype. While the cultivars 'Rambler' and 'RegenS' form embryos via a callus phase, 'Rangelander' exhibits almost direct embryogenesis. Exposure of 'Rangelander' protoplasts to a low voltage electric field induced direct embryogenesis in 100 per cent preparations as against 40 per cent preparations of the untreated controls. The number of embryos per plate was enhanced from 76 to 116. Even the cv. 'RegenS' protoplasts exposed to electric field showed direct embryogenesis.

Selective Subculture

Multicellular explants are generally heterogeneous in terms of the morphogenic potential of its constituent cells. Only a small proportion of these cells are able to express their cellular totipotency under a set of culture conditions. Therefore, the calli derived from such explants are also heterogeneous. Sometime the embryogenic/ organogenic portions of the callus are distinct from the non-morphogenic tissue on the basis of their morphological appearance and it is essential to make artistic subcultures to establish regenerating tissue cultures.

In cereals, irrespective of the explant used, two types of calli are formed: (1) white, off-white or pale yellow, compact and often nodular and (2) soft, granular and transluscent. Of these, only the first type of calli exhibit embryogenic differentiation. Finer classified the cotyledon callus of *Gossypium hirsutum* on the basis of their colour as green, yellow, white, brown and red. Only yellow callus yielded embryogenic cultures. Maintenance of embryogenic cultures of conifers involves subculture, at extended intervals, of carefully selected, morphogenically distinct embryogenic tissue. Selective or artistic subculture has also been practised in producing embryogenic cultures of *Daucus carota, Zea mays, Sorghum bicolor, Hordeum vulgare* and *Picea abies*. Calli derived from seedling explants of *Cucumis melo* developed green nodular structures after 3-4 weeks of culture. If subcultured along with the whole callus the nodules continued unorganized proliferation. However, if the nodular structures were carefully isolated from the rest of the callus and cultured separately they produced multiple shoot buds.

Other Factors

Brown *et al.* (1985) reported that high potassium (20 µM) is necessary for embryogenesis in wild carrot. This effect also occurred in barley. A promotion of somatic embryogenesis in maize was induced by ethylene action inhibitors, including $AgNO_3$. In a recalcitrant inbred line of maize treatment of ovules (3 DAP) *in vivo* or *in vitro* with high levels of dicamba increased embryogenic response from approx. 1 per cent to 30.

Tisserat and Murashige (1977) demonstrated that the ovules of monoembryonate *Citrus medica* synthesize and release certain volatile and non-volatile substances which can inhibit *in vitro* somatic embryogenesis in co-cultured nucellar tissue of polyembryonate *C. reticulata*. Ethanol is one of the volatile inhibitors. When applied at a concentration equal to that produced by the ovules of *C. medica*, it markedly inhibited embryogenesis in carrot cultures. The non-volatile component of the inhibitors has been identified with IAA, ABA, and GA_3.

Embryo Induction, Maturation and Plantlet Development

Induction

An auxin, particularly 2,4-D, is generally necessary to induce embryogenesis in plants such as carrot, alfalfa and cereals. However, the requirement of exogenous auxin for the induction of somatic embryogenesis depends on the nature of the explant used. For example, petiole explants, hypocotyl explants and single cells isolated from established suspension cultures of carrot required exposure to 2,4-D for 1,2 or 7 days, respectively, to acquire competence to form embryos on ED medium (devoid of 2,4-D). Microcalli of alfalfa required even a shorter pulse (a few minutes to a few hours) of relatively high concentration (100 µ M) of 2,4-D to produce embryos in ED, medium. An important phenomenon associated with the induction of somatic embryogenesis is the change of cellular polarity.

Polarity

Several observations support the hypothesis that plant growth regulators and other treatments employed for the induction of embryogenesis do this by altering the cell polarity and promotion of subsequent asymmetric division. On a cytokinin-containing medium the immature embryos of white clover develop several adventive embryos directly from the hypocotyl epidermis. The first cytological sign of the induction of embryogenic cells is a shift from the normal anticlinal divisions in the epidermis to periclinal or oblique divisions. Following stimulation by auxin, asymmetric cell divisions were frequently observed in leaf protoplast cultures of an embryogenic cultivar of alfalfa; the protoplasts from non-embryogenic cultivar divided symmetrically. In carrot, the first division of single suspension cells was asymmetric and only the small daughter cell ultimately developed into embryo. Since the root pole of the SE is always oriented towards the larger cell the polarity of the entire SE is already determined prior to the first division of an embryogenic cell. The positive effect of pH and electric field on the induction of embryogenesis appears to be due to their effect on cell polarity.

Maturation and Development

Germination of the somatic embryo can occur only when it is mature enough to have functional shoot and root apices capable of meristematic growth. Somatic embryos show poor germinable quality with respect to their convertibility into plants. This is because these embryos do not go through 'embryo maturation' phase which is characteristic of seed or zygotic embryos. During this phase, accumulation of embryo-specific reserve food materials and proteins imports desiccation tolerance to seed embryos and thereby promote their normal development for germination. Maturation of somatic embryos, however, could be accomplished with abscisic acid (ABA) which is known to increase desiccation tolerance in somatic embryos of carrot, celery, soybean, alfalfa and other plants.

High auxin levels can inhibit development and growth of the shoot meristem and often embryos mature when the embryogenic cell suspension is transferred to a medium lacking auxin. The addition of a low level of cytokinin (zeatin 0.1 μ M) in combination with ABA may prove beneficial for embryogenesis of low density cell cultures. Sometimes somatic embryogenesis may be repetitive and new centers of embryogeny arise from maturing embryos. Thus, the germination phase in somatic embryos is adversely affected by their highly asynchronous development. Therefore, it becomes essential to obtain some degree of uniformity in terms of the initial population by sieving the inoculum. A graded series of stainless steel or nylon mesh sieves is recommended for sieving the embryogenic cell suspension, followed by centrifugation in 16 per cent Ficoll solution containing 2 per cent sucrose. Replacing sieves with glass beads has also been effective. This ensures isolation of cell clusters, of 3-10 cells each, and as a result embryo development and maturation occurs more or less synchronously when these clusters are moved to an auxin-free medium. Alternatively, computer sorting methods could be applied to obtain synchronous embryos.

Various physical factors may also affect embryo maturation, depending on the requirement of the species. In such species which require cold treatment for seed germination, it may be necessary to chill the young or mature embryos for their normal development into plantlets. This procedure, which, mimics seed maturation *in vivo*, may be necessary to trigger metabolic processes needed for germination and seedling growth. Progressive increase in sucrose levels are also used to achieve maturation. Thus, embryo desiccation by a high-sucrose medium may be involved to develop plantlets. The maturation of somatic embryos proceeds more normally in complete darkness although light seems essential for somatic embryogenesis in some cultures (*e.g.*, tomato, egg-plant).

Somatic embryos germinate on agar medium without growth regulators. Single embryos may profit by the inclusion of low levels of zeatin (0.1 μ M) in the medium. Application of GA_3 is required for the development of root and shoot from embryos of *Citrus*. After a number of leaves have formed, the small plantlets are transferred to jiffy pots, or vermiculite, for subsequent growth and development.

Loss of Morphogenic Potential in Embryogenic Cultures

Callus or suspension cultures due to aging or prolonged subcultures often show a progressive decline and even complete loss of morphogenic ability. Three hypotheses have been proposed to explain this phenomenon.

Genetic and Molecular Aspects

Nuclear changes such as polyploidy, aneuploidy and chromosomal mutations in cultured cell may be responsible for the loss of organogenic or embryogenic potential in prolonged cultures. This loss is generally irreversible. Understanding the genetic mechanism of somatic embryo development, however, can be facilitated by identification of biochemical markers in the process of somatic embryogenesis. One of the approaches followed to determine biochemical markers is the use of drugs that can block specific stages of embryogenesis without affecting the proliferation of cells. Isolation and characterisation of drug-resistant mutants also enable a search for biochemical markers. Most interesting results may be cited from the characterisation of temperature-sensitive mutants in which the embryo development process is impaired. For example, phosphorylation seems to cause a defect in one or more peptides of 'heat shock' proteins induced at high temperature in carrot mutant cell line Ts59.

A number of extracellular proteins exuded by embryogenic cells in the medium are implicated in the induction and development of somatic embryos in carrot suspension cultures. Some of these include EP3, AGPs and 21D7 proteins. Some embryogenesis is a genetically control- process in orchardgrass and alfalfa, expressed as a dominant nuclear trait. Cytoplasmic factors have also been implicated in control of somatic embryogenesis. About 21 'embryo-specific' or 'embryo- enhanced' genes have been cloned and it is likely that some of these genes may be useful markers for early embryo development. AGL 15-specific antibodies found to accumulate in microspore embryos in oil seed rape and somatic embryos of alfalfa are found to participate in regulation of programmes active during the early stages of embryo development.

Physiological Hypothesis

Altered hormonal balance within the cells or tissues may also be associated with decline in the embryogenic potential. In such cases it may be possible to restore the potentiality of cells (differentiating organs and embryos) by modifying the growth constituents of the media. The loss of embryogenic potential could also be restored by adding 1-4 per cent activated charcoal in the auxin-free medium (*e.g.*, carrot cultures) or on giving cold treatment to the tissues. The induction of embryogenic tissue can also be aided by various stress factors, such as heat, cadmium and anaerobsis.

Competitive Hypothesis

In the complex multicellular explant only a few cells are able to give rise to embryogenic clumps while the remaining cells are non-totipotent. According to the competitive hypothesis, the non-embryogenic cells of the explant will increase under conditions favourable to their growth, resulting in a gradual loss of embryogenic component during repeated subcultures. Restoration of embryogenesis in such cultures is impossible, but if the culture carries few embryogenic cells which are not able to express their totipotency due to the inhibitory effect of the non- embryonic cells, it should be possible to restore the morphogenic potential of these cultures by altering the composition of the medium in such way that selective proliferation of the embryogenic totipotent cells occurs.

Practical Applications of Somatic Embryogenesis

Clonal Propagation

Somatic embryogenesis has a potential application in plant improvement. Since both the growth of embryogenic cells and subsequent development of somatic embryos can be carried out in a liquid medium, it is possible to combine somatic embryogenesis with engineering technology to create large-scale mechanised or automated culture systems. Such systems are capable of producing propagules (somatic embryos) repetitively with low labour inputs. In this process of repetitive somatic embryogenesis (also referred to as accessory, adventive, or secondary somatic embryogenesis) a cycle is initiated whereby somatic embryos proliferate from the previously existing somatic embryo in order to produce clones.

(i) Cloning Zygotic Embryos for Repetitive Somatic Embryogenesis

A wide range of soybean genotypes, have been tested for their ability to undergo auxin-stimulated somatic embryogenesis during cloning of zygotic embryos. All of them are reported to form somatic embryos provided appropriate nutrients are provided in the medium. Parrot *et al.* (1983) evaluated a diverse group of 33 soybean genotypes and observed that the genotype with the highest regeneration capacity had an average number of somatic embryos per explanted cotyledon (SE/COT) from immature zygotic embryos equivalent to 2.09. This figure was lOOx higher than the SE/COT of the worst responder. The role of genotypes in conferring regeneration capacity is further supported by studies on zygotic embryo cloning of wheat, rice and maize. Diallel analysis of various cultivars demonstrated that the regeneration capacity of these crops was directly affected by non-additive, additive and cytoplasmic factors, raising somaclonal variations in tree species.

Embryos formed directly from PEDCs appear to produce relatively uniform clonal material, whereas the indirect pathway involving lEDCs generates a high frequency of somaclonal variants. Mutation during adventive embryogenesis may give rise to a mutant embryo which on germination would form a new strain of plant. Nucellar embryos, like shoot tips, are free of virus and can be used for raising virus-free clones, especially from some tree species (*e.g.*, poly- embryonate *Citrus*) where shoot tip culture has not been successful. For clonal propagation of tree species, somatic embryogenesis from nucellar cells may offer the only rapid means of obtaining juvenile plants equivalent to seedlings with parental genotype. Clonal propagation through somatic embryogenesis has been reported in 60 species of woody trees representing 25 families.

Synthesis of Synthetic/Artificial Seeds

There is considerable worldwide interest in the development of methods for encapsulation of somatic embryos to enable them to be sown under field conditions as 'synthetic' or 'artificial' seeds. Research programmes on production of artificial seeds via somatic embryogenesis in respect of commercially important crops would not only contribute to increased agricultural production, but also add to our basic knowledge of the regulatory mechanisms which control plant growth and differentiation. Synthetic seeds, consisting of somatic embryos enclosed in a protective coating, have been proposed as a 'low-cost-high-volume' propagation system. The inherent advantages of synthetic seeds are the production of many somatic embryos and the use of conventional seed-handling techniques for embryo delivery. The objective is to produce clonal 'seeds' at a cost comparable to true seeds.

Source of Regenerable Protoplast System

Embryogenic callus, suspension cultures and somatic embryos have been employed as sources of protoplast isolation for a range of species. Cells or tissues in these systems have demonstrated the potentiality to regenerate in cultures and, therefore, yield protoplasts that are capable of forming whole plants. Embryogenic cultures are especially valuable in providing a source of regenerable protoplasts in the graminaceous, coniferous and *Citrus* species. Attempts to achieve regeneration of callus or even sustained divisions in mesophyll-derived protoplasts of Gramineae proved unsuccessful until Vasil and Vasil (1981) turned to embryogenic cultures obtained from immature embryos of pearl millet (*Pennisetum purpureum*) as the source of protoplasts. Protoplasts from these cultures were induced to divide to form a cell mass from which embryoids, and even plantlets, regenerated on a suitable nutrient medium. Similar success was subsequently reported by other workers with embryogenic suspension of *Panicum maximum, Pennisetum purpureum, Oryza sativa, Saccharum officinarum, Lolium perenne, Festuca arundinacea* and *Dactylis glomerata*.

Genetic Transformation

In seed embryogenesis, zygotic embryos are seated deep inside the nucellar tissue. They live in a protected environment besides being genetically heterogeneous. On the contrary, somatic embryos remain virtually unprotected and more or less give rise to genetically uniform plants. The advent of leaf-disc transformation systems has made it possible to successfully engineer species (*Nicotiana tabacum, Medicago sativa*) in which tissues are capable of regeneration *via* somatic embryogenesis. In these species, isolated single cells can be transformed in cultures and grown on a selection medium (nutrient medium containing an antibiotic, kanamycin) to callus colonies which eventually form somatic embryos on removal of auxin from this medium. Since the callus phase seems essential in this type of indirect somatic embryogenesis, the possibility of chimeric embryos arising from transformed and non-transformed tissues cannot be ruled out. There is also evidence to show that repetitive embryos originate from single epidermal or subepidermal cells which can be readily exposed to *Agrobacterium.* Thus, the transformation technique applied to a primary somatic embryo, instead of a zygotic embryo, should give rise to totally transgenosic somatic embryos. Repetitive embryogenesis is also ideally suited to particle gun mediated genetic transformation. Instead of relying on *Agrobacterium* to mediate the transfer of genes into plant cells, the particle gun literally shoots DNA that has been precipitated onto particles of a heavy metal, into the plant cells.

Synthesis of Metabolites

The repetitive embryogenesis system is of potential use in the synthesis of metabolites such as pharmaceuticals and oils. Borage (*Borage officinalis* L.) seeds contain high levels of γ-linolenic acid, used as a precursor of postglandins or in the treatment of atopic eczema. Somatic embryos of borage also produce this metabolite but through repetitive somatic embryogenesis a continuous supply of γ-linolenic acid is ensured, which otherwise would be limited to the growing season in the zygotic embryo. The same principle can be applied for production *in vitro* of industrial lubricants from jojoba (*Simmondsia chinensis*) and leo-palmi-tostearin (the major ingredient in cocoa butter) from Cacao (*Theobroma cacao*).

Conservation of Genetic Resources

Somatic embryos which originate from single cells and subsequently regenerate mostly genetically uniform plants are good candidates for genetic resource (germplasm) conservation. The totipotency in somatic embryos occurs because the process of somatic embryogenesis relies on reprogramming of gene expression patterns in a single cell and on triggering the cascade of structural embryogenic changes as in zygotic development. Embryogenic cultures as well as somatic embryos remain viable upon storage at ambient temperature, cold storage, or cryostorage.

Somatic Embryos versus Zygotic Embryos

In the classical embryology of angiosperms the early segmentation pattern of the zygote and its derivatives has been used as a key to the classification of embryogeny which, in turn, is employed as a taxonomic character. This is based on the assumption that the sequence of early divisions in the zygotic embryogeny is fixed for a plant. However, the published evidence for the sequence of early development of SEs corresponding to that followed during zygotic embryogeny does not appear very satisfactory. It should not be surprising if the SEs, which originate from superficial cells of calli or PEMs and develop under conditions very different from those experienced by zygotic embryos, do not follow a fixed pattern of early segmentation. In nature the adventive embryos do not follow the sequence of early divisions as strictly as do the zygotic embryos. Even the zygotic embryos often exhibit deviation from the normal pattern of development. Irrespective of the early mode of development, the zygotic embryos and SEs share similar gross ontogenies, with both typically passing through globular, torpedo and cotyledonary stages of dicots and gymnosperms, and globular, scutellar and coleoptilar stages of monocots. The SEs also accumulate seed-specific storage reserves and proteins characteristic of the same species, although in less amounts than the zygotic embryos.

Reports about the occurrence of a distinct suspensor in SEs are inconsistent. In some cases a few-celled suspensor- like structure, connecting the embryo proper to the parent callus tissue, has been reported but in most cases a distinct suspensor is not present. Even where present it may not be functional. Other abnormalities exhibited by somatic embryos are a double or triple vascular system caused by polar transport of auxins and non-development of shoot (rubber plant) or root (palms) meristems. Unlike the zygotic embryos, the SEs generally lack a dormant phase and often show

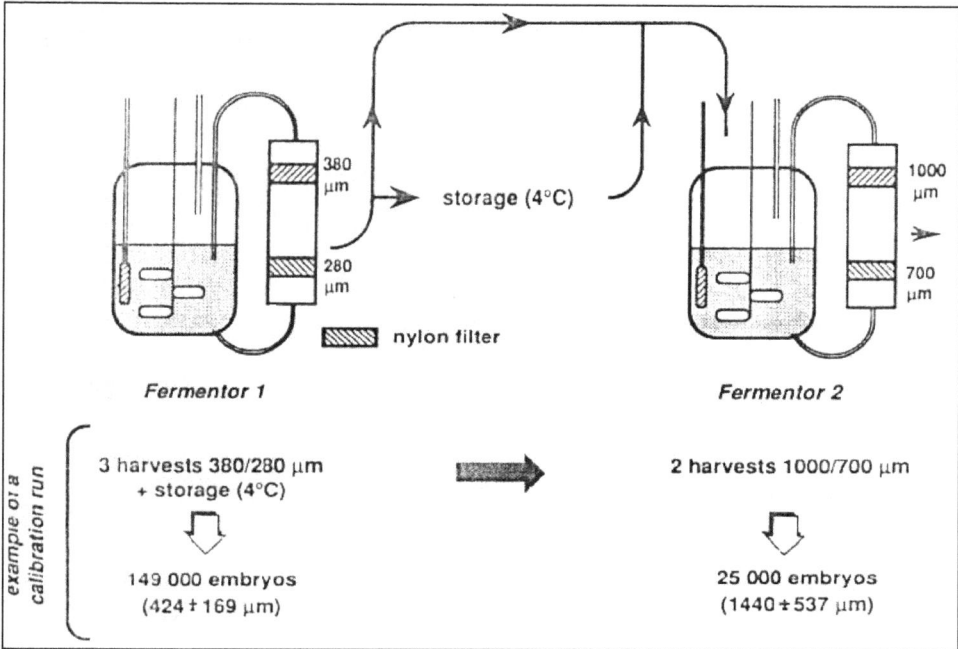

Figure 5.1: Schematic representation of a stirred-tank bioreactor (= fermentor) system designed for synchronized development of somatic embryos of carrot. Embryogenic cultures are initiated in Fermentor I and the suspension is regularly passed, by highly pressurized air, through the two nylon filters of different pore size (380 μm and 280 μm) fixed in the side arm. The filtrate of the 280 μm pore size filter, containing globular embryos or PEMs, are sent back into Fermentor 1 and the heart and young torpedo shaped embryos, collected on 28 μ m filter, are either used as such or transferred to Fermentor 2 for further synchronous development. Harvests from Fermentor 1 can also be stored at 4°C and transferred together to Fermentor 2 for synchronization of the maturation phase.

secondary embryogenesis and pluricotyledony. Some of these abnormalities can be corrected by the application of a low concentration (0.1-1.0 μ M) of ABA.

Large Scale Production of Somatic Embryos

As the multiplication of embryogenic cells and the subsequent development of SEs can occur in liquid medium, somatic embryogenesis offers a potential system for large scale plant propagation in automated bioreactors, with low labour inputs. For mass production of SEs in bioreactors, callus is initiated on a semi-solid medium. Pieces of undifferentiated or embryogenic callus are transferred to liquid medium in small flasks and agitated on a shaker. After a few cycles of multiplication in flasks, the embryogenic suspension may be filtered through a sieve of suitable pore size and PEMs or globular embryos transferred to the bioreactor flask. SEs being individual propagules, a 2-5 1 bioreactor with a production capacity of 10–100 X 10^{3} embryos should be sufficient for commercial micropropagation.

Most of the modern bioreactors are fitted with probes for measurement and control of temperature, agitator speed, pH, pO_2 and pCO_2, which not only allows cultivation of cells under highly controlled conditions but also enables precise analysis of interacting factors for cell growth and embryo development. Over the last 25 years different types of stirred-tank bioreactors, originally designed for microbial cultures, have been tested for plant cell cultures. However, a major problem in using such bioreactors for plant systems is the sheer damage caused to the cells which are relatively larger in size and possess a thinner wall than the microbes. Air-lift or bubble-column bioreactors reduce sheer damage but cause the undesirable formation of foam and callus growth above the surface of the medium. Vibration mixers, in which the medium is agitated by reciprocating vertical motion of a centrally mounted agitator shaft, provided with horizontally inserted discs, generate less sheer stress and minimize foam formation. Conical holes in the discs cause an upward or downward stream in various flow patterns.

For the production of poinsettia SEs, Preil used a round bottom 2 1 bioreactor, in which stirring was achieved by vibrating plates (vibro mixer plates of 55 mm diameter) and bubble-free O_2 was supplied through a stabilized silicorj tubing which was inserted as a spiral of 140 cm total tube length. A vibro stirring system proved to be suitable for gentle agitation that did not cause any cell damage. The poinsettia plants derived from the bioreactor-raised SEs exhibited high genetic stability (98 per cent). In bioreactors, embryogenesis is highly asynchronous.

Only a few articles on plant propagation in bioreactors have been published so far which does not allow a critical assessment of the progress achieved in the area. Different types of bioreactors have been used for the production of alfalfa SEs. In the air-lift bioreactor used by Chen *et al.* (1985) no embryos were formed. Stuart *et al.* (1985) achieved high yields of alfalfa SEs from air-lift bioreactor but the conversion rates of these embryos were disappointingly low (2-3 per cent). Bioreactor raised poinsettia embryos showed normal DNA content but the plantlets derived from them differed from the donor cultivar in various characters. Even the production of a large number of normally formed embryos in this system was not reproducible. Thus, the present success in bioreactor production of SEs is far from a practical application.

Synthetic Seeds

Somatic embryogenesis is expected to be the only clonal propagation system economically viable for crops currently propagated by seeds. However, it would require mechanical planting of SEs. Although suggestions have been made to use naked embryos for large scale planting, it would be desirable to convert them into 'synthetic seeds' or 'synseeds' by encapsulating in a protective covering. The coating material of synseeds should have several qualities: (a) it must be non-damaging to the embryo, (b) the coating should be mild enough to protect the embryo and allow germination but it must be sufficiently durable for rough handling during manufacture, storage, transportation and planting, (c) the coat must contain nutrients, growth regulators, and other components necessary for germination, and (d) the artificial seeds should be transplantable using existing farm machinery. The success of synthetic seed technology would also depend on the quality of the SEs; uniform stage with reversible arrested growth and showing high rates of conversion on planting.

Currently two types of synthetic seeds are being developed: (1) desiccated and (2) hydrated. Of these, desiccated synthetic seeds, of course, would be closer to true seeds and, therefore, have greater potential.

Desiccated Synthetic Seeds

The first synthetic seeds produced by Kitto and Janick (1985) involved encapsulation of multiple carrot SEs followed by their desiccation. Of the various compounds tested for encapsulation of celery embryos, Kitto and Janick (1985) selected polyoxyethylene (Polyox^r) which readily form a thin film, does not support growth of toxic to the embryos for encapsulation of carrot SEs with polyox suspension and a 5 per cent (w/v) solution. The suspension was dispensed as 0.2 ml drops from a pipette on to Teflon sheets (dried suspension sticks to glass plate) and dried to wafers in a laminar flow hood. The drying time is based on the ability of wafer to separate from teflon plate (about 5 h). Embryo survival and conversion of seeds are determined by redissolving the wafers in embryogenic medium and culturing the rehydrated embryos. With the considerable improvement made, during the past decade, in the desiccation of SEs and singulation of SEs by selecting them visually or by density separation, the survival of coated celery SEs could be improved from 35 per cent to 86 per cent. SEs of alfalfa desiccated to 10- 15 per cent could be stored at room temperature for 1 year without a decline in their germinability. However, efficient coating and encapsulation methods for desiccated embryos are yet to be developed.

Hydrated Synthetic Seeds

Since then encapsulation in hydrogel remains to be the most studied method of artificial seed production. Several methods have been examined to produce hydrated artificial seeds of which Ca-alginate encapsulation has been most widely used.

Concluding Remarks

During the last 10 years the list of species reported to regenerate plants in tissue cultures has considerably enlarged, and it now includes many such species which were once regarded recalcitrant. A major factor for this success has been the change of emphasis from medium manipulation to explant and genotype selection. Most of the cereals, grain legumes, cotton, tree species (including conifers) etc. express cellular totipotency only in the cultures of embryonic explants. Older tissues of these plants have remained recalcitrant. Probably in these species cells lose their competence to respond to the inductive conditions very early during development. Recent studies suggest that regeneration in tissue cultures is a three step process: (1) acquisition of competence, (2) induction and (3) development. Embryonic explants, which regenerate embryos directly without a callus phase, seem to carry the competent cells. In others the cells acquire competence on the induction medium. Rarely, as in some genotypes of *Convolvulus*, the first two steps require different treatments. We hardly know anything about the process of acquisition of competence which may be our handicap in achieving regeneration in hitherto recalcitrant taxa.

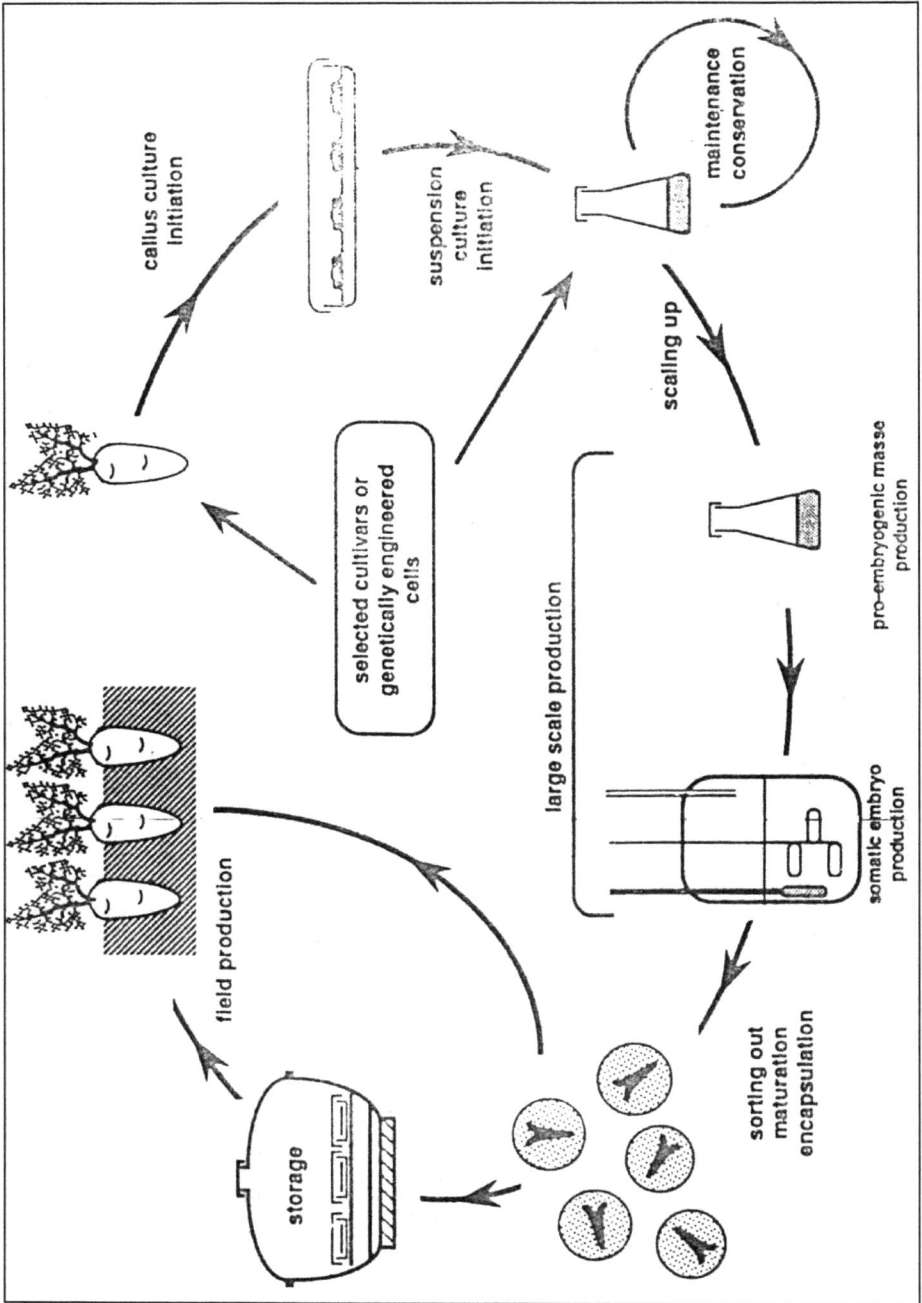

Figure 5.2: A Diagrammatic Scheme for the Production of Hydrated Synthetic Seeds of Carrot

The induction of somatic embryogenesis is being examined at the molecular level. Several 'embryo specific' or 'embryo enhanced' genes have been isolated from embryogenic cultures of carrot and some molecular markers to distinguish between embryogenic and non-embryogenic cultures have been identified. The information generated through such studies should enhance our manipulative power to induce somatic embryogenesis. For a long time, the literature on somatic embryogenesis dealt with only the production of embryos. However, since the potential application of this system in commercial propagation of plants has been recognized, a new problem has come to the fore. The somatic embryos show very poor germination because of their physiological and biochemical immaturity. Therefore, a new step of 'embryo maturation', requiring special treatment, has been introduced in the protocols of somatic embryogenesis. The prospects for commercialization of somatic embryogenesis also depends on the standardization of various parameters for large scale, synchronized production of SEs in bioreactors. Several novel bioreactors have been designed for embryogenic cultures but none has proved entirely satisfactory. The genetic/physiological instability of cultured cells leading to loss of morphogenic potential in long-term cultures continues to be a major handicap in commercial exploitation of cellular totipotency for plant propagation.

In spite of the importance attained by somatic embryogenesis and of the many studies that have been conducted on this developmental process, there are still many aspects that are not fully understood. Among those features, the involvement of plant hormones and plant growth regulators on determining the conversion of somatic onto embryogenic tissues, and on allowing progression and maturation of somatic embryos, are far away from being completely comprehended. Part of these difficulties relies on the frequent appearance of contradictory results when studying the effect of a particular stimulus over a specific stage in somatic embryogenesis. Recent progress achieved on understanding the interaction between exogenously added plant growth regulators over the concentration of endogenous hormones, together with the involvement of sensitivity of the tissues to particular hormone groups, might help clarifying the occurrence of divergent patterns in somatic embryogenesis, and in tissue culture in general. The aspects described above, emphasizing on the effect of the concentration of plant hormones and of the addition of plant growth regulators during the different phases of somatic embryogenesis, will be reviewed in this paper. Citations will be limited to review articles as much as possible and to individual articles only in those cases in which very specific or recent information is presented.

Somatic embryogenesis (SE) is the developmental pathway by which somatic cells develop into structures that resemble zygotic embryos (*i.e.*, bipolar and without vascular connection to the parental tissue) through an orderly series of characteristic embryological stages without fusion of gametes (Jimenez, 2001). From the time SE was described for the first time, almost 50 years ago (Steward *et al.* 1958; Reinert 1959), and due to the importance shown, it has been the subject of many studies. The connotation SE has acquired is the consequence of its usefulness as a tool to investigate zygotic embryogenesis, as well as an adequate system for mass-propagation of plants (Jimenez 2001). Jaganmohan Reddy *et al.* (2004) reported the effect of heavy metals on somatic embryojenesis in *Solanum melongena*. They studied the induction, maturation, germination and regeneration of induced embryos in relation to metals concentrations. Jaya Sree *et al.* (2001) reported and efficient procedure for inducing somatic embryognesis from leaf cultures of Potato CV. Jyothi. Vani Devi and Jaganmohan Reddy (1997) studied the interaction of different growth regulators and polyamine (SP) on dicotyledonary explant from their genotypes (ICCV-10 and ICCC-37) of Chick pea (*Cicer arietinum*)

SE has been traditionally divided in two main stages, namely induction and expression. In the former one, somatic cells acquire embryogenic (E) characteristics by means of a complete reorganization of the cellular state, including physiology, metabolism and gene expression (Feher *et al.*, 2002) it usually after a change in one or more culture conditions [*e.g.*, culture medium, composition of plant growth regulators (PGRs), carbohydrate source, osmotic potential, etc.] that the induced tissues or cells reach the expression stage, in which cells display their E competence and differentiate into somatic embryos. The importance of plant hormones and PGRs in both stages of SE has been widely documented during the last decades. In this review, 'plant hormones' will designate the endogenous compounds produced naturally by the tissues and cells, while 'PGRs' will account for those synthetic compounds added exogenously. Most studies on the participation of these compounds in SE has been conducted with the 'classical' groups [*i.e.*, auxins, cytokinins (CKs), gibberellins (GAs), abscisic acid (ABA) and ethylene], as the following pages make obvious. From the other substances that share some characteristics with the 'classical' hormones and that have been included more recently into this group (*i.e.*, jasmonates, salicylic acid, brassinosteroids, and polyamines), there is only evidence for polyamines to participate in SE. However, they will not be considered in this review. Further, information on the participation of polyamines in SE can be found elsewhere (*e.g.*, Minocha and Minocha 1995; Kakkar *et al.* 2000; Bais and Ravishankar 2002).

Most success achieved so far in understanding the mechanisms involved in determination and progress of SE in plants has been accomplished with model plant species, such as carrot, alfalfa and white spruce, and, more recently, hormone signaling pathways have been validated with the use of *Arabidopsis* mutants (Gazzarrini and McCourt, 2003). Very few reviews have been published dealing specifically with the role of hormones on initiation and development of SE in higher plants (*e.g.*, Jimenez, 2001). In most other reviews on SE, morphological and biochemical, and more recently, molecular aspects of this developmental process, are also included (Komamine *et al.*, 1992; Nomura and Komamine, 1995; Thorpe, 2000; Stasolla *et al.*, 2002).

Two main approaches have been followed to try to understand hormone regulation during SE. In the first and most widely used, PGRs have been added to the culture medium (evaluating different substances, concentrations and moments of application) to induce the desired developmental pattern, frequently in trial and error experiments. This approach is still being used and several new articles appear every year that propose the proper conditions, dose and combination of PGRs that permit efficient induction and progression of SE in recalcitrant or less studied species, genotypes or explant types (*e.g.*, Arunyanart and Chaitrayagun 2005; Park *et al.*, 2005; Stefanello *et al.*, 2005). Later on with the development of enough sensible techniques to analyze small molecules present in minute amounts in tissues, such as plant hormones, the role that these compounds might play in governing SE was investigated by means of determining and relating their endogenous concentrations to morphogenesis and development. In addition to the former two aspects, this review will also include other features that developed more recently, such as the interaction between added PGRs and endogenous hormones and the role played by sensitivity of the explants during SE.

The aim of this review is to summarize the vast amount of information published so far on the role of plant hormones and PGRs on the different phases of SE development. Those works that involve molecular and genetic studies of hormonal regulation will not be discussed here (further information on this particular topic can be found in Dudits *et al.* 1995; Dodeman *et al.* 1997; Dong and Dunstan 2000; Thorpe 2000; Stasolla *et al.* 2002; Phillips 2004; von Arnold *et al.* 2005, among other sources). Whenever possible, review' articles will be employed to avoid citing huge amounts of literature that could be found elsewhere. Individual articles will only be cited when they present very recent or specific information that will help to explain or illustrate a particular aspect.

Role Played by Endogenous Hormones on SE

Generalities

To study the role that endogenous hormone contents play on SE, different situations have been evaluated. In a first one, the endogenous hormones in responsive and non-responsive explants were compared to verify their involvement in initiating this developmental process. Also, the same aspect was evaluated in explants showing differences in E competence. Moreover, the evolution of the endogenous hormone levels along progression of SE was also the subject of intensive research. Finally, the use of inhibitors/antagonists of hormones that might give some light on the role of endogenous hormones was also analyzed. In this section, the evidence of the participation of endogenous hormones belonging to distinct groups on SE will be reviewed.

It is important to mention that initial explants, as well as E cultures derived from them, are composed of different kinds of cells, with varying degrees of competence. Analysis of such tissues and cell conglomerates would allow the global quantification of endogenous hormones in the tissues, but specific hormone levels in responsive or E cells could not be accounted, since they average with the levels in non-responsive or non embryogenic (NE) cells. It is only with the use of synchronized systems, such as the ones developed for carrot (Osuga *et al.*, 1999, and references therein) and alfalfa (reviewed by Feher *et al.*, 2002), that more exact contents have been measured.

Relationship Between Endogenous Hormone Contents in Explants Tissues and their E Competence

Level of endogenous hormones is considered to be one of the crucial factors determining E potential of explants (Feher *et al.*, 2003; Gaj, 2004). Explant tissues as source material to induce SE are very diverse, depending mainly on the species in study. There are very responsive plants, such as carrot, in which almost any part of the plant can be used to establish E cultures (Jimenez *et al.*, 2005 and references therein), and others more recalcitrant, in which only very specific, usually juvenile, explants are responsive. The latter is the case of most cereals and conifers (Bhaskaran and Smith, 1990; Stasolla *et al.*, 2002).

In some of the works, in which immature zygotic embryos were used as explants (mainly in cereals), the endogenous hormone contents were measured in the isolated embryos (Kopertekh and Butenko, 1995; Jimenez and Bangerth, 2001b, 2001c, d), while in some other works, those contents were analyzed in the whole grains (Carnes and Wright 1988; Hess and Carman 1998). The results in the latter studies may not reflect the situation very exactly, because it has been observed that the whole kernel hormone level poorly reveals the levels in the immature embryos. This is simply because the endosperm constitutes the majority of kernel dry matter, and the hormone levels in the endosperm might vary greatly to those of the immature embryos, as observed in wheat and maize (Jimenez and Bangerth, 2001 b,c; Hess *et al.*, 2002).

In the least responsive species, it is common to find genotypes that react more readily than others to a particular set of inductive conditions. One approach employed to study relationships between endogenous hormone levels and E competence has been the comparison of those contents in genotypes with differential E potential. There are several works in which such differences have been reported. For example, higher indole-3-acetic acid (IAA) and ABA and lower CK levels were found in zygotic embryos of the most competent wheat genotype evaluated by Kopertekh and Butenko (1995). In a similar experiment, but using another set of wheat genotypes, also differing in their E competence, Jimenez and Bangerth (2001b) found higher ABA levels in the competent one as the unique difference among them. This result was related to differences in the rate of precocious germination in the individual genotypes, which indirectly affects E callus formation, as reported by Qureshi *et al.* (1989). Furthermore, very recently, Tran Thi and Pleschka (2005) reported a positive relationship between the endogenous contents of ABA in petioles in some *Daucus* species and their capacity for SE.

Additionally, Centeno *et al.* (1997) found higher endogenous levels of N^6 (Δ^2-isopentenyl) adenine (iP) and lower of zeatin in E genotypes of *Corylus avellana*, even if the total CK contents did not vary among evaluated genotypes. Concerning a different hormone group, Hutchinson *et al.* (1997) reported that high endogenous GA contents play a negative role on induction of SE in geranium (*Pelargonium* x *hortorum*) hypocotyl explants. On the other hand, Limanton Grevet *et al.* (2000) and Jimenez and Bangerth (2001c) practically did not find any significant variation in the levels of several hormones between genotypes of asparagus and maize, differing in their E potential, pointing against a determinant role of endogenous hormone levels in the initial explants on their E competence.

Instead of comparing genotypes varying in their E competence, in some cases tissues or organ parts of the same genotype, but with different E ability, were evaluated. In several works, in which leaf sections from different parts of the leaves were analyzed higher IAA levels in E than in ME explants were found, *e.g.*, Rajasekaran *et al.* (1987a) in *Pennisetum purpureum*, Wenck *et al.* (1988) in *Dactylis glomerata*, Ivanova *et al.* (1994) in *Medicago falcata*. The same behavior was found for ABA in *Pennisetum purpureum*

(Rajasekaran *et al.*, 1987b) but the opposite was found for this hormone in *Medicago falcata* (Ivanova *et al.*, 1994). Wenck *et al.* (1988) and Rajasekaran *et al.* (1987a) reported lower endogenous levels of CKs in *Dactylis glomerata* leaves and in leaf regions of *Pennisetum purpureum*, respectively, with high E capacity. Even if in early works in this field it was postulated that endogenous hormone levels were the main difference between genotypes with various grades of competence (reviewed by Bhaskaran and Smith, 1990), the contrasting results cited above show that this is not always the case. Supporting this latter assertion, Jimenez and Bangerth (2001d) found that two barley genotypes, largely known for having broad differences in the contents of IAA and GAs ($GA_{1, 3, 20}$) (Mounla *et al.*, 1980), did not differ evidently in their E potential.

Involvement of Endogenous Hormones during Induction of SE

Auxin is considered to be the most important hormone in regulating SE *in vitro* (Cooke *et al.* 1993). This regulation probably occurs through the establishment of an auxin gradient during the induction phase, which is essential for initiating bilateral symmetry during embryogenesis in somatic and zygotic embryos in dicotyledonous and monocotyledonous (Schiavone and Cooke 1987; Liu *et al.* 1993; Fischer and Neuhaus 1996). For this gradient to be established, relatively high levels of IAA in the competent tissues may be necessary. It has been reported that the culture of explants in medium containing 2,4-dichlorophenoxyacetic acid (2,4-D), the classic induction treatment for many species, increases the endogenous auxin levels in the responsive explants (Michalczuk *et al.*, 1992a; Pasternak *et al.*, 2002), being this synthesis one of the crucial signals determining E fate of cultured cells (Thomas *et al.*, 2002).

Most studies on hormone contents during induction of SE evaluate only one single time-point during the process, either estimating the hormone levels on the explants prior to their culture (discussed in the previous section), or when E or NE cultures have already developed (at the end of the induction process). In the few works, in which endogenous hormone contents were followed-up during induction of SE, a transient increase in the endogenous IAA levels seems to be a common feature. Charriere *et al.* (1999) observed, it 24 h after increasing the concentration of sucrose from 3 to 12 per cent, one of the inducer treatments in immature zygotic embryos of sunflower (*Helianthus annuus*), while Thomas *et al.* (2002) detected it 14 h after including a CK into the culture medium, another inductor for the same model system. The latter pattern was shown to correlate in time with the reactivation of the cell division in the region of the explants where the morphogenic reaction took place. In another work, Grieb *et al.* (1997) evaluated changes in the levels of IAA, ABA and six CKs during the first 18 d of culture of carrot petiole explants with a supplement of auxin. They found a peak of IAA at day six, but also a continual decrease of ABA during the period evaluated, and an increase in the CKs up to day 10.

As stated previously in this review, explant tissues consist of cells with distinct capacity to respond to an induction treatment and therefore to become E. For that reason, after treating competent explants with an adequate induction treatment, many cells, sometimes majority of them do not acquire E capacity, resulting in a mixed population of cells that have to be selected. More than 10 years ago, Kiyosue *et al.* (1993) postulated that characterizing the differences between E and NE cells would help to elucidate the mechanisms involved in the induction and maintenance of E competence of somatic cells. However, even if characterization of endogenous hormone levels in E and NE cultures has been accomplished in several model systems during the past years, progress in understanding the mechanisms involved in SE has come primarily from other type of studies (*i.e.*, analysis of mutants in model plants and surveys on gene expression) (Gazzarrini and McCourt, 2003).

Examples of E cultures that have higher contents of endogenous auxins than their NE counterparts can be found in carrot (Li and Neumann, 1985; Sasaki *et al.*, 1994; Jimenez and Bangerth, 2001a), *Pennisetum purpureum* (Rajasekaran *et al.*, 1987b), *Medicago falcata* (Ivanova *et al.*, 1994), sugarcane (Guiderdoni *et al.*, 1995), *Prunus* spp. (Michalczuk and Druart, 1999), wheat (Jimenez and Bangerth, 2001b) and maize (Jimenez and Bangerth, 2001c). However, there is a couple of works in which no differences could be found between E and NE cultures (Besse *et al.*, 1992 in oil palm and Michalczuk *et al.*, 1992a in carrot). Further evidence that correlate high levels of endogenous auxins with E competence comes from works in which a reduction in the E capacity, observed after prolonged culture in inductive conditions, coincides with a reduction in the endogenous IAA, practically to the levels present in the NE lines, *e.g.*, Rajasekaran *et al.* (1987b) in *P. purpureum*, Kopertekh and Butenko (1995), Jimenez and Bangerth (2001b) in wheat, Jimenez and Bangerth (2001c) in maize.

Once the stimulus for the further development of the somatic embryos is given (*i.e.*, through reduction or removal of 2,4-D from the culture medium, an aspect that will be described in detail below), the endogenous IAA levels must be reduced to allow the establishment of the mentioned polar auxin gradient. Continuous growth in medium containing 2, 4-D does not allow reduction in endogenous auxin levels (Nissen and Minocha, 1993) resulting in inhibition of E development. Fischer-Iglesias *et al.* (2001) evidenced a bidirectional transport of auxins in wheat embryos growing on medium without auxin, and an altered distribution pattern of this hormone when external auxin is supplied. In very efficient and highly controlled E systems, such as the one described by Pasternak *et al.* (2002) in alfalfa, it has been even possible to compare evolution of endogenous hormone contents in E and NE cultures. These authors identified a peak in IAA synthesis in NE cultures, which showed a delay of a few days in appearance, when compared to the peak observed in E cells. In sunflower, tissues grown under the E conditions described earlier in this review (by modifying the sucrose content in the medium) showed a 4-fold increase in their IAA content, as compared to those tissues that followed the caulogenic pathway (Thomas *et al.* 2002). Evidence from these and other experiments suggests that temporal and spatial changes in endogenous auxin levels may be one of the first signals leading to SE (reviewed by Feher *et al.*, 2003).

Endogenous levels of ABA appear to be significant for initiation of E cultures, especially in some monocots (reviewed by Bhaskaran and Smith 1990), but also in carrot (Kiyosue *et al.*, 1992). Favoring this hypothesis, Rajasekaran *et al.* (1987b) in *Pennisetum purpureum*, Kiyosue *et al.* (1992) and Jimenez and Bangerth (2001a) in carrot, Guiderdoni *et al.* (1995) in sugarcane, Jimenez and Bangerth (2000) in grapevine and Nakagawa *et al.* (2001) in melon found higher ABA levels in E callus lines, when compared to NE ones. However, the opposite was found in *Hevea brasiliensis* (Etienne *et al.*, 1993) and in alfalfa (Ivanova *et al.*, 1994), whose E callus cultures accumulated lower levels of ABA than their NE counterparts did. Concerning GAs, the situation is less clear, because the few works in which endogenous contents of these hormones in E and NE cultures have been analyzed, show ambiguous data. For example, Jimenez and Bangerth (2001c) found higher GA (GA$_{1, 3, 20}$) levels in E maize lines, but Noma *et al.* (1982) found the contrary for polar GAs (probably GA) in carrot and anise. On the other side, Jimenez and Bangerth (2001a) in carrot, Jimenez and Bangerth (2001b) in wheat and Jimenez and Bangerth (2000) in grapevine did not find any difference in GA levels among cultures showing different E characteristics.

Results in anise (Ernst and Oesterhelt, 1985 and references therein) and in grapevine (Jimenez and Bangerth, 2000) indicate that CK levels seem to be more related to the growth of the callus cultures than to their E competence. However, several reports indicate a relationship between E capacity and endogenous CK contents. Rajasekaran *et al.* (1987a) and Pinto *et al.* (2002) found higher levels of CKs in NE than in E callus in *Pennisetum purpureum* and in *Medicago arborea*, respectively. Guiderdoni *et al.* (1995) reported higher levels of iP and N^6 (Δ^2-isopentenyl) adenosine (iPA) in E calluses than in the NE calluses of sugarcane, the opposite for zeatin, and no differences in the zeatin riboside levels. The individual role that the iP- and zeatin type CKs might play has been discussed elsewhere (Jimenez, 2001). Even if the relationship between auxin and CK levels in determining E properties has been postulated to be important in *Pennisetum* (Rajasekaran *et al.*, 1987a), this could not to be confirmed in other species (Jimenez and Bangerth, 2001a-c). Concerning ethylene, higher levels of this hormone have been found in NE than in E cultures of white spruce and carrot (Kumar *et al.*, 1989; Feirer and Simon, 1991). A further difference between E and NE cultures can be related to the uptake rates of exogenous PGRs, as reported in winter wheat cultures (Filek *et al.*, 2004).

Changes in Endogenous Hormone Contents during Expression of SE
The expression stage of SE can be divided in two substages, the first one elapsed from the time a stimulus to induce progression in embryo development is applied (*e.g.*, eliminating 2, 4-D from the culture medium), to the moment in which the first visible changes are observed. In the second phase embryos pass through the typical stages of zygotic embryogenesis, *i.e.*, globular, heart-shaped and torpedo shaped stages in dicots, globular, scutellar (transition), and coleoptilar stages in monocots, or globular, early cotyledonary and late cotyledonary embryos in conifers (Ciray *et al.*, 1995; Toonen and de Vries, 1996; Dong and Dunstan, 2000). Few works concentrate in the earliest of these sub stages, which usually comprises the first 7 days after the induction treatment. In one of these works, Michalczuk *et al.* (1992a) observed a rapid decline in both free and conjugated 2, 4-D metabolites while IAA levels expressed remained relatively steady within seven days after transfer from the auxin-containing medium. In a very early report also related to auxins, Fujimura and Komamine (1979a) did not find significant changes in IAA levels during the first 2 weeks after eliminating 2, 4-D in carrot suspension cultures.

Analyzing progression of ABA concentrations in carrot SE, Kamada and Harada (1981) found that the amount of endogenous ABA in cell clusters and embryos did not change and remained low during the first 7 days of culture. In a more recent work, Jimenez *et al.* (2005) analyzed evolution of endogenous IAA, ABA, GAs (GA$_{1, 3, 20}$) and several CKs, during the first 7 days after 2, 4-D was removed from the culture medium. They only found minor changes in concentration of the evaluated hormones along time, *i.e.*, a peak of IAA in callus 48 h after 2, 4-D removal, and some fluctuations in the CKs, making high levels of zeatin/zeatin riboside to coincide with low concentrations of iP/iPA. There are several plant model systems in which expression of SE is induced by stimuli different than a reduction in the content of 2, 4-D in the culture medium. For example, in E callus cultures of nucellar origin in several *Citrus* species, proembryos are induced to develop into embryos by a change in the energy source of the culture medium from sucrose to glycerol. Using this systerm, Jimenez *et al.* (2001) found that the treatment that stimulated the further development of the formed somatic embryos also stimulated auxin and CK accumulation within the first 5 days, maintaining the levels of ABA and GAs steady.

An advantage of analyzing endogenous hormones after the first morphological changes in SE had occurred (the second substage described above), is the fact that the cultures can be purified and synchronized by different procedures (reviewed by Osuga *et al.* 1999; Sharma 1999). This allows analysis of stage-specific embryo populations, and avoids the dilution effect observed when using the heterogeneous cell populations described above. Using this approach, Michalczuk *et al.* (1992a) reported that auxin levels decline steadily after the globular stage in all subsequent stages of embryo development. In another work, it has been observed that, after remaining low during the first 7 days of culture in absence of 2,4-D (see above), the concentrations of ABA increased during further development of carrot somatic embryos until day 10, and then decreased (Kamada and Harada, 1981). Similarly, Rajasekaran *et al.* (1982) found that ABA levels in hybrid grapevine somatic embryos decreased from the globular to the mature stage.

Application of a maturation treatment, such as chilling at 4 °C, induced partial desiccation of well-developed somatic embryos in some species. Desiccation considerably improves the germination frequency of somatic embryos by either reducing endogenous ABA content (Kermode *et al.*, 1989), such as chilling do (Rajasekaran *et al.*, 1982; Jimenez and Bangerth, 2000), or by changing the sensitivity to ABA (Finkelstein *et al.*, 1985). However, there is at least one example in which endogenous ABA levels in chilled somatic embryos of carrot were similar to those of non-chilled embryos (Spencer and Kitto, 1988). When evaluating evolution of endogenous ABA contents in *Quercus robur*, Prewein *et al.* (2004) found a reduction along germination of somatic embryos. Studies on development of endogenous levels of GAs during latter progression of SE are limited to, to the best of our knowledge, two works conducted early in the 1980s. In the earliest one, Noma *et al.* (1982) found carrot and anise developing somatic embryos to contain lower levels of polar GAs (GA$_1$ like), but 13-22 times higher levels of less polar GAs (GA$_{4/7}$ like) than embryos staggered in their development. One year later, Takeno *et al.* (1983) reported that free and highly water-soluble GA-like substances in a hybrid grape decreased on a dry weight basis during embryo development, but increased on an embryo basis.

Considering ethylene, Kong and Yeung (1994) studied evolution of endogenous levels of this hormone after transferring somatic embryos of white spruce into maturation medium. They found an initial rise at day one, followed by a decline and by a gradual raise in the latter half of the culture period (day 22nd). In a more recent investigation, El Meskaoui and Tremblay (2001) related supra optimal production of ethylene to a low maturation capacity of somatic embryos of a particular cell line of black spruce. They related the adequate maturation capacity in a different cell line in this species to an adequate ethylene production.

Use of Inhibitors/Antagonists as Tool to Determine the Role of Endogenous Hormones

An indirect approach to evaluate the effect of endogenous hormone concentrations on different morphogenetic processes in plants, SE among them, has been the use of inhibitors of or antagonists to the different hormone groups. The necessity for auxins on SE has been established in several plants by using polar auxin transport inhibitors and substances with anti auxin properties. When included in induction media, it has been observed that polar auxin transport inhibitors hampered embryo development in carrot (Schiavone and Cooke, 1987; Tokuji and Kuriyama, 2003), changed the morphogenic pathway from E to organogenic in sunflower (Char-riere and Hahne, 1998) and inhibited somatic embryo formation in ginseng (Choi *et al.*, 1997) and *Eleutherococcus senticosus* (Choi *et al.*, 2001). Experiments with the antiauxins (substances that inhibit biosynthesis of auxins), 2,4,6-trichloro-phenoxyacetic acid and *p*-chlorophenoxyisobutyric acid, supported the findings described above for polar auxin transport inhibitors, since addition of these compounds inhibited embryogenesis in carrot (Fujimura and Komamine, 1979b, Tokuji and Kuriyama, 2003).

Regarding CKs, it has recently been observed that purine riboside, an anti-CK, also inhibited direct SE in carrot (Tokuji and Kuriyama, 2003). Additionally, the use of triazine and carbamate type of anti CKs influenced SE in *Dactylis glomerata* by reducing the number of somatic embryos produced when high concentrations of the former were employed, while the opposite was found for the latter compound (Somleva *et al.*, 1995). As previously stated, several works (Rajasekaran *et al.*, 1987b; Kiyosue *et al.*, 1992; Guiderdoni *et al.*, 1995; Jimenez and Bangerth, 2000, 2001a) pointed out to a significant role of endogenous ABA during the induction phase of SE. Further experimental evidence for the contribution of this hormone was provided by Senger *et al.* (2001), by reducing endogenous ABA contents in *Nicotiana plumbaginifolia* by diverse means (*i.e.*, by producing a transgenic line constitutively expressing an anti-ABA single chain variable fragment antibody, by treating wild-type cultures with the ABA synthesis inhibitor flouridone, and by using two ABA-synthesis mutants). As a result, they observed disturbed morphogenesis at pre-globular formation of somatic embryos, which could be reverted by exogenous ABA application. Similarly, the loss in E capacity, caused by the application of flouridone in *Pennisetum purpureum*, was partially overcome by the addition of ABA (Rajasekaran *et al.* 1982). Also, inclusion of flouridone inhibited secondary embryogenesis from E cell clusters originated from carrot seed coats cultured in absence of PGRs, while this inhibition was counteracted by including ABA into the culture medium (Ogata *et al.* 2005).

Concerning GAs, there are contrasting reports about their involvement in SE, according to the results from experiments employing inhibitors of their biosynthesis. Rajasekaran *et al.* (1987a) observed a neutral effect, since neither paclobutrazol nor the reduced levels of GAs, which may have resulted from its application, altered E character of the *P. purpureum* explants evaluated. A negative effect was reported by Mitsuhashi *et al.* (2003) who found that uniconazole, another inhibitor of GA biosynthesis, induced shrunken embryos when applied during expression of SE in carrot. Similarly, the use of paclobutrazol in alfalfa significantly decreased the number of somatic embryos formed (Rudus *et al.*, 2002). Another effect of uniconazole is the aforementioned promotion of secondary SE in carrot (Tokuji and Kuriyama, 2003). Finally, Pullman *et al.* (2005b) recently found an improvement in initiation of SE in several conifers using paclobutrazol.

It has been observed that ethylene plays a negative role on induction of SE (reviewed by Buddendorf-joosten and Woltenng, 1994; Litz and Yurgalevitch, 1997). Supporting this statement, it has been reported that application of inhibitors of ethylene biosynthesis increased induction rate of SE, as observed in maize, white spruce and soybean with aminoethoxyvinyl-glycine (Vain *et al.* 1989; Kong and Yeung, 1994; Santos *et al.*, 1997), and in carrot with cobalt, nickel and salicylic acid (Roustan *et al.*, 1989, 1990 a). A similar effect was observed using inhibitors of ethylene action, such as silver nitrate in carrot, soybean and date palm (Roustan *et al.*, 1990b; Santos *et al.*, 1997; Al-Khayri and Al-Bahrany, 2001) and by using silver nitrate and silver thiosulfate in maize (Vain *et al.*, 1989). Several works reported an influence of the genotype in the response to inhibitors of ethylene biosynthesis and action (Litz and Yurgalevitch, 1997; El Meskaoui and Tremblay, 2001; Huang *et al.*, 2001; Al-Khayri and Al-Bahrany, 2004). Nissen (1994) worked with a carrot line in which addition of low concentrations of amino cyclopropane carboxylic acid, an ethylene precursor, or ethephon, a compound metabolized to ethylene by plant tissues, stimulated SE. In this line, inhibitors of ethylene biosynthesis caused a slight inhibition in SE. A similar effect was observed during direct SE in *Oncidium* (Orchidaceae) leaf cultures (Chen and Chang 2003). On the other side, there are some reports in which modulation of endogenous levels of ethylene by the use of inhibitors or modulation of the action of this compound by antagonists did not have an effect on maturation and conversion of somatic embryos into plants (*e.g.*, *Picea sitchensis* Selby *et al.*, 1996 and *P. mariana* El Meskaoui and Tremblay 1999).

Effect of PGRs on SE

On Induction of SE

Addition of PGRs into the culture medium is the preferred way to induce morphogenetic responses in *vitro* in most plant tissue culture systems evaluated, being SE no exception. It has even been observed that, depending on the PGR composition of a particular culture medium, either SE, organogenesis or axillary bud development can be induced, *e.g.*, in seedlings of *Arachis hypogaea* (Victor *et al.*, 1999) and in embryonic axes developed from mature seeds of *Juglans regia* (Fernandez *et al.*, 2000). In only less than 7 per cent of the protocols surveyed by Gaj (2004), SE was induced in culture media devoid of PGRs, being several new examples reported constantly, *e.g.*, *Eleutherococcus koreanum* (Park *et al.*, 2005). Lakshmanan and Taji (2000) pointed out that detailed study of those model systems in which addition of PGRs are not necessary to induce SE will be very valuable to elucidate early regulatory events in embryo development. In the majority of the species studied, in which addition of PGRs is necessary to induce SE, auxins and CKs are key factors in the determination of E response, probably because they strongly participate in cell cycle regulation and cell division (Francis and Sorrell 2001; Feher *et al.*, 2003; Gaj 2004). However, ABA, ethylene, GAs and other hormones have regulatory roles which must not be ignored in culture systems. Moreover, a new generation of PGRs, such as thidiazuron, a CK that belongs to the phenylureas, is emerging as successful alternative for high-frequency direct regeneration of somatic embryos, even from well differentiated explant tissues (Gairi and Rashid, 2004a,b; Panaia *et al.*, 2004; Zhang *et al.*, 2005).

Raemakers *et al.* (1995) and Gaj (2004) presented statistics about the number of species that respond to and of protocols that use different PGR groups and combinations to induce SE. While Raemakers *et al.* (1995) informed that 45 per cent of the species reported in the publications evaluated by them, responded to an auxin treatment for induction of SE, Gaj (2004) mentioned that in more than 80 per cent of the protocols studied, SE was induced in the presence of auxins alone, or in combination with CKs. Raemakers *et al.* (1995) also reported that about 48 per cent of the dicot species evaluated reacted to a combination of auxins and CKs for induction of SE. These latter authors also stated that, among auxins, the most frequently used was 2,4-D (49 per cent) followed by naphthalene acetic acid (NAA, 27 per cent), IAA (6 per cent), indole-3-butyric acid (6 per cent), Picloram (5 per cent) and Dicamba (5 per cent). Gaj (2004) pointed out the important role of 2,4-D, by mentioning that in more than 65 per cent of the recent protocols, this compound was applied alone or in combination with other PGRs.

Lakshmanan and Taji (2000), reviewed the response of legumes to different auxin sources during induction of SE. They pointed out that, even if most specie respond favorably to auxins, especially to 2,4-D this PGR was much less efficient than IAA in inducing somatic embryos in suspension cultures of *Chamaecytisus austriacus* (Greinwald and Czygan, 1991) and it completely inhibited the production of E callus in *Hardwickia binata* (Das *et al.*, 1995). They also mentioned that, for many legume species, use of 2, 4-D resulted in a high frequency of morphologically abnormal embryos, which failed to convert into plantlets later on. In an earlier review, Nomura and Komamine (1995) summarized the results of several works in carrot, in which the effect of different exogenous auxin sources on induction of SE were evaluated. 2, 4-D appears to act an effective stressor, being one of the triggers of E development in cultured plant cells (reviewed by Feher *et al.*, 2003). This mode of action should be also considered in those systems in which very high concentrations of exogenous auxins are necessary for induction of SE in some plant systems (*e.g.*, *Pisum sativum* Ozcan *et al.*, 1993 and *Serenoa repens* Gallo-Meagher and Green, 2002). NAA, being the second most frequently used auxin to induce SE, as reported by Raemakers *et al.* (1995, see above), has shown this outcome alone or in combination with CKs mainly in woody dicots (*e.g.*, Cuenca *et al.*, 1999; Pinto *et al.*, 2002; Hernandez *et al.*, 2003; Toribio *et al.*, 2004, and references therein).

Concerning the role that another group of PGRs, the CKs, has played on plant SE, Gaj (2004) reported that induction of SE by members of this group occurred in less than 14 per cent of the publications evaluated by her. For several species in the genus *Medicago*, in which SE occurs indirectly, the effect of CKs appears to be mostly on extensive cell proliferation prior to embryo differentiation (reviewed by Lakshmanan and Taji, 2000). Even in some cases, addition of CKs inhibited the induction of SE promoted by auxins, *e.g.*, direct SE in pea, soybean and *Coronilla varia*, (reviewed by Lakshmanan and Taji, 2000). Reports of species that respond to CKs as the sole source of PGRs include *Zoysia japonica* (Asano *et al.*, 1996), *Begonia gracilis* (Castillo and Smith 1997), six *Citrus* species (Carimi *et al.*, 1999) and *Oncidium* sp. (Chen and Chang, 2001). Among the protocols in which CKs were used as the sole PGR for induction of SE, Rasmakers *et al.* (1995) mentioned that N^6-benzylaminopurine (BAP was the most frequently employed (57 per cent), followed by kinetin (37 per cent), zeatin (3 per cent) and thidiazuron (3 per cent). Concerning this last product, it has been observed that it induces SE in *Cajanus cajan* and in peanut more efficiently than auxins, resulting in the development of one of the most competent genotype-independent peanut SE systems described to date (Saxena *et al.*, 1992; Murthy *et al.*, 1995;. Several review works, with detailed information and examples of plant systems in which exogenous CKs act alone inducing SE, are available (*e.g.*, Komamine *et al.*, 1992; Nomura and Komamine, 1995; Raemakers *et al.*, 1995; Lakshmanan and Taji, 2000; Gaj, 2004). In some plant systems, the couple of CKs with auxins has been more effective to induce SE than CKs alone (reviewed by Merkle *et al.*, 1995). For example, in conifers a low percentage of sucrose, in combination with auxins and CKs, is generally necessary for induction of this process (reviewed by Dong and Dunstan, 2000; Stasolla *et al.*, 2002).

In addition to auxins and CKs, supplement of other PGRs has been found to be necessary for induction of somatic embryos in some cases. There are several examples that evidence stimulation of SE by means of ABA. In one of them, an increase in the number of somatic embryos formed in explants of E genotypes of *Dactylis glomerata* was observed by inclusion of this PGR (Bell *et al.*, 1993). Also, seedlings of carrot cultured on medium containing ABA formed somatic embryos directly from the epidermal cells, being the number of embryos formed dependant of the concentration of this PGR (Nishiwaki *et al.*, 2000). Moreover, inclusion of ABA into a culture medium that normally induces organogenesis in sunflower immature zygotic embryos produced somatic embryos instead (Charriere and Hahne 1998). Induction of SE in hybrid bermudagrass also benefited from ABA supplement (Li and Qu, 2002).

The effect of exogenously applied GAs on SE is highly variable from one to another species or tissues. For example, when GAs was added to the culture medium, mainly as gibberellic acid (GA_3), they inhibited SE in carrot (Fujimura and Komamine, 1975; Tokuji and Kuriyama, 2003), citrus (Kochba *et al.*, 1978) and geranium (Hutchinson *et al.*, 1997). However, there are also some examples, in which exogenous GA_3 stimulated embryogenesis, such as in chickpea immature cotyledon cultures (Hita *et al.*, 1997) and in *Medicago sativa* petiole-derived tissue cultures (Rudus *et al.*,

2002). A similar effect (*i.e.,* inhibition of SE) was observed, in general, for ethylene (applied as eth-ephon or ethrel) (reviewed by Minocha and Minocha, 1995; Nomura and Komamine, 1995; Thorpe, 2000). However, in certain carrot cell lines, addition of low concentrations of ethephon or aminocyclopropane carboxylic acid stimulated SE (Minocha and Minocha, 1995).

Sometimes, a multi step protocol is necessary to induce SE in certain woody species. For example, Fernandez-Guijarro *et al.* (1995) could only induce SE in *Quercus suber* by reducing the high concentrations of BAP and NAA present in the first step (medium) to lower levels in the second one. A similar methodology was successfully employed by Hernandez *et al.* (2003) with the same species, but adding a preconditioning phase that consisted in placing the explants on medium devoid of PGRs.

On Secondary SE (Proliferation)

In some E systems (*e.g.,* the carrot system), SE is a recurrent process (*i.e.,* new somatic embryos are initiated from existing somatic embryos). The proliferative process has been termed secondary, recurrent or repetitive embryogenesis (Raemakers *et al.,* 1995). In some species, this proliferation may occur indefinitely (Merkle *et al.,* 1995; Thorpe, 2000). Usually, E callus is maintained and proliferated on a medium similar to that used for initiation, being the use of liquid cultures preferred for large-scale propagation (von Arnold *et al.,* 2002). For most species studied, auxin is the main factor associated with proliferation but, at the same time, with inhibition of development of pro embryogenic masses into somatic embryos, probably by inhibiting electric cellular polarity (Thorpe, 2000; von Arnold *et al.,* 2002; Feher, 2003) or by impairing establishment of an auxin gradient (discussed above). For example, repetitive SE in peanut requires the presence of 2, 4-D and the secondary embryos produced appeared to be arrested between the late globular and early torpedo stages of development (Durham and Parrott, 1992). Secondary SE is induced in cassava by another synthetic auxin, picloram, while further development of the somatic embryos required removal of this compound (Groll *et al.,* 2001).

However, there are some examples of species that deviate from the model described above. Primary somatic embryos of alfalfa induced on medium containing IAA, NAA and kinetin produced new somatic embryos directly, when transferred onto culture medium devoid of PGRs. Repetitive somatic E capacity of these cultures remained stable for 2 years (Parrott and Bailey, 1993). Similarly, the combination of a different auxin (2, 4-D) with the CK BAP, induced secondary SE in *Morus alba* (Agarwal *et al.,* 2004). Also, in some species, secondary SE occurs in absence of PGRs (*e.g.,* das Neves *et al.,* 1999; Koh and Loh, 2000; Puigderrajols *et al.,* 2000; Calic *et al.,* 2005). Additionally, in the case of the banana cultivar Dwarf Brazilian, addition of coconut milk induced secondary SE (Khalil *et al.,* 2002). Coconut milk has been known for some decades now to be a natural source of CKs (Amasino 2005). Synthetic CKs have also stimulated this phenomenon in several woody species, such as *Abies numidica,* cherry, coffee and mango (Vookova *et al.,* 2003; Fernandez-Da Silva and Menendez-Yuffa, 2003; Gutierrez and Rugini, 2004; Xiao *et al.,* 2004). In another example, Tokuji and Kuriyama (2003) observed that inhibition of GA synthesis promoted secondary embryogenesis from the primary embryo. Moreover, Mondal *et al*. (2001) reported that BAP, indole-3-butyric acid and glutamine are necessary to produce secondary somatic embryos in a synchronous manner.

In conifers, maintenance of E tissue occurs in a liquid or on solid medium of composition similar to the induction medium, but with a lower concentration of auxin and CK, and often a reduced amount of sucrose (Stasolla *et al*. 2002). It has also been observed that ABA reduces secondary embryogenesis in this plant group and in *Quercus ilex* (reviewed by von Arnold *et al.*, 2002; Mauri and Manzanera, 2004). Secondary SE, although rare in monocots, has also been reported in bermuda grass and was induced a few weeks after removal of ABA from the culture medium (Li and Qu, 2002). On the contrary, secondary, and even 'tertiary', SE has been recently described to be induced on E cultures of carrot when ABA was applied (Ogata *et al.*, 2005). Induction of secondary SE was also stimulated in *Rosa hybrida* cv. Carefree Beauty and in hybrid, larch by this PGR (Li *et al.*, 2002; Saly *et al.*, 2002). Additionally, in the latter work, it was observed that enrichment of the vessel atmosphere with ethylene, or addition of ethephon or aminocyclopropane carboxylic acid reduced induction of this process.

On Expression of SE

As practically every developmental process in plants, expression of SE might be triggered by different factors, depending on species, cultivar, and physiological conditions of the donor plant and so on. However, in those cases in which the exogenous application of auxins has proved to be the most efficient treatment to induce SE, further development of the existing somatic embryos has commonly been reached by reducing or removing auxin from the culture media, as mentioned previously. However, some cases that deviate from this general behavior have been recorded in the literature. For example, there are few examples in which somatic embryos continue their development in the same medium in which they formed, *e.g.*, *Arachis hypogaea* (Hazra *et al.*, 1989; Wetzstein and Baker, 1993), *Eleutherococcus koreanum* (Park *et al.*, 2005). In addition, a change in the auxin type, together with a reduction in its concentration induced formation of somatic embryos in *Psophocarpus tetragonolobus* (Ahmed *et al.*, 1996). In several instances, it was the addition of low levels of CK (*e.g.*, zeatin), together with a reduction in auxin levels, what was beneficial, such as in carrot (Fujimura and Komamine, 1975). Also, a positive effect of CKs on embryo progression was observed in *Corydalis yanhusuo* (Sagare *et al.*, 2000).

There are several examples of other PGRs exerting an effect on expression of SE in a number of model systems, sometimes with contradictory results. For example, ABA did not affect the number of carrot embryos in globular and early heart stages, but caused a decrease in the amount of embryos in heart and torpedo stages (Fujimura and Komamine, 1975). It has been also documented that this PGR caused a decrease in the total number of somatic embryos in carrot (Kamada and Harada, 1981). In most conifer species evaluated so far, somatic embryo development usually has to be stimulated by exogenous ABA, a treatment that concomitantly reduces cell proliferation, probably by affecting nucleotide biosynthesis (reviewed by Dong and Dunstan, 2000; Stasolla *et al.*, 2002). Very recently, a positive interaction of ABA with activated carbon in development and yield of somatic embryos was pointed out by Pullman *et al.* (2005a) for Norway spruce. Probably the main effect of exogenous ABA on progression of SE has been an improvement in embryo morphology, *e.g.*, in caraway, as described by Ammirato (1977), an event probably related to the effect of this PGR on maturation of the embryos, as will be described below.

Exogenous application of GA$_3$ inhibited development of somatic embryos in most species evaluated (Takeno *et al.*, 1983; Hutchinson *et al.*, 1997 and references therein). However, it has also been observed that the combination of L-glutamic acid ind GA$_3$ in cultures of *Hardwickia binata*, greatly improved the frequency of normal embryo differentiation (Das *et al.*, 1995), and that GA$_3$ strongly stimulated somatic embryo production in *Medicago sativa* (Rudus *et al.*, 2002). Concerning ethylene, Roustan *et al.* (1994) observed an inhibitory effect on embryo formation in carrot when this compound was applied at the beginning of the embryo developmental phase.

On Maturation of Somatic Embryos and Conversion of Somatic Embryos into Plants
Another important phase in zygotic, but also in somatic, embryo development is the process of maturation. During this phase embryos undergo various morphological and biochemical changes, which are evident by deposition of storage materials, repression of germination and acquisition of desiccation tolerance (the latter aspect mainly in species with Orthodox seeds) (Thomas, 1993; MoKersic and Brown, 1996). Nevertheless; there are several examples in the literature in which somatic cultured embryos do not develop normally, germinate, nor convert into normal plantlets. In other cases, embryo development and maturation are interrupted by precocious germination, leading to the occurrence of poorly developed plantlets. Great efforts have been devoted to circumvent these problems, especially by supplementing culture media with certain PGRs that allow latter phases of SE to progress similarly to those in zygotic embryogenesis.

Inclusion of ABA into the culture medium during the final phases of somatic embryo development, resembling in certain way the natural increase in endogenous hormones observed in several zygotic embryos, is necessary to stimulate maturation and, at the same time, to prevent precocious germination, especially, but not only in conifers (Mauri and Manzanera, 2004; Sharma *et al.*, 2004; Garcia-Martin *et al.*, 2005). Bozhkov *et al.* (2002) found that the yield of mature somatic embryos of Norway spruce on ABA-containing medium was increased up to 10-fold when a pre-treatment of 1-9 days with this PGR was applied. In legumes, a similar effect to the one observed in conifers was also noted. Moreover, partial desiccation or exposure to cold, heat, water and osmotic stresses have shown to enhance somatic embryo germination and conversion in many members of this family (reviewed by Lakshmanan and Taji, 2000). Very recently, Blochl *et al.* (2005) related the effect of ABA on maturation of alfalfa somatic embryos to an accumulation of raffinose oligosaccharides, such as it occurs during late seed development in orthodox seeds. In spite of the previous results, in peanut, application of ABA failed to improve somatic embryo maturation or conversion (Mhaske *et al.*, 1998).

Concerning individual effect of other PGR groups, addition of GA_3 to the regeneration medium of Bermuda grass, which usually contains BAP, accelerated germination/regeneration of the somatic embryos present (Li and Qu, 2002). With reference to ethylene, its application in form of ethephon during maturation has been related to an increase in morphological abnormalities in white spruce (Kong and Yeung, 1994), but did not show any apparent effect on *Picea sitchensis* (Selby *et al.*, 1996). Again in legumes maturation has also been stimulated by inclusion of a CK alone, or in combination with an auxin, into the culture medium, being relevant the particular substance added (reviewed by Lakshmanan and Taji, 2000). Nevertheless, there are some examples in this family, in which maturation and further development of somatic embryos occur only on growth regulator-free medium (Buchheim *et al.*, 1989; Durham and Parrott, 1992). Sreenivasu *et al.* (1998) explained such an event by suggesting that differentiated somatic embryos possibly acquire the ability to endogenously synthesize the hormones required to continue their development.

Even if obtaining high quantities of somatic embryos has not constituted a problem in several plant systems, a bottleneck encountered for massive propagation of certain species is the conversion of the somatic embryos into plants (Gaj, 2004). In some cases, somatic embryos develop into small plants on culture medium without PGRs, whereas there are several experiments in which addition of different PGRs, together with the use of a different or altered basal medium, were necessary (reviewed by von Arnold *et al.*, 2002). Among PGRs studied, CKs and auxins appear to have certain regulatory functions during somatic embryo germination and conversion, as demonstrated by the positive effect of the use of these PGRs separately or in combination (reviewed by Lakshmanan and Taji, 2000).

The positive effect of ABA on inhibition of precocious germination and stimulation of maturation stimulation (see above) extends well into conversion of somatic embryos into 'normal-shaped' plants. This fact is well documented in grapevine (Rajasekaran *et al.* 1982; Goebel-Tourand *et al.*, 1993) and *Brassica oleracea* (Hansen 2000). In *Medicago falcata* this was only evident when the treatment was performed at the torpedo stage (Kuklin *et al.*, 1994). Despite the stimulatory effect of ABA, prolonged exposure to this compound was reported by Bozhkov *et al.* (2002) to suppress the growth of the formed plants. Dormancy of zygotic embryos in several species is counteracted by chilling. This treatment has been related to an increase in endogenous GAs (Takeno *et al.*, 1983) and to a reduction in ABA endogenous contents (Rajasekarar *et al.*, 1982; Jimenez and Bangerth, 2000). Similarly, in somatic embryos of several species, especially in those which under dormancy germination and conversion of somatic embryos into plants were stimulated by inclusion of GA_3 in to the culture medium (reviewed by Gaj, 2004). There are also some cases in which particular treatments have a carry-on effect on later development of the explants. That is the case for the detrimental effect of the 2, 4-D used during induction of SE observed afterwards on the regeneration ability of the somatic embryos obtained (Ozcan *et al.*, 1993; Rodriguez and Wetzstein, 1998). The morphological abnormalities observed, such as multi-cotyledon or 'fan-shaped' embryos, have been related to disruption in polar auxin transport (Liuetal, 1993).

Interaction Between PGRs and Endogenous Hormones during SE

Several observations support the premise that PGRs added exogenously exert part of their effect by modifying the concentrations of endogenous hormones (Gaspar *et al.*, 1996, 2003). This mechanism has also been postulated to explain partially the regulation of SE by supplied PGRs (Neumann 1988; Carman 1990; Ribnicky *et al.*, 1996; Thorpe, 2000). Especially, the effect of added auxins and CKs has been related to an interaction with other endogenous plant hormones, such as ABA, ethylene, and GAs, producing, at the end, the conspicuous changes in development (Gasnar *et al.*, 1996; Lakshmanan and Taji, 2000). Modulation of endogenous hormones by exogenous PGRs may occur either directly (through enzyme synthesis) or indirectly (through effectors), as it was postulated for auxins by Gaspar *et al.* (1996).

There is evidence for PGRs modifying levels of endogenous hormones belonging, both to the same and to a different group, during SE. Examples of the former include the increase in the contents of IAA as a result of 2, 4-D supply into the culture medium in carrot E cultures (Michalczuk *et al.*, 1992a, b) and alfalfa leaf protoplasts (Pasternak *et al.*, 2002). Moreover, two variant carrot cell lines able to grow at very high concentrations of 2, 4-D (92 uM) increased their levels of endogenous IAA in response to this situation (Ceccarelli *et al.*, 2002). Furthermore, Liu *et al.* (1998) reported that NAA and indole-3-butyric acid treatments promote an increase in the endogenous IAA levels in soybean hypocotyl explants.

The other alternative, modulation of endogenous hormone levels by exogenous PGRs belonging to a different group, is exemplified in a very early work, in which Noma *et al.* (1982) observed that 2,4-D regulated the relationship among polar and less polar GAs in carrot and anise. More recently, Charriere *et al.* (1999) reported an increase in the endogenous contents of IAA in immature zygotic embryos of sunflower as a consequence of ABA application. This PGR, added in high quantities, also reduced ethylene contents during maturation of somatic embryos of white spruce (Stasolla *et al.*, 2002). In another report involving modulation of ethylene levels, an increase in the synthesis of this gaseous hormone was observed in response to application of high amounts of 2,4-D, which impaired embryo development (Minocha and Minocha, 1995). A further example of the interaction among PGRs and endogenous plant hormones has been proposed to be the mechanism by which thidiazuron induces SE in peanut. This CK apparently modulates endogenous levels of auxins and CKs, which caused the observed effect (Murthy *et al.*, 1995). This is supported by the reduction in the endogenous contents of IAA and BAP caused by thidiazuron in callus cultures of *Scutellaria baicalensis* (Zhang *et al.*, 2005)

Sensitivity as a Factor Regulating SE

Trewavas (1981) raised, more than twenty years ago, the point that sensitivity to plant hormones has an important role in the way hormones modulate several processes in plants. He postulated that the sensitivity of the tissues to a change in the hormone concentration (probably perceived by particular receptors) is more important than the change in the concentration itself. Involvement of sensitivity during induction of SE could be evidenced by the fact that only responsive tissues react to the PGR contents in culture media (Bell *et al.*, 1993; Somleva *et al.*, 1995). Sensitivity to auxins might explain, at least partially, differences in response between plant species, genotypes or cells in the same explants or in explants with different origin, in their capability to become E (Dudits *et al.*, 1995). Divergences in sensitivity are probably the consequence of variation in the ability of certain explants to produce the proper receptors, and thus continuing with the E developmental pattern (Guzzo *et al.*, 1994). There is evidence that E lines are more sensitive than their NE counterparts to particular PGRs.

This was observed by Bogre *et al.* (1990), in protoplast-derived cells or root explants from alfalfa. It has also been detected that 2, 4-D can modulate the level of auxin-binding proteins in the membranes of carrot cell suspension cultures (reviewed by Lo Schiavo, 1995), being this a mechanism by which sensitivity might be affected. Additionally, loss in E competence in sweet potato after prolonged time in culture was related, by Padmanabhan *et al.* (2001), to a decrease in auxin-responsiveness. Also, variation in CK requirements for optimal expression of SE in different species of *Medicago* and *Trifolium* were explained by Lakshmanan and Taji (2000), to be the consequence, in addition to genetic variability, of differential sensitivity to CKs. Moreover, since auxin- and CK-autonomy of habituated tissues could not be explained simply by an overproduction of these hormones (Jimenez, 2001), it is high probable that sensitivity plays a role in development of this phenomenon (Gaspar *et al.*, 2003).

In spite of the large amount of research conducted during the last years, knowledge is still vague in regards to the mechanisms by which plant hormones are involved in regulation of SE. There is a pattern well defined for highly responsive species and genotypes, in which the auxin 2, 4-D plays a positive role for induction, while withdrawal of this compound triggers expression, allowing development of somatic embryos. However, this scenery is not so clear for the more recalcitrant genotypes, in which the requirements can vary greatly. Evaluating endogenous hormones in explants varying in their degree of competence, as well as along development of SE, was proposed by the mid-1990s to be an approach that would improve induction and expression of this developmental process in recalcitrant genotypes (Merkle *et al.*, 1995). However, use of custom-designed culture media that counteracts the deficiency of a particular hormone in an explant, by supplying the corresponding PGR, has not been employed frequently. The former has occurred despite the relatively large number of studies in which endogenous hormones were analyzed in the explants and tissue cultures. Moreover, to date the usual strategy to develop adequate 'recipes' to culture, multiply and regenerate plant through SE still involves addition of PGRs in a trial and error basis.

Findings summarized in this review give a clear indication on the absence of a unifying mechanism for induction, development and expression of SE in the different species and genotypes in which studies have been conducted. It has been postulated that the use of distinct methodologies to purify plant extracts and to quantify plant hormones is, at least, partially, responsible for the differences reported in various works (Jimenez, 2001). However, nowadays the methodologies employed are highly reliable and very large differences among their results are not to be expected (reviewed by Ljung *et al.*, 2004). It is more feasible that the encountered differences are the result of genotypic diversity among cultivars and species or of physiological determination of the explants.

The absence of tight relationships between endogenous hormones and E competence, as well as the large variability in the requirements of PGRs to promote and govern SE, as described in this review, point out to the participation of additional factors. Interaction between PGRs and endogenous hormones seems to be one of the mechanisms involved that might explain the absence of a common pattern of hormonal regulation in this process, and has to be studied in more detail. An aspect that is gaining importance is the differential sensitivity of particular tissues/genotypes to specific factors (*i.e.*, PGR and endogenous hormones). Recent advances in identifying the molecular receptors for some hormonal groups (reviewed by Napier, 2004), together with comprehending the responses that plant hormones and PGRs induce at the level of gene expression might bring new insights to the subject (Dong and Dunstan, 1997; Shakirova *et al.*, 2002) and might help to gain a better understanding of the actors involved.

Somatic Embryogenesis in Medicinal Plants

Centella asiatica L. is a member of Apiaceae family, popularly known as 'Mandukaparni'. The plant is a prostrate, stoloniferous perennial herb grown in marshy areas. *C. asiatica* grows in India to an altitude of 600 m above sea level. The key constituents are 'asiaticosides' which have effect on skin diseases and psychotropic diseases (Srivastava *et al.*, 1998). *Centella asiatica* leaves are rich in carotenoids, vitamin B and C. The plant also used as neutraceutical in the form of brahmigritha a medicated ghee and as syrup by the brand name 'Mentat'. In 1990, the estimated annual requirement of *C. asiatica* was around 12 700 tonnes of dry mass valued at Rs.1.5 billion (Ahmad, 1993). The entire ever growing demand is met from the natural populations. Because of large scale and unrestricted exploitation of this natural resource coupled with limited cultivation and insufficient attempts for its replenishment, the wild stock of this species has been markedly depleted and now it is listed as threatened species by the International Union for Conservation of Nature and National Resources (IUCN) (Pandey *et al.*, 1993) and an endangered species (Singh, 1989; Sharma and Kumar, 1998) According to the report prepared by the Export and Import bank of India *C. asiatica* is one of the important medicinal plant in the International market of medicinal plant trade (Kameshwara Rao, 2000).

Hence, there is an urgent need to conserve this valuable germplasm. Somatic embryogenesis has proved to be useful for micropropagation and the production of mutants, artificial seeds and materials for use plant genetic engineering (Tanaka *et al.*, 2000). There has been no report to date on somatic embryogenesis in this species. However, shoot regeneration has been reported from leaf derived callus (Patra *et al.*, 1998; Banerjee *et al.*, 1999), stem (Patra *et al.*, 1998) and nodal (Tiwari *et al.*, 2000) segments of *C. asiatica*. In the present investigation a protocol has been developed to initiate somatic embryogenesis from the callus raised from leaf explants. Young leaves were collected from *C. asiatica* plants maintained in our botanical garden washed thoroughly with distilled water, then surface sterilized with mercuric chloride (0.1 per cent for 3 min) and washed thoroughly again with sterile distilled water for several times under asceptic conditions. One centimeter diameter explants with midrib were used as explants and the explants were placed on the medium abaxially.

Callus Induction and Embryogenic Medium

The explants were inoculated on to basal MS (Murashige and Skoog, 1962) medium supplemented with various combinations and concentrations of plant growth regulators such as IAA, NAA, 2, 4-D, BAP and Kinetin. Sucrose (3 per cent) was used as carbon source and the medium was solidified with 0.8 per cent agar (Difco or Bacteriological). The pH of the medium was adjusted to 5.6 before autoclaving at 102 kpa for 15 min. The cultures were incubated at 25 ± 2 °C under a 16-h photoperiod of 50 mmol $m^{-2} S^{-1}$ irradiance provided by cool, white, fluorescent tubes (Phillips, India). Twenty culture tubes were used per treatment and each experiment was repeated three times. All the cultures were examined periodically and visual observations were recorded.

Embryo Maturation and Germination Medium

The embryogenic callus was produced on MS medium containing BAP (4.4 and 8.87 µM) or kinetin in combination with 0.45 and 2.26 µM 2, 4-D. However, embryo maturation was slow. In order to enhance the frequency of maturation of somatic embryos to plantlets MS medium with basal salts and either 2.32 µM kinetin or 2.22 µM BAP were used in combination with various concentrations of GA_3 (0.29, 1.44 and 2.89 µM). All other culture conditions were maintained as the same.

Hardening

Matured plantlets were washed with sterile distilled water and transferred into a mixture of soil and vermiculite (1:1) in 6-cm plastic cups and covered with polythene bags for 2 weeks under culture room conditions. They were later gradually exposed to low humidity by removing the polythene cover and transferred to a shade house. Percentage of survival was calculated at the end of 6th week after transplantation. To ascertain the embryogenic nature of differentiating structures, cultured tissues were subjected to a histological study. Callus bearing somatic embryos at different developmental stages was fixed in acetic acid alcohol (1:3) dehydrated in ethanol-xylol series, embedded in paraffin wax sectioned at 10 µM thickness and stained with haematoxylin and basic fuschcin.

Profuse, friable callusing was observed on explants cultured on MS medium supplemented with either BAP (2.22 µM) in combination with 2.26 µM 2, 4-D or 2.69 µM NAA or kinetin (2.32 µM) in combination with 2.26 µM 2, 4-D. Lower concentrations of BAP or kinetin either in combination with IAA (0.57 and 2.85 µM) or NAA (0.54 and 2.69 µM) resulted in medium to poor growth of callus (Figure 6.1). Higher concentrations of BAP (4.44 and 8.87 µM) or kinetin (4.65 and 9.29 µM) in combination either with IAA (0.57 and 2.85 µM) or NAA (0.54 and 2.69 µM) resulted in shoot organogenesis.

Higher concentrations of BAP (4.44 and 8.87 µM) or Kinetin (4.65 and 9.29 µM) in combination with 2, 4-D (0.45 and 2.26 µM) resulted in embryogenic calli. Although, the embryos differentiated from the callus in the presence of higher concentrations of

BAP/kinetin with 2, 4-D at 0.45 and 2.26 µM, the most effective concentration of BAP was 8.87 µM in combination with 2.26 µM 2, 4-D resulted in 84 per cent of induction and with an average of 37.3 somatic embryos per explant. Cytokinins have been reported to induce somatic embryogenesis in *Echinochloa colona* (Samataray *et al.*, 1997), *Dendanthema grandiflorum* (Tanaka *et al.*, 2000), *Bacopa monniera* (Tiwari *et al.*, 1998) solely or in combination with an auxin. In the first week of cutute highly granular and shiny masses of calli were produced in small groups from the abaxial surface of the leaf. Later they turned into globular structures in the second. During the third week of culture these structures started to turn green. At first, embryo-like, cell clusters appeared from the surface cells and grew into heart shaped structure as evidenced from the histological studies. In the fourth week of culture some heart shaped embryos differentiated into cotyledonary stage and secondary embryogenesis was also observed. At this stage these structures could be easily separated into individual embryos.

To promote development of somatic embryos in *C. asiatica* MS medium containing BAP (2.22 µM)or kinetin (2.32 µM) in combination with GA_3 (0.29, 1.44 and 2.89 µM) were tested. Maturation of somatic embryos was noticed in all other combinations except in growth regulator free medium. 2.22 µM BAP in combination with 2.89 µM GA_3 showed 58 per cent cultures with unipolar conversion, 57 per cent of bipolar conversion and only 49 per cent of the cultures survived with a high rate of mortality. However, in MS medium supplemented with 2.32 µM kinetin and 2.89 µM GA_3 a maximum of 83 per cent of cultures showed unipolar conversion, 76 per cent of bipolar conversion and 77 per cent of the cultures showed survival rate. Promotion of germination of somatic embryos by GA_3 has been also observed in *Pimpinella* (Prakash *et al.*, 2001). In the present investigation 2.32 µM Kinetin in combination with GA_3 (2.89 µM) was more effective than BAP for somatic embryo maturation, where a maximum of 76 per cent of somatic embryos were successfully regenerated into plantlets. Of the 259 plantlets transferred to soil 256 plantlets survived. Regenerated plantlets through somatic embryos showed normal growth in 95 per cent of cultures. The method described here is useful for producing large scale propagation and conservation of this important neutracutical herb.

References

Agarwal, S., Kanwar, K. and Sharma, D.R. 2004. Factors affecting secondary somatic embryogenesis and embryo maturation in *Morus alba* L. Sci. Hort. 102: 359-368.

Ahmad Ru. 1993. Medicinal plants used in ISM–their procurement, cultivation, regeneration and import/export aspects: a review. In: Govil, J.N., Singh, V.K. & Hashmi, S. (eds) Medicinal Plants: New Vistas of Research. Part I (pp 221-258). Today and Tomorrow printers and publishers, New Delhi.

Ahmed, R., Gupta, S.D. and De, D.N. 1996. Somatic embryogenesis and plant regeneration from leaf derived callus of winged bean (*Psophocarpus tetragonolobus* L). Plant Cell Rep. 15: 531 -535.

Al-Khayri, J.M. and Al-Bahrany, A.M. 2001. Silver nitrate and 2- isopentenyladenine promote somatic embryogenesis in date palm (*Phoenix dactylifera* L.). Sci. Hort. 89: 291-298.

Al-Khayri, J.M. and Al-Bahrany, A.M. 2004. Genotype-dependent *in vitro* response of date palm (*Phoenix dactylifera* L.) cultivars to silver nitrate. Sci. Hort. 99: 153-162.

Amasino R. 2005. 1955. Kinetin arrives. The 50th anniversary of a new plant hormone. Plant Physiol. 138: 1177-1184.

Ammirato, P.V. 1977. Hormonal control of somatic embryo development from cultured cells of caraway. Plant Physiol. 59: 579-586.

Arunyanart, S. and Chaitrayagun, M. 2005. Induction of somatic embryogenesis in lotus (*Nelumbo nucifea* Geartn.). Sci. Hort. 105: 411-420.

Asano, Y., Katsumoto, H., Inokuma, C, Kaneko, S., Ito, Y. and Fujiie, A. 1996. Cytokinin and thiamine requirements and stimulative effects of riboflavin and α-ketoglutaric acid on embryogenic callus induction from the seeds of *Zoysia japonica* Steud. J. Plant Physiol. 149: 412- 417.

Bais, H.P. and Ravishankar, G.A. 2002. Role of polyamines in the ontogeny of plants and their biotechnological applications. Plant Cell Tiss. Org. Cult. 69: 1-34.

Banerjee, S, Zehra, M. and Kumar, S. 1999. *In vitro* multiplication of *Centella asiatica*, a medicinal herb from leaf explants. Curr. Sci. 76:147-148.

Bell, L.M., Trigiano, R.N. and Conger, B.V. 1993. Relationship of abscisic acid to somatic embryogenesis in *Dactylis glomerata*. Environ. Exp. Bot. 33: 495-499.

Besse, I., Verdeil, J.L., Duval, Y., Sotta, B., Maldiney, R. and Miginiac, E. 1992. Oil palm (*Elaeis guinensis* Jacq.) clonal fidelity: endogenous cytokinins and indoleacetic acid in embryogenic callus cultures. J. Exp. Bot. 43: 983-989.

Bhaskaran, S. and Smith, R.H. 1990. Regeneration in cereal tissue culture: a review. Crop Sci. 30: 1328-1337.

Blochl, A., Grenier-de March, G., Sourdioux, M., Peterbauer, T. and Richter, A. 2005. Induction of raffinose oligosaccharide biosynthesis by abscisic acid in somatic embryos of alfalfa (*Medicago sativa* L.). Plant Sci. 168: 1075-1082.

Bogre, L., Stefanov, I., Abraham, M., Somogyi, I. and Dudits, D. 1990. Differences in response to 2, 4-dichlorophenoxy acetic acid (2,4-D) treatment between E and NE lines of alfalfa. In: Nijkamp, H.J.J., Van Der Plas, L.H.M. and Van Aartrijk, J. (eds), Progress in Plant Cellular and Molecular Biology. Kluwer Academic Publishers, Dordrecht, pp. 427-436.

Bozhkov, P.V., Filonova, L.H. and von Arnold, S. 2002. A key developmental switch during Norway spruce somatic embryogenesis is induced by withdrawal of growth regulators and is associated with cell death and extracellular acidification. Biotechnol. Bioeng. 77: 658-667.

Brown, D.C.W. and Atanassav, A. 1985. Role of genetic back ground in somatic embryogenesis in *Medicago*. Plant Cell Tiss. Org. Cult. 4: 111-122.

Buchheim, J.A., Colburn, S.M. and Ranch, J.P. 1989. Maturation of soybean somatic embryos and the transition to plantlet growth. Plant Physiol. 89: 768-775.

Buddendorf-Joosten, J.M.C. and Woltering, E.J. 1994. Components of the gaseous environment and their effects on plant growth and development *in vitro*. Phnt Growth Regulat. 15: 1-16.

Calic, D., Zdravkovic-Korac, S. and Radojevic, L. 2005. Secondary embryogenesis in androgenic embryo cultures of *Aesculus hippocastanum* L. Biol. Plant. 49: 435-438.

Carimi, F., De Pasquale, F. and Crescimanno, F.G.1999. Somatic embryogenesis and plant regeneration from pistil thin cell layers of *Citrus*. Plant Cell Rep 18: 935-940.

Carman, J.G. 1990. Embryogenic cells in plant tissue cultures: occurrence and behavior. *In Vitro* Cell. Dev. Biol. 26:746-753.

Carnes, M.G. and Wright, M.S. 1988. Endogenous hormone levels of immature corn kernels of A188, Missouri-17, and Dekalb XL-12. Plant Sci. 57: 95-203.

Castillo, B. and Smith, M.A.L. 1997. Direct somatic embryogenesis from *Begonia gracilis* explants. Plant Cell Rep. 16: 385-388.

Ceccarelli, N., Mondin, A., Lorenzi, R., Picciarelli, P. and Lo Schiavo, F. 2002. The metabolic basis for 2, 4-D resistance in two variant cell lines of carrot. Funct. Plant Biol. 29: 575-583.

Centeno, M.L., Rodriguez, R., Berros, B. and Rodriguez, A. 1997. Endogenous hormonal content and somatic embryogenic capacity of *Corylus avellana* L. cotyledons. Plant Cell Rep. 17: 139-144.

Charriere, F. and Hahne, G. 1998. Induction of embryogenesis versus caulogcnesis on *in vitro* cultured sunflower (*Helianthus annuus* L.) immature zygotic embryos: role of plant growth regulators. Plant Sci. 137: 63-71.

Charriere, F. Sotta, B., Miginiac, E. and Hahne, G.1999. Induction of adventitious shoots or somatic embryos on *in vitro* cultured zygotic embryos of *Helianthus annuus:* variation of endogenous hormone levels. Plant Physiol. Biochem. 37: 751-757.

Chen, J.-T. and Chang, W.-C. 2001. Effects of auxins and cytokinins on direct somatic embryogenesis on leaf explants of *Oncidium* 'Gower Ramsey'. Plant Growth Regulat. 34: 229-232.

Chen, J.-T. and Chang, W.-C. 2003. 1-Aminocyclopropane-l-carboxylic acid enhanced direct somatic embryogenesis from *Oncidium* leaf cultures. Biol. Plant. 46: 455-458.

Chen *et al.*, 1985. Somatic embryogenesis and plant regeneration from cultured inflorescence of *Oryza sativa* L. (rice). Plant Cell Tiss. Org. Cult., 4: 51-54.

Choi, Y.E., Kim, H.S., Soh, W.Y. and Yang, D.C. 1997. Development and structural aspects of somatic embryos formed on medium containing 2, 3, 5-triiodobenzoic acid. Plant Cell Rep. 16: 738-744.

Choi, Y.E., Katsumi, M. and Sano, H. 2001. Triiodobenzoic acid, an auxin polar transport inhibitor, suppresses somatic embryo formation and postembryonic shoot/root development in *Eleutherococcus senticosus*. Plant Sci. 160: 1183-1190.

Cooke, T.J., Racusen, R.H. and Cohen, J.D. 1993. The role of auxin in plant embryogenesis. Plant Cell 5: 1494-1495.

Cuenca, B., San-Jose, M.C., Martinez, M.T., Ballestcr, A. and Vieitez, A.M. 1999. Somatic embryogenesis from stem and leaf explants of *Quercus robur* L. Plant Cell Rep. 18: 538-543.

Das, A.B., Rout, G.R. and Das, P. 1995. *In vitro* somatic embryogenesis from callus culture of the timber yielding tree *Hardwickia binata* Roxb. Plant Cell Rep. 15: 147-149.

das Neves, L.O., Duque, S.R.L., de Almeida, J.S. and Fevereiro, P.S. 1999. Repetitive somatic embryogenesis in *Medicago truncatula* ssp. *narbonensis* and *M. iruncaiula* Gaertn cv. Jemalong. Plant Cell Rep. 18: 398-^05.

Dodeman, V.L., Docreux, G. and Kreis, M. 1997. Zygotic embryogenesis versus somatic embryogenesis. J. Exp. Bot. 48: 1493-1509.

Dong, J.-Z. and Dunstan, D.I. 1997. Characterization of cDNAs representing five abscisic acid-responsive genes associated with somatic embryogenesis in *Picea glauca,* and their responses to abscisic acid stereostructure. Planta 203: 448-453.

Dong, J.-Z. and Dunstan, D.I. 2000. Molecular biology of somatic embryogenesis in conifers. In: Jain S.M. and Minocha S.C. (eds), Molecular Biology of Woody Plants, Vol.1. Kluwer Academic Publishers, Dordrecht, 51-87.

Dudits, D., Gyorgyey, J., Bogie, L. and Bako, L. 1995. Molecular biology of somatic embryogenesis. In: Thorpe, T.A. (ed.), *In Vitro* Embryogenesis in Plants. Kluwer Academic Publishers, Dordrecht, pp. 276-308.

Durham, R.E. and Parrott, W.A. 1992. Repetitive somatic embryogenesis from peanut cultures in liquid medium. Plant Cell Rep. 11: 122-125.

El Meskaoui, A. and Tremblay, F.M. 1999. Effects of sealed and vented gaseous microenvironments on the maturation of somatic embryos of black spruce with special emphasis on ethylene. Plant Cell Tiss. Org. Cult. 56: 201-209.

El Meskaoui, A.. and Tremblay, F.M. 2001. Involvement of ethylene in the maturation of black spruce embryogenic cell lines with different maturation capacities. J. Exp. Bot. 52: 761-769.

Ernst, D. and Oesterhelt, D. 1985. Changes of cytokinin nucleotides in an anise cell culture (*Pimpinella anisum* L.) during growth and embryogenesis. Plant Cell Rep. 4: 140-143.

Etienne, H., Sotta, B., Montoro, P, Miginiac, T. and Carron, M.P. 1993. Relation between exogenous growth regulators and endogenous indoIe-3-acetic acid and abscisic acid in the expression of somatic embryogenesis in *Hevea brasiliensis* Mull.-Arg. Plant Sci. 88: 91-96.

Feher, A., Pasternak, T., Otvos, K., Miskolczi, P. and Dudits, D. 2002. Induction of embryogenic competence in somatic plant cells: a review. Biologia (Bratislava) 57: 5-12.

Feher, A., Pasternak, T.P. and Dudits, D. 2003. Transition of somatic plant cells to an embryogenic state. Plant Cell Tiss. Org. Cult. 74: 201-228.

Feirer, R.P. and Simon, P.W. 1991. Biochemical differences between carrot inbreds differing in plant regeneration potential. Plant Cell Rep. 10: 152-155.

Fernandez, H., Perez, C. and Sanchez-Tames, R. 2000. Modulation of the morphogenic potential of the embryonic axis of *Juglans regia* by cultural conditions. Plant Growth Regulat. 30: 125-131.

Fernandez-Da Silva, R. and Menendez-Yuffa, A. 2003. Transient gene expression in secondary somatic embryos from coffee tissues electroporated with the genes *gus* and *bar*. Electr. J. Biotechnol. 6: 29-38.

Fernandez-Guijarro, B., Celestino, C. and Toribio, M. 1995. Influence of external factors on secondary embryogenesis and germination in somatic embryos from leaves of *Quercus suber* L. Plant Cell Tiss. Org. Cult. 41: 99-106.

Filek, M., Biesaga-Koscielniak, J., Marcinska, I., Machackova, I. and Krekule, J. 2004. The influence of growth regulators on membrane permeability in cultures of winter wheat cells. Z. Naturforsch. 59c: 673-678.

Finkelstein, R.R., Tenbarge, K.M., Shumway, J.E. and Crouch, M.L. 1985. Role of ABA in maturation of rapeseed embryos. Plant Physiol. 78: 630-636.

Fischer, C. and Neuhaus, G. 1996. Influence of auxin on the establishment of bilateral symmetry in monocots. Plant J. 9: 659-669.

Fischer-Iglesias, C, Sundberg, B., Neuhaus, G. and Jones, A.M. 2001. Auxin distribution and transport during embryonic pattern formation in wheat. Plant J. 26: 115-129.

Francis, D. and Sorrell, D.A. 2001. The interface between the cell cycle and plant growth regulator: a mini review. Plant Growth Regulat. 33: 1-2.

Fujimura, T. and Komamine, A. 1975. Effects of various growth regulators on the embryogenesis in a carrot cell suspension culture. Plant Sci. Lett. 5: 359-364.

Fujimura, T. and Komamine, A. 1979a. Involvement of endogenous auxin in somatic embryogenesis in a carrot cell suspension culture. Z. Pflanzenphysiol. 95: 13-19.

Fujimura, T. and Komamine, A. 1979b. Synchronization of somatic embryogenesis in a carrot cell suspension culture. Plant Physiol. 64: 162- 164.

Gairi, A. and Rashid, A. 2004a. Direct differentiation of somatic embryos on different regions of intact seedlings of *Azadirachta* in response to thidiazuron. J. Plant Physiol. 161: 1073-1077.

Gairi, A. and Rashid, A. 2004b. TDZ-induced somatic embryogenesis in non-responsive caryopses of rice using a short treatment with 2, 4-D. Plant Cell Tiss. Org. Cult. 76: 29-33.

Gaj, M.D. 2004. Factors influencing somatic embryogenesis induction and plant regeneration with particular reference to *Arabidopsis thaliana* (L.) Heynh. Plant Growth Regulat. 43: 27-47.

Gallo-Meagher, M. and Green, J. 2002. Somatic embryogenesis and plant regeneration from immature embryos of saw palmetto, an important landscape and medicinal plant. Plant Cell Tiss. Org. Cult. 68: 253-256.

Garcia-Martin, G., Manzanera, J.A. and Gonzalez-Benito, M.E. 2005. Effect of exogenous ABA on embryo maturation and quantification of endogenous levels of ABA and IAA in *Quercus suber* somatic embryos. Plant Cell Tiss. Org. Cult. 80: 171-177.

Gaspar, T., Kevers, C, Perel, C., Greppin, H., Reid, D.M. and Thorpe, T.A. 1996. Plant hormones and plant growth regulators in plant tissue culture. *In Vitro* Cell. Dev. Biol. Plant 32: 272-289.

Gaspar, T., Kevers, C, Faivre-Rampant, O., Crevecoeur, M., Penel, C, Greppin, H. and Dommes, J. 2003. Changing concepts in plant hormone action. *In Vitro* Cell. Dev. Biol. Plant 39: 85-105.

Gazzarrini, S. and McCourt, P. 2003. Cross-talk in plant hormone signaling: what *Arabidopsis* mutants are telling us? Ann. Bot. 91: 605-612.

Goebel-Tourand, I., Mauro, M.-C, Sossountzov, L., Miginiac, E. and Deloire, A. 1993. Arrest of somatic embryo development in grapevine: histological characterization and the effect of ABA, BAP and zeatin in stimulating plantlet development. Plant Cell Tiss. Org. Cult. 33: 91-103.

Gray, D.J., Compton, M.E., Harrell, R.C. and Cantliffe, D.J. 1995. Somatic embryogenesis and the technology of synthetic seed. In: Bajaj, Y.P.S. (ed.), Somatic Embryogenesis and Synthetic Seed II. Biotechnology in Agriculture and Forestry, Vol. 30. Springer, Berlin, pp. 126-151.

Greinwald, R. and Czygan, F.-C. 1991. Regeneration of plantlets from callus cultures of *Chamaecytisus purpureus* and *Chamaecytisus austriacus* (Leguminosae). Bot. Acta 104: 64-67.

Grieb, B., Schafer, F., Imani, J., Mashayekhi, K.N., Arnholdt-Schmitt, B. and Neumann, K.H. 1997. Changes in soluble proteins and phytohormone concentrations of cultured carrot petiole explants during induction of somatic embryogenesis (*Daucus carota* L.). J. Appl. Bot. 71: 94-103.

Groll, J., Mycock, D.J., Gray, V.M. and Laminski, S. 2001. Secondary somatic embryogenesis of cassava on picloram supplemented media. Plant Cell Tiss. Org. Cult. 65: 201-210.

Guiderdoni, E., Merot, B., Ek-omtramage, T., Fjlet, T., Teldmann, P. and Glaszmann, J.C. 1995. Somatic embryogenesis in sugarcane (*Saccharum* species). In: Bajaj, Y.P.S. (Ed.). Somatic embryogenesis and synthetic seed I. Biotechnology in Agriculture and Forestry, Vol. 31. Springer, Berlin, pp. 92-113.

Gutierrez, P. and Rugini, E. 2004. Influence of plant growth regulators, carbon sources and iron on the cyclic secondary somatic embryogenesis and plant regeneration of transgenic cherry rootstock 'Colt' (*Prunus avium x P. pseudocerasus*). Plant Cell Tiss. Org. Cult. 79: 223-232.

Guzzo, F., Baldan, B., Mariani, P., Lo Schiavo, F. and Terzi, M. 1994. Studies on the origin of totipotent cells in explants of *Daucus carota* L. J. Exp. Bot. 45: 1427-1432.

Halperin, W. 1970. Embryos from somatic plant cells. In: Padykula (ed.), Control mechanisms in the expression of cellular phenotypes. Academic Press, New York, pp. 169-191.

Hansen, M. 2000. ABA treatment and desiccation of microspore-derived embryos of cabbage (*Brassica oleracea* ssp. *capitata* L.) improves plant development. J. Plant Physiol. 156: 164-167.

Hazra, S., Sathaye, S.S. and Mascarenhas, A.F. 1989. Direct somatic embryogenesis in peanut (*Arachis hypogaea*). Bio Technology 7: 949-951.

Hernandez, I., Celestino, C. and Toribio, M. 2003. Vegetative propagation of *Quercus suber* L. by somatic embryogenesis. I. Factors affecting the induction in leaves from mature cork oak trees. Plant Cell Rep. 21: 759- 764.

Hess, J.R. and Carman, J.G. 1998. Embryogenic competence of immature wheat embryos: genotype, donor plant, environment, and endogenous hormone levels. Crop. Sci. 38:249-253.

Hess, J.R., Carman, J.G. and Banowetz, G.M. 2002. Hormones in wheat kernels during embryony. J. Plant Physiol. 159:379-386.

Hita, O., Lafarga, C. and Guerra, H. 1997. Somatic embryogenesis from chickpea (*Cicer arietinum* L.) immature cotyledons: the effect of zeatin, gibberellic acid and indole-3-butyric acid. Acta Physiol. Plant. 19: 333-338.

Huang, X.-L., Li, X.-J., Li, Y. and Huang, L.-Z. 2001. The effect of AOA on ethylene and polyamine metabolism during early phases of somatic embryogenesis in *Medicago sativa*. Physiol. Plant. 113:424-429.

Hutchinson, M.J., Krishna Raj, S. and Saxena, P.K. 1997. Inhibitory effect of GA_3 on the development of thidiazuron-induced somatic embryogenesis in geranium (*Pelargonium x hortorum* Bailey) hypocotyl cultures. Plant Cell Rep. 16: 435-438.

Ivanova, A., Velcheva, M., Denchev, P., Atanassov, A. and Van Onckelen, H.A.1994. Endogenous hormone levels during direct somatic embryogenesis in *Medicago falcata*. Physiol. Plant 92: 85-89.

Jaganmohan Reddy, K., Neelima and Chandra Shekar, Ch. 2004. The effect of heavy metals on somatic embryogenesis in *Solanum melongena*. Plant cell biotechnology and Molecular biology, 5(3 & 4): 129-134.

Jaya Sree, T, Pavan, V, Ramesh, M., Rao, A.V. Jaganmohan Reddy, K and Sadanandam, A. 2001. Efficient procedure for inducing somatic embryognesis from leaf cultures of Potato cv. Jyothi. Plant Cell Tiss. and Organ Cult. 64: 13-17.

Jimenez, V.M. 2001. Regulation of *in vitro* somatic embryogenesis with emphasis on the role of endogenous hormones. Rev. Bras. Fisiol. Veg. 13: 196-223.

Jimenez, V.M. and Bangerth, F. 2000. Relationship between endogenous hormone levels in grapevine callus cultures and their morphogenetic behaviour. Vitis 39: 151-157.

Jimenez, V.M. and Bangerth, F. 2001a.Endogenous hormone levels in explants and in embryogenic and non-embryogenic cultures of carrot. Physiol. Plant 111: 389-395.

Jimenez, V.M. and Bangerth, F. 2001b. Endogenous hormone levels in initial explants and in embryogenic and non-embryogenic callus cultures of competent and non-competent wheat genotypes. Plant Cell Tiss. Org. Cult. 67: 37-46.

Jimenez, V.M. and Bangerth, F. 2001c. Hormonal status of maize initial explants and of the embryogenic 2nd non-embryogenic callus cultures derived from them as related to morphogenesis *in vitro*. Plant Sci. 160: 247–257.

Jimenez, V.M. and Bangerth, F. 2001d. *In vitro* culture and endogenous hormone levels in immature zygotic embryos, endosperm and callus cultures of normal and high-lysine barley genotypes. J. Appl. Bot. 75: 1-7.

Jimenez, V.M., Guevara, E., Herrera, J. and Bangerth, F. 2001. Endogenous hormone levels in habituated nucellar *Citrus* callus during the initial stages of regeneration. Plant Cell Rep. 20: 92-100.

Jimenez, V.M., Guevara, E., Herrera J. and Bangerth, F. 2005. Evolution of endogenous hormone concentration in embryogenic cultures of carrot during early expression of somatic embryogenesis. Plant Cell Rep. 23: 567- 572.

Kakkar, R.K., Nagar, P.K., Ahuja, P.S. and Kai, V.K. 2000. Polyamines and plant morphogenesis. Biol. Plant 43: 1-11.

Kamada, H and Harada, H. 1979. Studies on the organogenesis in carrot tissue culture. I. Effects of growth regulators on somatic embryogenesis and root formation. Z. Pflanzen Physiol., 91: 255-266.

Kamada, H. and Harada, H. 1981. Changes in the endogenous level and effects of abscisic acid during somatic embryogenesis *of Daucus carota* L. Plant Cell Physiol. 22: 1423-1429.

Kameshwara Rao, C. (ed.) 2000. In: Material for the Database of Medicinal Plants. Karnataka State Council for Science and Technology for the Dept. of Forests, Environment and Ecology Govt, of Karnataka. 90 pp.

Kermode, A.R., Dumbroff, E.B. and Bewley, J.D. 1989. The role of maturation drying in the transition from seed development to germination. VII. Effects of partial and complete desiccation on abscisic acid levels and sensitivity in *Ricinus communis* L. seeds. J. Exp. Bot. 40: 303-313.

Kessell, R.H.J., Goodwin, C., Philip, J. and Fowler, W. 1977. The relationship between dissolved oxygen concentration, ATP and embryogenesis in carrot (*Daucus carota*) tissue cultures. Plant Sci. Lett., 10: 265-274.

Khalil, S.M., Cheah, K.T., Perez,E.A., Gaskill, D.A. and Hu, J.S. 2002. Regeneration of banana (*Musa* spp. AAB cv. Dwarf Brazilian) via secondary somatic embryogenesis. Plant Cell Rep. 20: 1128-1134.

Kitto, S.L. and Janick, J. 1985. Production of synthetic seeds by encapsulating asexual embryos of carrot. J. Am. Soc. Hortic–Sci., 110: 277-282.

Kiyosue, T., Nakajima, M., Yamaguchi, I., Satoh, S., Kamada, H. and Harada, H. 1992. Endogenous levels of abscisic acid in embryogenic cells, non- embryogenic cells and somatic embryos of carrot (*Daucus carota* L.). Biochem. Physiol. Pflanzen 188: 343-347.

Kiyosue, T., Satoh, S., Kamada, H. and Harada, H. 1993. Somatic embryogenesis in higher plants. J. Plant Res. (Special Issue) 3: 75-82.

Kochba, J., Spiegel-Roy, P., Neumann, H. and Saad, S. 1978. Stimulation of embryogenesis in *Citrus* ovular callus by ABA, Ethephon, CCC and Alar and its suppression by GA_3. Z. Pflanzenphysiol. 89: 427-432.

Koh, W.L. and Loh, C.S. 2000. Direct somatic embryogenesis, plant regeneration and *in vitro* flowering in rapid-cycling *Brassica napus*. Plant Cell Rep. 19: 1177-1183.

Kohlenbach, H.W. 1978. Comparative somatic embryogenesis In: T.A. Thorpe (ed.), Frontiers of plant Tissue Culture 1978. Univ. Calgary Press, Canada. pp. 59-66.

Komamine, A., Kawahara, R., Matsumoto, M., Sunabori, S., Toya, T., Fujiwara, A., Tsukuhara, M., Smith, J., Ito, M., Fukuda, H., Nomura, K. and Fujimura, T. 1992. Mechanisms of somatic embryogenesis in cell cultures: physiology, biochemistry, and molecular biology. *In Vitro* Cell. Dev. Biol. 28: 1-14.

Kong, L. and Yeung, E.C. 1994. Effects of ethylene and ethylene inhibitors on white spruce somatic embryo maturation. Plant Sci. 104: 71-80.

Kopertekh, L.G. and Butenko, R.G. 1995. Naturally occurring phytohormones in wheat explants as related to wheat morphogenesis *in vitro*. Russ. J. Plant Physiol. 42: 488-491.

Kuklin, A.I., Denchev, P.D., Atanassov, A.I. and Scragg, A.H. 1954. Alfalfa embryo production in airlift vessels via direct somatic embryogenesis. Plant Cell Tiss. Org. Cult. 38: 19-23.

Kumar, P.P., Joy, R.W. IV and Thorpe, T.A. 1989. Ethylene and carbon dioxide accumulation, and growth of cell suspension cultures of *Picea glauca* (white spruce). J. Plant Physiol 135: 592-596.

Lakshmanan, P. and Taji, A. 2000. Somatic embryogenesis in leguminous plants. Plant Biol. 2: 136-148.

Li, L. and Qu, R. 2002. *In vitro* somatic embryogenesis in turf-type bermudagrass: roles of abscisic acid and gibberellic acid, and occurrence of secondary somatic embryogenesis. Plant Breed. 121: 155-158.

Li, T. and Neumann, K.-H. 1985. Embryogenesis and endogenous hormone content of cell cultures of some carrot varieties (*Daucus carota* L.). Ber. Deutsch. Bot. Ges. 98: 227-235.

Li, X.Q., Krasnyanski, S.F. and Korban, S.S. 2002. Somatic embryogenesis, secondary somatic embryogenesis, and shoot organogenesis in *Rosa*. J. Plant Physiol. 159: 313-319.

Limanton-Grevet, A., Sotta, B., Brown, S. and Jullien, M. 2000. Analysis of habituated embryogenic lines in *Asparagus officinalis* L.: growth characteristics, hormone content and ploidy level of cam and regenerated plants. Plant Sci. 160: 15-26.

Litz, R.E. and Yurgalevitch, C. 1997. Effects of 1-aminocy/clopropane-1-carboxylic acid, aminoethoxyvinylglycine, methylglyoxal fcu- (guanylhydrazone) and dicyclohexylammonium sulphate on induction of embryogenic competence of mango nucellar explants. Plant Cell Tiss. Org. Cult. 51: 171-176.

Liu, C.-M., Xu, Z.H. and Chua, N.-H. 1993. Auxin polar transport is essential for the establishment of bilateral symmetry during early plant embryogenesis. Plant Cell 5: 621-630.

Liu, Z.-H., Wang, W.-C. and Yen, Y.-S. 1998. Effect of hormone treatment on root formation and endogenous indole-3-acetic acid and polyamine levels of *Glycine max* cultivated *in vitro*. Bot. Bull. Acad. Sin. 39: 113-118.

Ljung, K., Sandberg, G. and Moritz, T. 2004. Hormone analysis. Methods of plant hormone analysis. In: Davies P.J. (ed.), Plant Hormones. Biosynthesis, Signal Transduction, Action Kluwer, Dordrecht, pp. 671-694.

Lo Schiavo, F. 1995. Early events in embryogenesis. In: Bajaj Y.P.S. (ed.), Somatic Embryogenesis and Synthetic Seed I. Biotechnology in Agriculture and Forestry, Vol. 30. Springer, Berlin, pp. 20-29.

Mauri, P.V. and Manzanera, J.A. 2004. Effect of abscisic acid and stratification on somatic embryo maturation and germination of holm oak (*Quercus ilex* L.). *In Vitro* Cell Dev. Biol. Plant 40: 495-498.

McKersie, B.D. and Brown, D.C.W. 1996. Somatic embryogenesis and artificial seeds in forage legumes. Seed Sci. Res. 6: 109-126.

Meijer, E.G.M and Brown, D.C.W. 1985. Screening of diploid *Medicago sativa* germ plasm for somatic embryogenesis. Plant. Cell. Rep., 4: 285- 288.

Merkle, S.A., Parrott, W.A. and Flinn, B.S. 1995. Morphogenic aspects of somatic embryogenesis. In: Thorpe, T.A. (ed.), *In Vitro* Embryogenesis in Plants. Kluwer Academic Publishers, Dordrecht, pp. 155-203.

Mhaske, V.B., Chengalrayan, K. and Hazra, S. 1998. Influence of osmotica and abscisic acid on triglyceride accumulation in peanut somatic embryos. Plant Cell Rep. 17: 742-746.

Michalczuk, L. and Druart, P. 1999. Indole-3-acetic acid metabolism in hormone-autotrophic, embryogenic callus of Inmil cherry rootstock (*Prunus incisa x serrula* 'GM 9') and in hormone-dependent, nonembryogenic calli of *Prunus incisa x serrula* and *Prunus domestica*. Physiol. Plant 107: 426-432.

Michalczuk, L., Cooke, T.J. and Cohen, J.D. 1992a. Auxin levels at different stages of carrot embryogenesis. Phytochemistry 31: 1097-1103.

Michalczuk, L., Ribnicky, D.M., Cooke, T.J. and Cohen, J.D. 1992b. Regulation of indole-3-acetic acid biosynthetic pathways in carrot cell cultures. Plant Physiol. 100: 1346-1353.

Minocha, S.C. and Minocha, R. 1995. Role of polyamines in somatic embryogenesis. In: Bajaj Y.P.S. (ed.), Somatic embryogenesis and Synthetic Seed I. Biotechnology in Agriculture and Forestry, Vol. 30. Springer, Berlin, pp. 53-70.

Mitsuhashi, W., Toyomasu, T., Masui,H., Katho, T., Nakami-nami, K., Kashiwagi, Y., Akutsu, M., Kenmoku, H., Sassa, T., Yamaguchi, S., Kamiya, Y. and Kamada, H. 2003. Gibberellin is essentially required for carrot (*Daucus carota* L.) somatic embryogenesis: dynamic regulation of gibberellin 3-oxidase gene expression. Biosci. Biotechnol. Biochem. 67: 2438-2447.

Mondal, T.K., Bhattacharya, A. and Ahuja, P.S. 2001. Induction of synchronous secondary somatic embryogenesis in *Camellia sinensis* (L.) O. Kuntze. J. Plant Physiol. 158: 945-951.

Mounla, M.A.K., Bangerth, F. and Stoy, V. 1980. Gibberellin-like substances and indole type auxins in developing grains of normal- and high-lysine genotypes of barley. Physiol. Plant 48: 568-573.

Murashige, T. and Skoog, F. 1962. A revised medium for rapid growth and bioassays with tobacco tissue culture. Physiol. Plant 15: 473-497.

Murthy, B.N.S., Murch, S.J. and Saxena, P.K. 1995. Thidiazuron-induced somatic embryogenesis in intact seedlings of peanut (*Arachis hypogaea*): endogenous growth regulator levels and significance of cotyledons. Physiol. Plant 94: 268-276.

Nakagawa, H., Saijyo, T., Yamauchi, N., Shigyo, M., Kako, S. and Ito, A. 2001. Effects of sugars and abscisic acid on somatic embryogenesis from melon (*Cucumis melo* L.) expanded cotyledon. Sci. Hort. 90: 85-92.

Napier R. 2004. Plant hormone binding sites. Ann. Bot. 93: 227-233.

Neumann, K.-H. 1988.Phytohormones in cell and tissue cultures. In: Constabel F. and Vasil I.K. (eds), Plant Cell Cultures. Cell Culture and Somatic Cell Genetics of Plants, Vol. 5. Academic Press, San Diego, pp. 587-599.

Nishiwaki, M., Fujino, K., Koda, Y., Masuda, K. and Kikuta, Y. 2000. Somatic embryogenesis induced by the simple application of abscisic acid to carrot (*Daucus carota* L.) seedlings in culture. Planta 211: 756-759.

Nissen, P. 1994. Stimulation of somatic embryogenesis in carrot by ethylene: effects of modulators of ethylene biosynthesis and action. Physiol. Plant 92: 397-403.

Nissen, P. and Minocha, S.C. 1993. Inhibition by 2, 4-D of somatic embryogenesis in carrot as explored by its reversal by difluoromethylornithine. Physiol. Plant 89: 673-680.

Noma, M., Huber, J., Ernst, D. and Pharis, R.P. 1982. Quantitation of gibberellins and the metabolism of [^{3}H] gibberellin A_1, during somatic embryogenesis in carrot and anise cell cultures. Planta 155: 369-376.

Nomura, K. and Komamine, A. 1995. Physiological and biochemical aspects of somatic embryogenesis. In: Thorpe, T.A. (ed.), *In Vitro* Embryogenesis in Plants. Kluwer Academic Publishers, Dordrecht, pp. 249-265.

Ogata, Y., Iizuka, M., Nakayama, D., Ikeda, M., Kamada, H. and Koshiba, T. 2005. Possible involvement of abscisic acid in the induction of secondary somatic embryogenesis on seed coat derived carrot somatic embryos. Planta 221: 417-423.

Osuga, K. Masuda, H. and Komamine, A. 1999. Synchronization of somatic embryogenesis at high frequency using carrot suspension cultures: model systems and application in plant development. Meth. Cell Sci. 21: 129-140.

Ozcan, S., Barghchi, M., Firek, S. and Draper, J. 1993. Efficient adventitious shoot regeneration and somatic embryogenesis in pea. Plant Cell Tiss. Org. Cult. 34: 271-277.

Padmanabhan, K., Cantliffe, D.J. and Koch, K.E. 2001. Auxin-regulated gene expression and embryogenic competence in callus cultures of sweet potato, *Ipomoea batatas* (L.) Lam. Plant Cell Rep. 20: 187-192.

Panaia, M., Senaratna, T., Dixon, K.W. and Sivasithamparam K. 2004. The role of cytokinins and thidiazuron in the stimulation of somatic embryogenesis in key members of the Restionaceae. Aust. J. Bot. 52: 257-267.

Pandey, N.K., Tewari, KC, Tewari, RN, Joshi, G.C, Pande, V.N. and Pandey, G. 1993.Medicinal plants of Kumaon Himalaya strategies for conservation. In: Dhar U (ed) Himalayan Biodiversity Conservation Strategies, No. 3 (pp 293-302).Himavikas Publications, Nanital.

Park, S.-Y., Ahn, J.-K., Lee, W.-Y., Murthy, H.N. and Paek, K.-Y. 2005. Mass production of *Eleutherococcus koreanum* plantlets via somatic embryogenesis from root cultures and accumulation of eleutherosides regenerants. Plant Sci. 168: 1221-1225.

Parrott, W.A., Dryden, G., Vogt, S., Hildebrabd, D.F., Collins, G.B., Williams, E.G. 1988. Optimization of somatic embryogensis and embryo germination in soybean. *In vitro* Cell Dev. Bio. Plant 24: 817-820

Parrott, W.A and Collins, G.B. 1983. Callus and Shoot tip culture of eight *Triphlium* species *in vitro* with regeneration via somatic embryogenesis of *T. rubens*. Plant. Sci. Lett., 28: 189-194.

Pasternak, T.P., Prinsen, E., Ayaydin, F., Miskolczi, P., Potters, G., Asard, H., Van Onckelen, H.A., Dudits, D. and Feher, A. 2002. The role of auxin, pH, and stress in the activation of embryogenic cell division in leaf protoplast- derived cells of alfalfa. Plant Physiol. 129: 1807-1819.

Patra, A, Rai, B, Rout, G.R and Das, P. 1998. Successful plant regeneration from callus cultures of *Centella asiatica* (Linn.) Urban. Plant Growth Regul. 24:13-16.

Phillips, G.C. 2004. *In vitro* morphogenesis in plants–recent advances. *In Vitro* Cell. Dev. Biol. Plant. 40: 342-345.

Pinto, G., Santos, C, Neves, L. and Araiijo, C. 2002. Somatic embryogenesis and plant regeneration in *Eucalyptus globulus* Labill. Plant Cell Rep. 21: 208-213.

Pintos, B,, Martin, J.P., Centeno, M.L., Villal bos, N., Guerra, H. and Martin, L. 2002. Endogenous cytokinin levels in embryogenic and non-embryogenic calli of *Medicago arborea* L. Plant Sci. 163: 955-960.

Prakash, E, Shavalli Khan and Rao K.R. 2001. Somatic embryogenesis in *Pimpinella tirupatiensis* Bal. et Subr., an endangered medicinal plant of Tirumala hills. Curr. Sci. 81(9): 1239-1242.

Prewein, C, Vagner, M. and Wilhelm, E. 2004. Changes in water status and proline and abscisic acid concentrations in developing somatic embryos of pedunculate oak (*Quercus robur*) during maturation and germination. Tree Physiol. 24: 1251-1257.

Puigderrajols, P., Celestino, C, Suils, M., Toribio, M. and Molinas, M. 2000. Histology of organogenic and embryogenic responses in cotyledons of somatic embryos of *Quercus suber* L. Int. J. Plant Sci. 161: 353-362.

Pullman, G.S., Gupta, P.K., Timmis, R., Carpenter, C, Kreitinger, M. and Welty, E. 2005a. Improved Norway spruce somatic embryo development through the use of abscisic acid combined with activated carbon. Plant Cell Rep. 24: 271-279.

Pullman, G.S., Mein, J., Johnson, S. and Zhang, Y. 2005b. Gibberellin inhibitors improve embryogenic tissue initiation in conifers. Plant Cell Rep. 23: 596-605.

Qureshi, J.A., Karatha, K.K., Abrams, S.R. and Steinhauer, L. 1989. Modulation of somatic embryogenesis in early and late-stage embryos of wheat (*Triticum aestivum* L.) under the influence of (±)-abscisic acid and its analogs. Plant Cell Tiss. Org. Cult. 18: 55-69.

Raemakers C.J.J.M., Jacobsen E. and Visser R.G.F. 1995. Secondary somatic embryogenesis and applications in plant breeding. Euphytica 81: 93-107.

Rajasekaran, K., Vine, J. and Mullins, M.G. 1982. Dormancy in somatic embryos and seeds of *Vitis:* changes in endogenous abscisic acid during embryogeny and germination. Planta 154: 139-144.

Rajasekaran, K., Hein, M.B. and Vasil, I.K. 1987a. Endogenous abscisic acid and indole-3-acetic acid and somatic embryogenesis in cultured leaf explants of *Pennisetum purpureum* Schum. Plant Physiol. 84: 47-51.

Rajasekaran, K, Hein, M.B., Davis, G.C, Carnes, M.G. and Vasil, I.K. 1987b. Endogenous growth regulators in leaves and tissue cultures of *Pennisetum purpureum* Schum. J. Plant Physiol. 130: 12-25.

Reinert and J. 1959. Untersuchungen fiber die Morphogenese an Gewebekulturen. Ber. Deutsch. Bot. Ges. 71: 15.

Ribnicky, D.M., Ilic, N., Cohen, J.D. and Cooke, T.J. 1996. The effects of exogenous auxins on endogenous indole-3-acetic acid metabolism. Plant Physiol. 112: 549-558.

Rodriguez, A.P.M.and Wetzstein, H.Y. 1998.A morphological and histological comparison of the initiation and development of pecan (*Carya illinoinensis*) somatic embryogenic cultures induced with napththaleneacetic acid or 2, 4- dichlorophenoxyacetic acid. Protoplasma 204: 71-83.

Roustan, J.-P., Latche, A. and Failot, J. 1989. Stimulation of *Daucus carota* somatic embryogenesis by inhibitors of ethylene synthesis: cobalt and nickel. Plant Cell Rep. 8: 182-185.

Roustan, J.-P., Latche, A. and Fallot, J. 1990a. Inhibition of ethylene production and stimulation of carrot somatic embryogenesis by salicylic acid. Biol. Plant 32: 273-275.

Roustan, J.-P., Latche, A. and Fallot, J. 1990b. Control of carrot somatic embryogenesis by $AgNO_3$, an inhibitor of ethylene action: effect on arginine decarboxylase activity. Plant Sci. 67: 89-95.

Roustan, J.-P., Latche, A. and Fallot, J.1994. Role of ethylene on induction and expression of carrot somatic embryogenesis: relationship with polyamine metabolism. Plant Sci. 103: 223-229.

Rudus, I., Kepczynska, E. and Kepczynski, J. 2002. Regulation of *Medicago sativa* L. somatic embryogenesis by gibberellins. Plant Growth Regulat. 36: 91-95.

Sagare, A.P., Lee, Y.L., Lin, T.C., Chen, C.C. and Tsay, U.S. 2000. Cytokinin- induced somatic embryogenesis and plant regeneration in *Corydalis yanhusuo* (Fumariaceae)–a medicinal plant. Plant Sci. 160: 139-147.

Saly, S., Joseph, C, Corbineau, F., Lelu, M.A. and Come, D.2002. Induction of secondary somatic embryogenesis in hybrid larch (*Larix leptoeuropaea*) as related to ethylene. Plant Growth Regulat. 37: 287-294.

Samantaray, S, Rout, R. and Das, P. 1997. Regeneration of plants via somatic embryogenesis from leaf base and leaf tip segments of *Echinochloa colona*. Plant Cell Tiss. Org. Cult. 47: 119-125.

Santos, K.G.B., Mudstock, E. and Bodanese-Zanettini, M.H. 1997. Genotype- specific normalization of soybean somatic embryogenesis through the use of an ethylene inhibitor. Plant Cell Rep. 16: 859-864.

Sasaki, K., Shimomura, K., Kamada, H. and Harada, H. 1994. IAA metabolism in embryogenic and non-embryogenic carrot cells. Plant Cell Physiol. 35: 1159-1164.

Saxena, P.K., Malik, K.A. and Gill, R. 1992. Induction by thidiazuron of somatic embryogenesis in intact seedlings of peanut. Planta 187: 421-424.

Schiavone, F.M. and Cooke, T.J. 1987. Unusual patterns of somatic embryogenesis in the domesticated carrot: developmental effects of exogenous auxins and auxin transport inhibitors. Cell Differ. 21: 53-62.

Selby, C, McRoberts, W-.C, Hamilton, J.T.G. and Harvey, B.M.R. 1996. The influence of culture vessel head-space volatiles on somatic embryo maturation in Sitka spruce [*Picea sitchensis* (Bong.) Carr.]. Plant Growth Regulat. 20: 37-42.

Senger, S., Mock, H.-P., Conrad, U. and Manteuffel, R. 2001. Immunomodulation of ABA function affects early events in somatic embryo development. Plant Cell Rep. 20: 112-120.

Shakirova, F.M., Aval'baev, A.M., Chemeris, A.V. and Vakhitov, V.A. 2002. Hormonal transcription regulation in plants. Mol. Biol. 36- 456-461.

Sharma, A.K. 1999.Synchronization in plant cells–an introduction. Meth. Cell Sci. 21: 73-78.

Sharma, B.L. and Kumar, A. 1998. Biodiversity of Medicinal plants of Trigugi Narain (Garhwal Himalaya) and their conservation. National Conference on Recent

Trends in spices and medicinal plant research, A-78 (2-4) April, Calcutta, WB, India.

Sharma, P., Pandey, S., Bhattacharya, A., Nagar, P.K. and Ahuja, P.S. 2004. ABA associated biochemical changes during somatic embryo development in *Camellia sinensis* (L.) O. Kuntze. J. Plant Physiol. 161: 1269-1276.

Sharp, W.R., Evans, D.A and Sondahl, M.R. 1982. Application of somatic embryogenesis to crop improvement. In: A. Fujimura (ed.), Plant Tissue Culture. Jpn. Assoc Plant Tissue Cult., Tokyo, pp. 759-762.

Singh, H.C. 1989.Himalayan herbs and drugs importance and extinction threat. J. Sci. Res. Plants Med. 10: 47-52.

Smith, S.M and Street, H.E., 1974. The decline of embryogenic potential as callus and suspension cultures of carrot (*Daucus carota*) are serially sub cultured. Ann. Bot. (Lond.), 38: 233-241.

Somleva, M.M., Kapchina, V., Alexieva, V. and Golovinsky, E. 1995. Anticytokinin effects on *in vitro* response of embryogenic and nonembryogenic genotypes of *Dactylis glomerata* L. Plant Growth Regulat. 16: 109-112.

Spencer, T.M. and Kitto, S.L. 1988. Measurement of endogenous ABA levels in chilled somatic embryos of carrot by immunoassay. Plant Cell Rep. 7: 352-355.

Sreenivasu, K., Malik, S.K., Kuma,r P.A. and Sharma, R.P. 1998. Plant regeneration via somatic embryogenesis in pigeonpea (*Cajanus cajan* L. Millsp). Plant Cell Rep. 17: 294-297.

Srivastava, R., Shukia, Y.N. and Kumar, S. 1998 Chemistry and pharmacology of *Centella asiatica:* a review. J. Med. Arom. Plant Sci. 19: 1049-1056.

Stasolla, C, Kong, L., Yeung, E.C. and Thorpe, T.A. 2002. Maturation of somatic embryos in conifers morphogenesis, physiology, biochemistry, and molecular biology. *In Vitro* Cell. Dev. Biol. Plant 38: 93-105.

Stefanello S., Dal Vesco L.L., Ducroquet J.P.H.J., Nodari R.O. and Guerra M.P. 2005. Somatic embryogenesis from floral tissues of feijoa (*Feijoa sellowiana* Berg). Sci. Hort. 106: 117-126.

Steward, F.C., Mapes, M.O and Meaia, K. 1958. Growth and organized development if cultured cells. II: organization in cultures grown from freely suspended cells. Am. J. Bot. 45: 705-708.

Stuart, D.A., Nelsen, J., Strickland, S.G. and Nichol, J.W. 1985. Factors affecting developmental processes in alfalfa cell cultures. In: R.R. Herke *et al.* (eds.), Tissue Culture in forestry and Agriculture. Plenum Press, New York, pp. 59-73.

Takeno, K., Koshioka, M., Pharis, R.P., Rajasekaran, K. and Mullins, M.G. 1983. Endogenous gibberellin-like substances in somatic embryos of grape (*Vitis vinifera x Vitis rupestris*)in relation to embryogenesis and the chilling requirement for subsequent development of mature embryos. Plant Physiol. 73: 803-808.

Tanaka, K, Kanno, Y. Kudo, S. and Suzuki, M 2000 Somatic embryogenesis and plant regeneration in Chrysanthemum (*Dendanihemu grandiflorum* (Rami.) Kitamura). Plant Cell Rep. 19:946-953.

Thomas, C, Bronner, R., Molinier, J., Prinsen, E., van Onckelen, H.and Hahne, G. 2002. Immuno-cytochemical localization of indole-3-acetic acid during induction of somatic embryogenesis in cultured sunflower embryos. Planta 215: 577-583.

Thomas, T.L. 1993. Gene expression during plant embryogenesis and germination: an overview. Plant Cell 5: 1401-1410.

Thorpe, T.A. 2000. Somatic embryogenesis: morphogenesis, physiology, biochemistry and molecular biology. Korean. J. Plant Tiss. Cult. 27: 245- 258.

Tisserat, B. and Murasighe, T. 1977. Probable identity of substances in *Citrus* that repress asexual embryogenesis. *In vitro* 13: 785-789.

Tiwari, V, Singh, B.D. and Tiwari, K.N. 1998. Shoot regeneration and somatic embryogenesis from different explants of brahmi (*Bacopa monnieri* L. Wettst). Plant Cell Rep. 17: 538-543.

Tiwari, K.N, Sharma, N.C. Tivari, V. and Singh, B.D. 2000. Micropropagation of *Centella asiatica* L., a valuable medicinal herb. Plant Cell Tiss. Org. Cult. 63: 179-185.

Tokuji, Y. and Kuriyama, K. 2003. Involvement of gibberellin and cytokinin in the formation of embryogenic cell clumps in carrot (*Daucus carota*). J. Plant -Physiol. 160: 131-141.

Toonen, M.A.J, and De Vries, S.C. 1996. Initiation of somatic embryos from single cells. In: Wang T.L. and Cuming A. (eds), Embryogenesis: the Generation of a Plant. Bios Scientific Publishers, Oxford, pp. 173-189.

Toribio, M., Fernandez, C., Celestino, C., Martinez, M.T., San-Jose, N.L.C. and Vieitez A.M. 2004. Somatic embryogenesis in mature *Quercus robur* trees. Plant Cell Tiss. Org. Cult. 76: 283-237.

Tran Thi, L. and Pleschka, E. 2005. Somatic embryogenesis of some *Daucus* species influenced by ABA. J. Appl. Bot. Food Qual. 79: 1-4.

Trewavas, A. 1981.How do plant growth substances work? Plant Cell Environ. 4: 203-228.

Vain, P., Flament, P. and Soudain, P. 1989. Role of ethylene in embryogenic callus initiation and regeneration in *Zea mays* L. J. Plant Physiol. 135: 537- 540.

Vani Devi and Jaganmohan Reddy, K. 1997. *In vitro* induction somatic embryogenesis in *Cicer arietinum*. Acta Botanica indica 25: 207-209.

Vasil, V. and Vasil, I.K., 1981. Somatic embryogenesis and plant regeneration from suspension cultures of Pearl Millet (*Pennisetum americanum*). Ann. Bot. (Lond)., 47: 669-678.

Victor, J.M.R., Murch, S.J., Krishna Raj, S. and Saxena, P.K. 1999. Somatic embryogenesis and organogenesis in peanut: the role of thidiazuron and N[6]-benzylaminopurine in the induction of plant morphogenesis. Plant Growth Regulat.28:9-15.

von Arnold, S., Sabala, I., Bozhkov, P., Kyachok, J. and Filonova, L. 2002. Developmental pathways of somatic embryogenesis. Plant Cell Tiss. Org. Cult. 69: 233-249.

von Arnold, S,, Bozhkov, P., Clapham, D., Dyachok, J., Filonova, L., Hogberg, K.-A., Ingouff, M. and Wiweger, M. 2005. Propagation of Norway spruce via somatic embryogenesis. Plant Cell Tiss. Org. Cult. 81: 323-329.

Vookova, B., Matusova, R. and Kormutak, A. 2003. Secondary somatic embryogenesis in *Abies numidica*. Biol. Plant 46: 513-517.

Wenck, A.R., Conger, B.V., Trigiano, R.N. and Sams, C.E. 1988. Inhibition of somatic embryogenesis in orchardgrass by endogenous cytokinins. Plant Physiol. 88: 990-992.

Wetzstein, H.Y. and Baker,. CM. 1993. The relationship between somatic embryo morphology and conversion in peanut (*Arachis hypogaea* L.). Plant Sci. 92: 81-89.

Xiao, J.-N., Huang, X.-L., Wu, Y.-J., Li, X.-J., Zhou, M.-D. and Engelmann F. 2004. Direct somatic embryogenesis induced from cotyledons of mango immature zygotic embryos. *In Vitro* Cell. Dev. Biol. Plant 40: 196-199.

Zhang, C.G., Li, W., Mao, Y.F., Zhao, D.L., Dong, W. and Guo, G.Q. 2005. Endogenous hormonal levels in *Scutellaria baicalensis* calli induced by thidiazuron. Russ. J. Plant Physiol. 52: 345-351.

Chapter 6

Bioelicitation of Gymnemic Acid in Suspension Cultures of *Gymnema sylvestre* (Retz.) Schultes

Subbanarasiman Balasubramanya[1],
Lingaiah Rajanna[1] and Maniyam Anuradha[2]

[1]*Department of Botany, Bangalore University, Bangalore – 560 056*
[2]*Rishi Foundation, # 234, 10 'C' Main, 1st Block, Jayanagar,*
Bangalore – 560 011

ABSTRACT

Gymnema sylvestre (Retz.) Schultes belonging to family Aclepiadaceae is medicinally important due to presence of anti-diabetic compounds Gymnemic acid. Establishment of suspension cultures were done by adding friable callus to the B_5 liquid media supplemented with 5 mg/l NAA and 0.5 mg/l BAP. Suspension cultures subjected to biotic elicitors *Aspergillus niger, Rhizophus oryzae, Fusarium oxysprorum* and Yeast extract induced increase in gymnemic acid content. Among all the elicitors tried Yeast extract proved to be best elicitor at 3 gm/l resulting in 541.67 mg/Kg and 959 mg/Kg of gymnemic acid after 12 hours and 24 hrs of incubation respectively.

Introduction

There are many popular herbs used in traditional medicinal systems to cure diabetes. Among such plants, *Gymnema sylvestre* (Figure 6.1) has an important place. The sugar destroying property is so effective that after chewing one or two leaves, one

Figure 6.1: *Gymnema sylvestre*

is unable to detect the sweet taste. The sugar crystals taste like sand. The taste suppression is believed to involve direct interaction of the active principles with the sweet taste buds (Liu *et al.*, 1992). This property has led to names like Gurmar and Madhunasini (meaning destroyer of sweet taste). The active principle of this plant also suppresses the sweet taste of sucrose, sodium saccharin, cyclamate, glycine, d-alanine, d-tryptophan, d-leucine, beryllium chloride and lead acetate but not that of chloroform (Warren and Pfaffmann, 1959; Kurihara, 1969; Meiselman and Halpern, 1970). The ancient ayurvedic texts apart from eulogising its anti-diabetic property has also mentioned it as bitter, astringent, acrid, thermogenic, anti-inflammatory, anodyne, digestive, liver tonic, emetic, diuretic, stomachic, stimulant, anthelmentics, laxative, cardiotonic, renal and vesical calculi, cardiopathy, dyspepsia, constipation, asthma, bronchitis, amenorrhoea and leucoderma (Chopra *et al.*,1992; Nadkarni, 1993, Vaidyaratnam, 1995). *Gymnema sylvestre* is also found useful in obesity and weight management (Preuss, 1998; Shigematsu *et al.*, 2001a, 2001b; Woodgate and Conquer, 2003).

Before discovery of insulin in 1920s, severe cases of diabetes resulted in death. Today, diabetes is not quite fatal, but evidence of cases of "insulin resistance" and reports of side effects from prolonged administration of synthetic drugs like metaformin, sulphonylureas, biguanidines and acarbose have generated renewed search for safer alternatives. NIDDM patients have been treated by a variety of plants and its extracts in many countries by indigenous system of medicine. Natural product industries in these countries have made available number of commercial, over-the-counter herbal products containing various amounts of *Gymnema sylvestre*. In order to meet the market requirements, the plant is wild harvested indiscriminately resulting in natural strands disappearing fast leading to its extinction (Choudhury, 1988). The medicinal properties attributed to presence of oleanane type triterpenoid saponins, known as gymnemic acid shows variations in its content depending on eco-climatic zones (Yokota *et al.*, 1994, Thamburaj *et al.*, 1996). This is compounded by low seed viability, poor rooting ability of vegetative cuttings and low multiplying rate in tissue cultured plants which is hindering all efforts towards conservation and sustained utilisation for commercial purposes.

In vitro production of bioactive gymnemic acid with increased focus towards yield enhancement can provide new means of obtaining large quantity of gymnemic acid in shorter period of time. In the present study an attempt to enhance the yield of gymnemic acid in *in vitro* using fungal elicitors is carried out.

Materials and Methods

Leaf explants were collected from full grown plants maintained at Rishi Foundation Herbal Garden to establish callus cultures. The leaves were treated with detergent (tween 20) solution for 10 min and thoroughly washed under a jet flow of tap water for 45 mins. Leaves were then disinfected with 100 ppm Bavistin (BASF India Ltd. Thane, India) for 45 mins followed by 70 per cent ethyl alcohol for 3 mins and rinsed thoroughly with sterile distilled water for several times. Finally, the leaves were surface sterilized with 0.05 per cent aqueous mercuric chloride (W/V) for 10 mins followed by 0.1 per cent of the same and washed with sterile distilled water several times. Leaf discs were placed on B_5 medium supplemented with 5 mg/l NAA and 0.5 mg/l BAP, 3 per cent sucrose and gelled with 0.7 per cent agar-agar (Himedia, Mumbai,). pH of the media was adjusted to 5.7–5.8 prior to the addition of agar. All growth regulators were added prior to autoclaving. The media was autoclaved at 15-psi, 121°C for 20 mins. The cultures were incubated at $25 \pm 2°C$ temperature under dark for callus proliferation.

Initiation and Establishment of Suspension Cultures

B_5 liquid medium supplemented with 5 mg/l NAA and 0.5 mg/l BAP with pH 5.7 was used. About 75 ml of liquid medium was dispensed into 250 ml Erlenmeyer flasks and friable callus weighing approximately 5 gms was transferred to the flasks under aseptic conditions. Cultures were incubated on a rotary shaker at 120 rpm in darkness and temperature was maintained at $25 \pm 2°$ C. After 15 days of incubation cell suspensions were filtered through appropriate sterile sieves to obtain single cells and few celled aggregates which are used as inoculum for subculture. Fine cell suspensions were obtained by repeated subculturing of callus and removing small clumps at every stage. The cell suspension cultures were maintained on the same medium by subculturing at 2-week interval.

Bio-elicitors Used

Aspergillus niger, Rhizopus oryzae and *Fusarium oxysporum* cultures were procured from the Padmashree Institute of Management and Sciences, Bangalore. Yeast extract was procured from Himedia Laboratories, Mumbai.

Elicitor Preparation

Potatoes (200gms) were cut into slices and boiled with 200 ml of distilled water and then filtered through muslin cloth. Twenty grams of dextrose was added to the filtrate and the volume was made up to 1000 ml. The pH of the medium was adjusted to 6.5. Into each 250 ml conical flask 50 ml of the media was dispensed and autoclaved. Loop full of fungal inoculum grown on solid medium was inoculated into medium. The flasks were incubated and allowed to grow. After 20 days, during sporulation flasks with the cultures were autoclaved. The fungal mat was separated by filtration and then washed several times with distilled water. Aqueous extract was made by homogenizing the mat in a mortar and pestle using acid washed neutralized sand. The homogenized mat was filtered through muslin cloth and the volume was made equal to that of spent medium. The mat extracts and filtrates were used as elicitors.

Four concentrations of cell filtrate (at 5, 10, 15 and 20 per cent v/v) and four concentrations of cell extract (at 5, 10, 15 and 20 per cent v/v) of all fungi added to the established cultures to study their effect on active principle production. Control cultures were supplemented with volumes and concentrations of the microbial culture medium equivalent to that of elicitor.

Preparation of Yeast Elicitor

Ten grams of yeast extract was added to 100ml of double distilled water and ethanol was added up to 80 per cent (v/v) and they were kept at 4°C for 3 days for precipitation. The supernatant was decanted and the precipitate was re-dissolved in 10 ml of distilled water, autoclaved and used as elicitor.

Results and Discussion

Only two recent reports of elicitation of gymnemic acid have been evidenced. Devi and Srinivasan (2009) and Ahmed *et al.* (2009) have reported elicitation of gymnemic acid using biotic (*Aspergillus niger*) and abiotic (light) elicitor respectively. The literature available and the present study establish the fact that cell cultures of *Gymnema sylvestre* produces gymnemic acid. But to be cost effective and commercially viable, increasing the production rate of gymnemic acid seems to be inevitable. Various biotechnological methods have been adopted by many researchers in various other plants to increase the production of secondary metabolites. Treatment of plant cells with biotic and/or abiotic elicitors has been considered as a useful strategy to enhance secondary metabolite production in cell cultures (Karuppusamy, 2009).

In the present study cell suspension cultures of *Gymnema sylvestre* were known to accumulate gymnemic acids. Established suspension cultures when treated with *Aspergillus niger, Rhizophus oryzae, Fusarium oxysprorum* and Yeast extract at different concentrations effected gymnemic acid production. Figures 6.2 and 6.3 illustrates the

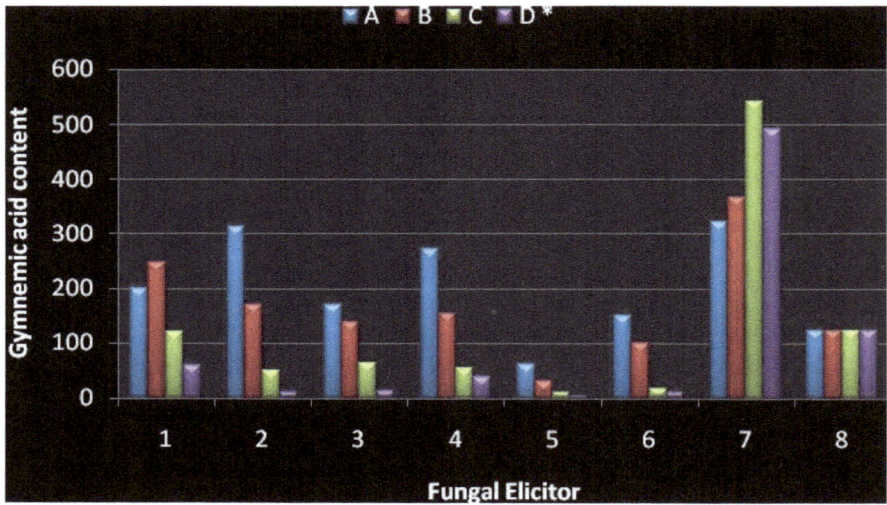

Sl.No.	Fungal Elicitor	A	B	C	D*
1.	*Aspergillus niger* (mat extract)	199.88	248.35	121.16	59.76
2.	*Aspergillus niger* (mat filtrate)	313.43	169.47	50.49	10.87
3.	*Rhizopus oryzae* (mat extract)	169.53	139.17	64.28	14.94
4.	*Rhizopus oryzae* (mat filtrate)	271.17	155.47	54.19	39.45
5.	*Fusarium oxysporum* (mat extract)	61.67	32.50	11.09	5.53
6.	*Fusarium oxysporum* (mat filtrate)	151.00	99.00	18.00	11.00
7.	Yeast Extract	321.33	366.67	541.67	491.67
8.	Control	123.00	123.00	123.00	123.00

Figure 6.2: Gymnemic Acid Content Estimated in Fungal Elicited Cultures of *Gymnema sylvestre* after 12 hrs of Incubation

accumulation pattern of gymnemic acid noticed after 12 and 24 hrs incubation on addition of fungal elicitors respectively.

Among all the elicitors tried Yeast extract treated suspension exhibited maximum amount of gymnemic acid after 24 hrs of incubation. Of all the concentrations of yeast extract, 3 gm/l elicited better response by production of 541.67 mg/Kg and 959 mg/Kg of gymnemic acid after 12 hours and 24 hrs of incubation respectively. Further, increase in concentration of yeast extract to 4 gm/l resulted in decreased gymnemic acid production both at 12 hrs and 24 hrs incubated cultures. The second best fungal elicitor was *Aspergillus niger* mat filterate at 5 per cent level which induced 361.20 mg/Kg gymnemic acid production after 24 hrs of incubation. The relative response of *Rhizophus oryzae* and *Fusarium oxysprorum* was not encouraging and at certain level these two fungal elicitors induced lesser production of gymnemic acid than controlled cultures.

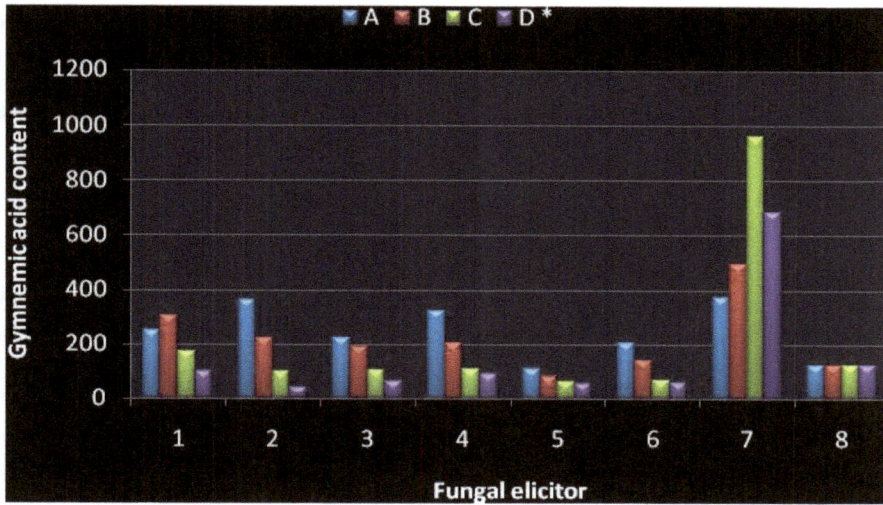

Sl. No.	Fungal Elicitor	A	B	C	D*
1.	*Aspergillus niger* (mat extract)	251.85	302.29	172.11	101.16
2.	*Aspergillus niger* (mat filtrate)	361.20	221.67	100.50	41.00
3.	*Rhizopus oryzae* (mat extract)	221.00	190.17	104.32	65.53
4.	*Rhizopus oryzae* (mat filtrate)	320.87	206.50	108.04	91.00
5.	*Fusarium oxysporum* (mat extract)	111.25	82.92	61.97	55.53
6.	*Fusarium oxysporum* (mat filtrate)	202.70	139.57	69.32	61.00
7.	Yeast Extract	372.00	494.67	959.00	684.00
8.	Control	124.67	124.67	124.67	124.67

Figure 6.3: Gymnemic Acid Content Estimated in Fungal Elicited Cultures of *Gymnema sylvestre* after 24 hrs of Incubation

* Treatment concentrations are 5 per cent (A) 10 per cent (B) 15 per cent (C) and 20 per cent (D) except in yeast extract which was used at the concentrations of 1 (A), 2 (B), 3 (C) and 4 (D) gms/l.

Devi and Srinivasan (2009) studied the effect of *Aspergillus niger* cell extract on gymnemic acid yield. They reported 8 – 9 folds increase of gymnemic acid when compared to control cultures. Maximum 900 mg/l of gymnemic acid was achieved. However, in present investigation yeast extract was found more effective than *Aspergillus niger*. The positive effect of yeast extract in secondary metabolites production is very well evidenced by many published literature. Berberine production in *Thalictrum rugosum* (Marques and Brodelius, 1998) was enhanced 4-folds when elicited with yeast extract. Similarly, increase in plumbagin content in suspension cultures of *Plumbago rosea* cultures was seen when yeast extract was used as elicitor (Komaraiah *et al.*, 2002).

The results of the present investigations clearly demonstrate the increase in gymnemic acid content by 7-folds after the addition of yeast extract. In conclusion, the method presented here could be an important clue for the increased production of anti-diabetic principles by the herbal drug industry, however subject to economic considerations.

References

Ahmed, A.B.A., A.S. Rao and M.V. Rao, 2009. *In vitro* production of gymnemic acid from *Gymnema sylvestre* (Retz.) R. Br. ex Roemer and Schultes through callus culture under abiotic stress conditions. In: Methods in molecular biology. Protocols for *in vitro* cultures and secondary metabolites analysis of aromatic and medicinal plants. S.M. Jain and P.K. Sexena, eds., Vol. 547, Springer Protocols, Humana Press, pp. 93 – 105.

Chopra, R.N., S.L. Nayar and I.C. Chopra, 1992. Glossary of Indian Medicinal Plants, CSIR, 3rd Edition, New Delhi, 319 – 322.

Choudhury, B.P., 1988. Assessment and conservation of medicinal plants of Bhubaneshwar and its neighbourhood. In: Indigenous medicinal plants, Today and Tomorrow's Printers and Publishers, New Delhi, India,pp. 211 – 219.

Devi, C.S. and V.M. Srinivasan, 2009. Bioelicitation of secondary metabolites in *in vitro* cultures of *Gymenma sylvestre* R. Br. New Biotechnol., 255: 3.2.10.

Karuppusamy, S. 2009. A review on trends in production of secondary metabolites from higher plants by *in vitro* tissue, organ and cell cultures. J. Med. Plant. Res., 3(13): 1222 – 1239.

Komaraiah, P., R. N. Amrutha, P.B. Kavi Kishor and S.V. Ramakrishna, 2002. Elicitor enhanced production of plumbagin in suspension cultures of *Plumbago rosea* L. Enzyme Microbial Technol., 31: 634–639.

Kurihara, Y., 1969. Anti-sweet activity of gymnemic acid A1 and its derivatives. Life. Sci., 8: 537 – 543.

Liu, H.M., K. Fumiyuki and T. Yoshisuke, 1992. Isolation and structure elucidation of Gymnemic Acids, anti-sweet principles *Gymnema sylvestre*. Chem. Pharm. Bull., 40(6): 1366–1375.

Marques, I.A. and P.E. Brodelius, 1998. Elicitor-induced L-tyrosine decarboxylase from plant cell suspension cultures induction and purification. Plant Physiology, 88: 46–51.

Meiselman, H. and B.P. Halpern, 1970. Effects of *Gymnema sylvestre* on complex tastes elicited by amino acids and sucrose. Physiol Behav., 5(12): 1379–1384.

Nadkarni, K.M., 1993. Indian Materia Medica, Popular Prakashan Pvt. Ltd., Bombay, Vol.1: 596–599.

Preuss, H.G., S.T. Jarrell, R. Scheckenbach, S. Lieberman and R.A. Anderson, 1998. Comparative effects of chromium, vanadium and *Gymnema sylvestre* on sugar-induced blood pressure elevations in SHR. J. Am. Coll. Nutr., 17(2): 116 – 123.

Shigematsu N, R. Asano, M. Shimosaka, and M. Okazaki. 2001b. Effect of long-term administration with *Gymnema sylvestre* R. Br on plasma and liver lipid in rats. Biol. Pharm. Bull., 24: 643–649.

Shigematsu N., R. Asano, M. Shimosaka and M. Okazaki, 2001a. Effect of administration with the extract of *Gymnema sylvestre* R.Br leaves on lipid metabolism in rats. Biol. Pharm. Bull., 24: 713–717.

Thamburaj, S., D. Subbaraj, S. Kasturi and M. Vijayakumar, 1996. Evaluation of germplasm accessions of *Gymnema sylvestre* R.Br. S. Indian Hort., 44: 174 – 176.

Vaidyaratnam, P., 1995. Indian Medicinal Plants, Vol. 3, Orient Longman Publisher, Madras–600107.

Warren, R.M. and C. Pfaffmann, 1959. Suppression of sweet sensitivity by potassium gymnemate. J. Appl. Physiol., 14: 40 – 42.

Woodgate, D. E. and J.A. Conquer, 2003. Effects of a stimulant-free dietary supplement on body weight and fat loss in obese adults: a six-week exploratory study. Curr. Ther. Res., 64: 248–262.

Yokota, T., K. Mizutani, K. Okada and O. Tanak, 1994. Quantitative analysis of gymnemic acids by high performance liquid chromatography. J. Jpn Soc. Food Sci. Technol., 41: 202 – 205.

Chapter 7

Multiple Shoot Regeneration from Leaf Callus of *Phyllanthus acidus* (L.) Skeels – Star Gooseberry

R. Usha Nagalakshmi, B. Mallikarjuna,
*D. Gayathri and G.Rama Gopal**

Department of Botany, Sri Venkateswara University,
Tirupati – 517 502, Andhra Pradesh

ABSTRACT

An efficient and reproducible protocol for multiple shoot regeneration through leaf callus cultures of *Phyllanthus acidus* L. has been reported. Luxuriant mass of green coloured compact nodular callus with high morphogenetic ability was produced from the tender leaf bits on Murashige and Skoog's (MS) medium supplemented with BA and NAA. The effects of different concentrations and combinations of growth regulators on shoot organogenesis from leaf callus were investigated. MS medium supplemented with 2 mg/l TDZ + 0.5 mg/l NAA was found to be effective in producing maximum number of multiple shoots (24.24) than with 3mg/l BAP + 0.5 mg/l NAA (17.48). Though TDZ was found to be effective in producing maximum number of multiple shoots, BAP produces more or less equal number of shoots and the regenerated shoots were normal. TDZ regenerated shoots were found to be hyperhydric, glossy with slightly stunted

* Corresponding Author: E-mail: ghantargopal@yahoo.co.in

growth. Further, the regenerated shoots were elongated and proficiently rooted on ½ MS medium supplemented with 1 mg/l IBA. Rooted plantlets were well acclimatized to the natural conditions.

Keywords: Phyllanthus acidus, Star gooseberry, Leaf callus, Multiple shoots.

Introduction

Phyllanthus acidus L., also called star goose berry, Malay goose berry, Otaheite goose berry, is a member of the Euphorbiaceae family that bear edible fruits. It is a small tropical tree bearing clusters of pale- yellow waxy fruits, with high content of Ascorbic acid. The fruits are eaten fresh, used in making jams, jellies, pickles etc. In Philippines, the fruits are used to make vinegar. The ripe fruits are served as a relish, syrup or sweet preserve. Several parts of *P. acidus* have been used in folk medicine. Early reports revealed that plants of Euphorbiaceae contain many kinds of secondary metabolites, such as triterpenoids, lipids (Teresa *et al.*, 1987), diterpenoids (Madureira *et al.*, 2004) and steroids (Biesboer *et al.*, 1982).

Roots are used in the treatment of psoriasis of the feet. The latex from *P. acidus* is credited with emetic and purgative activity (Lemmens *et al.*, 1999). The root bark contains saponins, gallic acid and tannins. Fruit is used as a laxative in Myanmar. In India fruits are taken as a liver tonic and to enrich blood. Methanolic extracts of *P. acidus* possess strong *in vitro* antimicrobial activity (Melendez *et al.*, 2006). *P. acidus* extract and its isolated compounds induce airway chloride secretion, a potential treatment for cystic fibrosis (Sousa *et al.*, 2006). Rats fed with extracts from *P. acidus* leaf showed hepatoprotective and anti-oxidant effect against acute liver damage induced by carbon tetrachloride by normalizing the liver enzymes (Lee *et al.*, 2006). The plant is also used as an antidote for snake bite (Muhammad Ishtiaq *et al.*, 2006).

The root extracts have long been used in the rehabilitation program for alcoholics in Thailand. Isolated compound Phyllanthusol A from the root of *P. acidus* has been proposed as a possible antitumor agent (Vongvanich *et al.*, 2000; Mahidol *et al.*, 2002).

The plant is propagated by seeds and also multiplied by budding, cuttings or air layers. Severe anthropogenic pressure, habitat destruction, scarce natural regeneration and low seed germination causes its decline in the wild. Considering the limitations of natural regeneration in its endemic locality, there is an urgent need for regeneration for *ex-situ* conservation of this plant species. Micropropagation is a most useful method for large scale multiplication and preservation of important medicinal plants (Mahantesh *et al.*, 2005). Duangporn and Siripong (2009) studied the effect of auxin and cytokinin on Phyllanthusol A production by callus cultures of *Phyllanthus acidus*. However multiple shoot induction through leaf callus cultures has not been reported in this plant. The present study reports an efficient and reproducible regeneration protocol in *Phyllanthus acidus* using mature leaf explants.

Materials and Methods

Young leaves from healthy plants of *Phyllanthus acidus* were collected from Sri Venkateswara University Campus, Tirupati. Explants were initially washed under running tap water with teepol solution (5 per cent v/v) for 5 min. Leaf explants were surface sterilized with 70 per cent ethanol for 30 sec and then with 0.1 per cent aqueous mercuric chloride solution for 5 minutes followed by three washings with sterile distilled water to remove the traces of excess sterilant in the surface of the explants. These explants were aseptically cut into 1 cm bits and transferred to MS (Murashige and Skoog, 1969) callus induction medium, supplemented with various growth regulators *viz.,* BAP, KN, 2,4-D and NAA (Table 7.1). 3 per cent sucrose was added before adjusting the pH between 5.7 and 5.8 and the media were jelled with 0.8 per cent (w/v) agar (regular grade, SRL, Bombay, India). 20 ml of medium was dispensed in tubes (15×2.5cm) prior to autoclaving at 121°C at 105kpa for 20 min. The cultures were illuminated by white fluorescent light with an intensity of 50 $\mu molm^{-2}s^{-1}$ and maintained at 25±2°C under 16:8 light dark regime. Sub culturing of the callus was conducted by transferring small bits of callus on to the same fresh medium. The morphological differences were recorded at the end of each subculture.

Table 7.1: Effect of Different Concentrations and Combinations of Plant Growth Regulators on the Callus Induction of *Phyllanthus acidus* L

Plant Growth Regulators (mg l^{-1})				Frequency of Callus Regeneration (per cent)	Nature of Callus	Response
2,4-D	NAA	BA	KN			
0.5				35.67 ± 0.88[a]	Light Cream col. compact	+
1				61.63 ± 0.18[i]	Pale yellow col. compact hard	-
2				48.22 ± 0.62[f]	Brown compact hard	-
	0.5			51.53 ± 0.24[g]	Light green col. compact	++
	1			47.67 ± 0.73[f]	Green col. compact nodular	++
	2			68.77 ± 0.14[k]	Green col. compact nodular	+++
0.5			1	35.57 ± 0.72[a]	Creamy white col. compact	+
0.5			2	42.63 ± 0.35[d]	Light green col. compact	++
1			0.5	55.17 ± 0.60[h]	Light cream col. Loose fragile and fast growing	-
	0.5		1	43.53 ± 0.61[d]	Light green col. compact	+
	1		0.5	45.60 ± 0.23[e]	Greenish white col. compact	++
	1		1	37.93 ± 0.35[b]	Greenish white col. compact	++
	2		1	55.67 ± 0.88[h]	Green col. compact nodular	+++
	1		2	63.83 ± 0.60[j]	Green col. compact nodular	+++
1		0.5		40.50 ± 0.87[c]	Light creamy col. compact	++
1		1		51.83 ± 0.33[g]	Light green col. compact nodular	++
	0.5	1		55.50 ± 0.87[h]	Green col. compact nodular	+++
	1	1		67.67 ± 0.88[k]	Green col. compact nodular	++++
	2	1		70.67 ± 0.60[l]	Green col. compact nodular	++++

No. of + indicates the degree of shoot regeneration.

The morphogenic calli obtained from the leaf explants were transferred to the shoot regeneration medium containing MS basal salts supplemented with different concentrations and combinations of TDZ, BAP and NAA (Table 7.2). The frequency of shoot regeneration, number of shoots per explant and shoot length of each explant were recorded 35 days after subculture. The regenerated shoots were transferred to the rooting medium (MS full, half strength with varying concentrations of IBA).

Table 7.2: Effect of Different Combinations of Growth Regulators on Shoot Morphogenesis from Callus Obtained from Leaf Explants of *Phyllanthus acidus* L.

Plant Growth Regulators (mg l⁻¹)			Frequency (per cent)	No. of Shoots/ Explant Mean ± S.E	Shoot Length (cm) Mean ± S.E
TDZ	BAP	NAA			
	0.5		52.33 ± 0.44[c]	2.04 ± 0.03[c a]	2.02 ± 0.06[fg]
	1		66.50 ± 0.76[f]	5.06 ± 0.12[c]	1.83 ± 0.02[ef]
	2		62.67 ± 0.44[e]	10.48 ± 0.12[f]	1.20 ± 0.08[bcd]
	3		83.17 ± 0.44[h]	15.71 ± 0.20[i]	0.81 ± 0.12[ab]
	5		53.83 ± 0.73[c]	6.10 ± 0.07[d]	1.40 ± 0.15[cde]
	2	0.5	47.10 ± 0.70[b]	12.53 ± 0.18[g]	2.12 ± 0.16[fg]
	3	0.5	78.17 ± 0.44[g]	17.48 ± 0.21[j]	3.43 ± 0.27[h]
0.5			55.90 ± 0.58[d]	3.25 ± 0.16[b]	1.63 ± 0.19[def]
1			42.83 ± 0.60[a]	7.43 ± 0.13[e]	1.17 ± 0.16[bcd]
2			46.37 ± 0.41[b]	16.07 ± 0.07[i]	0.87 ± 0.11[abc]
3			65.27 ± 0.43[f]	3.24 ± 0.15[b]	0.59 ± 0.21[a]
1		0.5	78.50 ± 0.56[g]	15.17 ± 0.20[h]	1.77 ± 0.22[ef]
2		0.5	88.37 ± 0.78[i]	24.24 ± 0.21[k]	2.43 ± 0.24[g]

All experiments were repeated thrice using twelve explants. The cultures were examined periodically. All the data were statistically analyzed using a completely randomized block design and means were evaluated at the $P = 0.05$ level of significance using Duncan's multiple range test.

Results and Discussion

Preliminary cultures produced different types of calli from leaf discs on MS medium containing different concentrations and combinations of NAA, 2,4-D, Kn and BAP (Table 7.1). Chitra *et al.* (2009) obtained callus from leaf bit explants of *Phyllanthus amarus*, on MS medium supplemented with various concentrations of NAA and 2, 4-D. Haicour (1974) also obtained callus from leaf bits of *Phyllanthus urinaria*.

Calli began to form on the cut edges of leaf explants 2 weeks after inoculation (Figure 7.1 A). MS medium supplemented with 2, 4-D alone was found to be sufficient for induction of callus. Light cream to pale yellow coloured compact calli were obtained with lower concentrations of 2, 4-D (<2mg/l) while higher concentrations yielded

Figure 7.1: Indirect Multiple Shoot Regeneration of *P. acidus* via Leaf Callus

A: Induction of callus from the margins of mature leaf explants (Bar = 3.6 mm); B: Green cultured callus from the mature leaf explants on MS medium supplemented with BA 1 mg/l +NAA 2 mg/l (Bar = 3.6 mm); C: Regeneration of shoots on MS medium with TDZ 2 mg/l +NAA 0.5 mg/l (Bar = 6.4 mm); D: Regeneration of shoots on MS medium with BA 3 mg/l +NAA 0.5 mg /l (Bar = 8.3 mm); E: Shoot elongation on MS basal medium (Bar = 8.3 mm); F: Rooting of *in vitro* regenerated shoot on ½ MS medium supplemented with IBA (1 mg/l) (Bar = 7.4 mm).

brown coloured callus. Most of the 2, 4-D derived calli upon subculturing in fresh 2, 4-D containing medium, turned dark brown and degenerated after 4 weeks of subculture. MS medium with either BA/Kn + 2, 4-D resulted in creamy to light green coloured compact calli. Light cream coloured, fragile and fast growing non morphogenetic callus was obtained on MS medium fortified with 1 mg/l 2, 4-D + 0.5 mg/l Kn. In the majority of the plant species, the synthetic hormone 2,4 D interacted with endogenous hormones of the explants and stimulated the cells to proliferate into callus mass (Narayanaswamy, 1994).

Green coloured compact calli were obtained from the leaf explants on MS medium supplemented with NAA. Morphogenetic callus was obtained at higher concentrations of NAA (2 mg/l). Green coloured compact nodular morphogenic calli were obtained on MS medium combined with Kn and NAA. Green coloured compact nodular calli with high morphogenetic ability was obtained on MS medium supplemented with BA and NAA (Figure 7.1 B). All combinations of growth regulators induced callus growth without an organogenesis response for over 42 days of cultivation. Frequent sub culturing of the callus into the same medium increases the amount of callus.

The possibility of callus regeneration that could be brought through the manipulation of hormone balance is a known factor (Rossini, 1969). The level of auxin and cytokinin alone or in combination decides the ability of callus for organogenesis (Kohlenbach, 1977). Regeneration of shoots from the callus was obtained when subcultured onto the regeneration medium.

BAP alone was found to be efficient in triggering the shoot bud formation from the callus tissues. Increase in the concentration of BAP up to 3mg/l increases the formation of multiple shoots. Further increase in the concentration of BAP decreases the number of shoots. MS medium supplemented with 2mg/l TDZ results in the formation of 16.07 multiple shoots, but at higher concentrations leads to the formation of deformed shoots. The effective and efficient role played by the BAP in the shoot proliferation of various medicinal plants was emphasized by Harikrishnan *et al.* (1999) in *Acorus calamus*, Arockiasamy *et al.* (1999) in *Datura metel*, Al-Wasel (1999) in *Atropa belladonna*, Smitha *et al.* (2000) in *Plumbago indica*, Selvakumar and Balakumar (2000) in *Acalypha fruticosa* and Muthuram *et al.* (2000) in *Scoparia dulcis*. Chitra *et al.* (2009) showed better regeneration capacity of shoots / plantlets from leaf and internode callus of *Phyllanthus amarus*, on MS medium at an effective concentration of cytokinin (2.0 mg/l BAP) along with GA_3 (0.5 mg/l).

Synergistic effect of the cytokinins used (BAP, TDZ) along with the auxin NAA in MS medium was evidenced by the increase in the number of multiple shoots. 3mg/l BA + 0.5 mg/l NAA was the most effective combination for shoot multiplication resulting in 17.48 shoots. The interaction of high levels of BAP with lower levels of NAA in provoking the caulogenic potency of the callus have been reported in many of the medicinally important species (Saini and Jaiwal, 2000; Ramulu *et al.*, 2002). High frequency of shoot regeneration from callus was achieved with BAP along with low levels of NAA (0.5 mg/l) and this is in correlation with the reports on *Dioscorea zingiberensis* (Chen *et al.*, 2003), *Morus alba* (Vijayan *et al.*, 2000), and *Entada pursaetha*

(Sai Vishnu Priya, 2003). In *Embellia ribes* leaf callus cultures, high frequency shoot bud differentiation was noticed with 2.5 mg/l BAP and 0.5 mg/l NAA (Shankaramurthy *et al.*, 2004). In the present study maximum number of multiple shoots (24.24) was observed on MS medium supplemented with 2 mg/l TDZ + 0.5 mg/l NAA (Table 7.2, Figure 7.1 C, D). Though TDZ was found to be effective in producing maximum number of multiple shoots, BAP produced more or less equal number of shoots and the regenerated shoots were normal. Shoots regenerated on TDZ containing medium were found to be hyperhydric, glossy with slightly stunted growth.

Considerable elongation of shoots was found when the individual shoots excised from 4-6 weeks cultures were subcultured on MS basal medium (Figure 7.1 E). Further sub culturing of these elongated shoots in ½ MS medium containing IBA showed rooting after 10 days of transfer. Best rooting was achieved on ½ MS medium supplemented with 1 mg/l IBA (Figure 7.1 F). Manickam *et al.* (2000) obtained maximum rooting on MS medium with 9.84 µM IBA. The rooted shoots were washed with distilled water to remove the traces of agar and transferred to pots for hardening. After hardening, plantlets with fully expanded leaves and well developed roots were eventually established in green house. The regenerated plants showed 90 per cent of survival rate.

References

Al-Wasel, A.S. 1999. *In vitro* multiplication of *Atropa bellad*onna L. using nodal segments. Alexandria J. Agric. Res., 44: 263-274.

Arockiasamy, D.I., Muthukumar, B. and Britto, S.J. 1999. *In vitro* plant regeneration from internodal segments of *Datura metel* L. Adv. Plant Sci., 12: 227-231.

Biesboer, D.D., Amour, P.D., Wilson, S.R., Mahlberg, P. 1982.Sterols and triterpenols in latex and cultured tissues of *Euphorbia pulcherrima*. Phytochemistry, 21(5): 1115-1118.

Chen,..Y., Fan Luo, Z. and Fu, Y. 2003. Rapid clonal propagation of *Dioscorea zingiberensis*. Plant Cell Tiss. Org. Cult., 73: 75-80.

Chitra, R., Rajamani, K. and Vadivel, E. 2009. Regeneration of plantlets from leaf and internode explants of *Phyllanthus amarus* Schum. & Thonn. African Journal of Biotechnology, 8(10): 2209-2211.

Duangporn, P. and Siripong, P. 2009. Effect of auxin and cytokinin on phyllanthusol A production by callus cultures of *Phyllanthus acidus* Skeels. American-Eurasian J. Agric. & Environ. Sci., 5(2): 258-263.

Haicour, R. 1974. Comparasion chez *Phyllanthus urinaria* L. de l'activite antibact erienne desd'ecoctions de diverses portions de laplante *et al.*, de cultures de tissue qui en proviennent. CR Seances Acad Sci Paris S'er D, 278: 3323 – 3325.

Harikrishnan, K.N. and Moly Hariharan 1999. *In vitro* clonal propagation of sweet flag (*Acorus calamus*) – a medicinal plant. In: Plant tissue culture and biotechnology emerging trends. Jan: 29-31.

Kohlenbach, H.W. 1977. Basic Aspects of Differentiation and Plant Regeneration from Cell and Tissue Culture. In: Plant Tissue Culture and Its Biotechnological Application, Barg W, Reinhard E and Zenk MH (eds.), Springer Verlag, New York, p. 355- 368

Lee, C., Peng, Y., Cheng, W.H., Cheng, H.Y., Lai, F.N.M.T. and Chiu, T.H. 2006. Hepatoprotective Effect of *Phyllanthus* in Taiwan on Acute Liver Damage Induced by Carbon Tetrachloride. Am. J. Chin. Med., 30(3): 471-482.

Lemmens, R.H., Bunyapraphatsara, M.J. and Padua de, L.S.N. 1999. Plant Resources of South-East Asia No 12(1) Medicinal and Poisonous Plants 1. Prosea Foundation, Bogor, Indonesia, pp: 386-387.

Madureira, A.M., Duarte, M.T., Piedade, M.F.M., Ascenso, J.R., Ferreira, M.J.U. 2004. Isoprenoid compounds from *Euphorbia portlandica* X-ray structure of lupeportlandol, a new lupane triterpene. J. Braz. Chem. Soc., 15(5): 742-747

Mahantesh, K., Shashidhara, S. and Rajasekharan, P.E. 2005. *In vitro* multiplication of *Wedelia chinensis* (Osbeck) Merr. Plant Cell Biotechnology and Molecular Biology, 6(3&4): 147-150.

Mahidol, C., Prawat, H., Prachyawarakorn, V. and Ruchirawat, S. 2002. Investigation of some bioactive Thai medicinal plants. Phytochemistry Reviews, 1: 278-297.

Manickam, V.S., Elango, M.R. and Antonisamy, R. 2000. Regeneration of Indian ginseng plantlets from stem callus. Plant Cell Tiss. Org. Cult., 62: 181-185.

Melendez, P.A. and Capriles, V.A. 2006. Antibacterial properties of tropical plants from Puerto Rico. Phytomedicine, 13(4): 272-6.

Muhammad Ishtiaq, Ch., Khan, M.A. and Wajahat Hanif 2006. Ethnoveterinary medicinal uses of plants from Samahni valley Dist. Bhimber, Pakistan. Asian Journal of Plant Sciences, 5(2): 390-396.

Murashige, T. and Skoog, F. 1962. A revised medium for rapid growth and bioassays with tobacco tissue cultures. Physiol. Plant., 15: 437-497.

Muthuram, G., Brindha, P. and Lokeswari, T.S. 2000. *In vitro* propagation of *Scoparia dulcis* Linn. A medicinal herb. In: Recent Trends in Spices and Medicinal Plants Research. Amit Krishna De (eds.)., Associated Publishing Co., New Delhi.

Narayanaswamy, S. 1994. Plant Micropropagation, In: Plant Cell and Tissue Culture, Tata McGraw Hill Publishing Company Limited, New Delhi, pp.194-235.

Ramulu, D.R., Murthy, K.S.R. and Pullaiah, T. 2002. *In vitro* propagation of *Cynanchum callialatum.* J. Trop. Med. Plants, 3:233-238.

Rossini 1969. One nogevelle method de culture *in vitro* de cellules parenchymatessus separess de fevilles de *Calystegia sepium* L. CR.Acad. Sci., Paris, In : Plant tissue culture an dits biotechnology application. W.Berg, E. Rainard, M.H. Zenk, (Eds.). Springer Venlarg, Berlin.

Sai Vishnu Priya, K. 2003. Studies on *in vitro* culture of endangered medicinal plant *Entada pursaetha* DC. Ph.D. thesis, S. V. University, Tirupati, India.

Saini, R. and Jaiwal, P.K. 2000. *In vitro* multiplication of *Peganum harmala* an important medicinal plant. Indian J. Exp. Biol., 38: 499–503.

Selvakumar, V. and Balakumar, T. 2000. *In vitro* induction of axillary branching in the medicinal plant *Acalypha fruticosa* Forsk. In: Recent Trends in Spices and Medicinal Plants Research. Amit Krishna De (eds.). Associated publishing Co., New Delhi: 96-97.

Shankaramurthy, K., Krishna, V., Maruthi, K.R., Nagaraja, Y.P. and Rahiman, B.A. 2004. High frequency plant regeneration from leaf callus cultures of *Embelia ribes* Burm. A threatened medicinal plant. Plant Cell Biotechnology and Molecular Biology, 5(3&4): 115-120.

Smitha Chetia and Handique, P.J. 2000. High frequency *in vitro* shoots multiplication of *Plumbago indica*, a rare medicinal plant. Curr. Sci., 78: 1187-1188.

Sousa, M., Ousingsawat, J., Seitz, R., Puntheeranurak, S., Regalado, A., Schmidt, A., Grego, T., Jansakul, C., Amaral, M.D., Schreiber, R. and Karl, K.K. 2007. An extract from the medicinal plant *Phyllanthus acidus* and its isolated compounds induce airway secretion: A potential treatment for cystric fibrosis. *Mol. Pharmacol.*, 7(1): 366-376.

Teresa, J.D.P., Urones, J.G., Marcos, I.S., Basabe, P., Cuadrado, M.J.S. and Moro, R.F. 1987. Triterpenes from *Euphorbia broteri*. Phytochem., 26(6): 1767-1776.

Vijayan, K., Chakraborti, S.P. and Roy, B.N. 2000. Plant regeneration from leaf explants of mulberry. Influence of sugar, genotype and 6-benzyl ademine. Ind. J. Exp. Biol., 38: 504-508

Vongvanich, N., Kittakoop, P., Kramyu, J., Tanticharoen, M. and Thebtaranonth, Y. 2000. Phyllanthusols A and B, Cytotoxic Norbisabolane Glycosides from *Phyllanthus acidus* Skeels. J. Org. Chem., 65: 5420-5423.

Chapter 8

An Efficient Method of Regeneration of *Dysophylla myosuroides* (Roth) Benth. by *in vitro* Culture of Nodal Explants

N. Savithramma, A. Sasikala, Beena Prabha and S.K.M. Basha

Department of Botany, Sri Venkateswara University, Tirupati, A.P.

ABSTRACT

An attempt was made to develop a protocol for rapid *In-vitro* multiplication of *Dysophylla myosuroides* which is a medicinally important perennial herb. Nodal segments were cultured on Murashige and Skoog's (MS) medium supplemented with different concentrations and combinations of auxins like Indole acetic acid (IAA), α-naphthalene acetic acid (NAA) and 3-Indole butyric acid (IBA), and cytokinins like Benzyl adenine (BA) and Kinetin (Kn). Maximum number of shoots obtained on BA at 1.0 mg l^{-1} followed by 1.0 mg l^{-1} BA + 0.01 mg l^{-1} NAA. *In-vitro* shoots were rooted onto the half strength MS medium supplemented with different concentrations of IBA, NAA and IAA. Highest frequency of root induction was achieved on MS medium fortified with 1.0 mg l^{-1} IBA. Regenerated plants were successfully acclimatized and 60 plantlets survived under e*x-vitro* conditions.

Keywords: Regeneration, Dysophylla myosuroides, Nodal explant, Growth regulators.

Introduction

Medicinal plants are now under great pressure of extinction due to uncontrolled exploitation for product isolation (IUCN, 2004). To meet the demand of plant, tissue culture technique has been progressing to over come the limitations of natural propagation of plant species like slow rate of multiplication. Edaphic and climatic factors, seasonal dormancy, low percentage of seed set and germination and clonal uniformity etc. Multiple shoot formation from nodal explant has been reported from several medicinally important plant taxa like *Ruta graveolens* (Bohidar *et al.,* 2008), *Trichosanthes cucumeriana* (Kawale and Choudhari, 2009), *Talinum cuneifolium* (Saradvathi *et al.,* 2009) and *Plantago ovata* (Mahato *et al.,* 2009) etc.

Dysophylla myosuroides, a member of the family Lamiaceae, is a much branched perennial herb with woody root stock on open rock crevices on hilly slopes. From the foregoing review of research carried out on this plant reveals that until now there is no published report on its economic and medicinal significances except traditional use of this plant. The local healers and tribal people are using the plant extract for relieving anxiety and stimulation of brain. As the herb is uprooted in the wild the number has been reduced on the hill slopes. Hence the present study was conducted to develop protocol to obtain plants at a quicker rate through the multiple shoots using the nodal explant.

Materials and Methods

Explant Source

Nodal segments were collected from wild grown mature plants on Tirumala hill slopes and washed under running tap water for 15-20 min., treated with 5 per cent teepol detergent for 15 min. In order to remove traces of detergent, all nodal segments were washed thoroughly with distilled water. Disinfection was performed under aseptic conditions in a laminar air flow cabinet by keeping them in 70 per cent alcohol for 60 sec. followed by rinsing for 3 times in sterile distilled water. Finally the explants were immersed in 0.1 per cent mercuric chloride solution for 3 min. followed by 5-6 rinses in sterile distilled water for surface sterilization before inoculation onto the sterilized medium (Murashige and Skoog, 1962) in culture tubes for shoot induction.

Shoot Multiplication

MS basal medium containing 3 per cent sucrose and 0.8 per cent agar and supplemented plant growth regulators was used for regeneration. The pH of the medium was adjusted to 5.8. The explants were implanted in the medium augmented with BA (0.5–3.0 mg l^{-1}) and Kn (1.0–3.0 mg l^{-1}) singly and in combination with auxins (IAA and NAA) for shoot proliferation. The cultures were maintained and incubated in 16 h light and 8 h dark photoperiod under light intensity 3000-4000 lux at 25 ± 2°C with 55 ± 5 per cent relative humidity. Six replicates for treatment were raised and the experiment was repeated thrice. The cultures were examined periodically.

Rooting of Shoots and Plant Establishment

In vitro raised shoots measuring 5 to 6 cm were excised and cultured on half strength MS medium supplemented with different concentrations of various auxins IAA, IBA, NAA (0.1-1.0 mg l⁻¹). Plantlets with well developed roots were taken out from culture tubes, washed with distilled water to remove adhering culture medium and were transferred to poly pots containing soil rite and soil (1:1). Pottted plants were incubated in an acclimatization chamber at 28 ± 1°C. After 3 weeks of acclimatization, they were transferred to poly bags containing soil : sand : farm yard manure (1:2:1) for a period of 2 weeks and then transferred to field. Statistical analysis was carried out using SPSS packages on IBM personal computer for different parameters.

Results and Discussion

Various growth regulators viz BA (0.5–3.0 mg l⁻¹), Kn (1.0-3.0 mg l⁻¹), IAA (0.01–0.20 mg l⁻¹) and NAA (0.01 – 0.2 mg l⁻¹) were tried individually and in combinations to obtain the most suitable growth regulator level for the proliferation of shoots

Table 8.1: Effect of Different PGR in Half Strength MS Medium on Shoot Multiplication from Mature Nodal Explants of *Dysophylla myosuroides*

Sl.No.	Plant Growth Regulators (mg l⁻¹)				Frequency of Shoot Regeneration	Mean No. of Shoots/ Explant	Mean Length of the Shoot (cm)
	BA	Kn	IAA	NAA			
1.	0.5				72.42 ± 0.06^n	6.7 ± 0.07^e	1.33 ± 0.03^h
2.	1.0				88.52 ± 0.02^p	22.61 ± 0.11^p	1.27 ± 0.02^{fg}
3.	2.0				74.72 ± 0.05^q	14.42 ± 0.08^n	0.53 ± 0.02^{bc}
4.	3.0				66.58 ± 0.14^g	8.5 ± 0.1^g	0.46 ± 0.02^b
5.		1			62.78 ± 0.03^d	1.75 ± 0.01^a	1.76 ± 0.01^i
6.		2			64.2 ± 0.12^e	5.16 ± 0.05^b	0.62 ± 0.02^d
7.		3			68.74 ± 0.05^i	7.53 ± 0.03^f	0.56 ± 0.03^{cd}
8.	0.5	1			71.34 ± 0.1^m	11.65 ± 0.04^i	0.62 ± 0.03^d
9.	0.5	2			70.8 ± 0.01^l	12.53 ± 0.04^k	0.3 ± 0.03^a
10.	1	1			72.63 ± 0.08^n	13.75 ± 0.08^l	1.2 ± 0.02^f
11.	1	2			67.35 ± 0.07^h	14.22 ± 0.03^m	0.92 ± 0.03^e
12.	2	1			58.47 ± 0.02^c	9.76 ± 0.07^h	0.51 ± 0.02^{bc}
13.	1			0.01	70.34 ± 0.04^k	14.67 ± 0.05^q	0.63 ± 0.01^d
14.	1			0.1	67.43 ± 0.05^h	12.13 ± 0.07^j	0.58 ± 0.02^{cd}
15.	1			0.2	69.56 ± 0.04^j	13.68 ± 0.1^l	0.33 ± 0.03^a
16.	1		0.01		66.26 ± 0.04^f	7.35 ± 0.06^f	0.61 ± 0.02^d
17.	1		0.1		55.84 ± 0.04^b	6.48 ± 0.05^d	0.50 ± 0.03^{bc}
18.	1		0.2		52.47 ± 0.04^a	6.15 ± 0.02^c	0.30 ± 0.01^a

Values represented above are the mean values of 12 replicates. '±' indicates the standard error. Observations after 6 weeks of culture. Mean values having the same letter in each column don't differ significantly at P=0.05 (Duncans Test).

fromnodal explant. The effect of BA alone and its interaction with NAA found to be significant on *in vitro* multiplication rate (Table 8.1). BA at 1.0 mg l⁻¹ recorded maximum multiplication rate *i.e.*, 23 mean number of shoots per nodal explant with a highest frequency of shoot generation followed by 15 shoots at the combination of BA and NAA (1.0 and 0.01 mg l⁻¹) respectively.

Similar morphogenic response with BA was encountered by Zebakhan *et al.* (2009). Addition of BA in the culture medium increased number of multiple shoots in *Plumbago zeylanica* (Sahoo and Debata, 1998), *Desmodium gangeticum* (Behera and Thirunavoukkarasu, 2006), *Rouwolfia serpentina* (Singh and Guru, 2007) and *Ruta graveolens* (Bohidar *et al.*, 2008). As concentration of BA increases the number of multiple shoots decline. Slight decrease and increase in strength of growth regulators gave varied results during present work.

Table 8.2: Effect of Different Auxins on the Root Induction of *in vitro* Raised Shoots of *Dysophylla myosuroides* on MS Half Strength Medium

Plant Growth Regulators (mg l⁻¹)			Frequency of Regeneration	Mean Number of Roots	Mean Length of Roots
NAA	IBA	IAA			(in cm)
–	–	–	46.05 ± 0.03ᵇ	2.63 ± 0.03ᵇ	0.67 ± 0.03ᵃ
0.1	–	–	66.71 ± 0.01ᵍ	4.21 ± 0.02ᶠ	3.28 ± 0.02ᶠ
0.5	–	–	58.22 ± 0.03ᵉ	3.17 ± 0.03ᵈ	2.76 ± 0.04ᵉ
–	–	0.1	52.61 ± 0.01ᶜ	3.66 ± 0.01ᵉ	1.82 ± 0.02ᶜ
–	–	0.5	43.47 ± 0.02ᵃ	2.52 ± 0.05ᵃ	1.47 ± 0.01ᵇ
–	0.01	–	57.56 ± 0.03ᵈ	2.84 ± 0.04ᶜ	1.83 ± 0.02ᶜ
–	0.1	–	65.32 ± 0.01ᶠ	4.61 ± 0.02ᵍ	2.23 ± 0.01ᵈ
–	0.5	–	69.42 ± 0.02ʰ	7.42 ± 0.03ʰ	3.73 ± 0.03ᵍ
–	1.0	–	76.38 ± 0.03ⁱ	12.75 ± 0.04ⁱ	3.84 ± 0.02ʰ

Values represented above are the mean value of 12 replicates. '±' indicates the standard error. Observations after 4 weeks of culture. Mean values having the same letter in each column don't differ significantly at P=0.05 (Duncans Test).

The well developed elongated shoots were excised from shoot clump and transferred to half strength MS medium containing IAA (0.1 – 0.5 mg l⁻¹), IBA (0.01–1.0 mg l⁻¹) and NAA (0.1–0.5 mg l⁻¹) (Figure 8.1). Auxin concentration in tissue cause root induction (Narayanaswamy, 1994) and phenolics and peroxidase activity play key role for induction of adventitious roots. Induction of roots was observed on basal MS media containing IBA and NAA. On a medium containing IAA explants did not respond properly. More number of roots with high frequency were found on a medium containing IBA 1.0 mg l⁻¹ followed by 0.5 mg l⁻¹. NAA also induced roots but the response is not like IBA. The frequency of the root induction increased with concentration of IBA from 0.01 to 1.0 mg l⁻¹. In case of NAA the roots were induced at frequency rate of 66.7 per cent on 0.1 mg l⁻¹ the increase concentration reduced the number of roots.

Figure 8.1: Micropropagation of *Dysophylla myosuroides*
(a) Nodal explant, (b) multiple shoots on MS + BA 1.0 mg l⁻¹, (c) elongated
shoot on rooting medium, (d) *In-vitro* shoot rooted on ½ MS + IBA 1.0 mg l⁻¹,
(e) Plantlet during hardening.

However in case of IAA 0.1 mg l^{-1} induced roots at frequency of 52 per cent while increasing concentration reduced the roots (Table 8.2). Similar results were obtained in *Camellia sinensis* (Rout, 2006), *Trichosanthes cucumeriana* (Kawale and Choudhary, 2009) and *Ruta graveolens* (Bohidar *et al.*, 2008 and Reddy *et al.*, 2009). Rooted plantlets were transferred to acclimatization in lab conditions and transferred to field. In this study 40 per cent plants transferred to field, did not survive during the acclimatization process. The reason could be that *In-vitro* derived roots are often delicate and easily damaged during transplantation and often show limited physiological functioning when in contact with the soil (Gangopadhyay *et al.*, 2002). 60 per cent of plants survived and well adjusted to natural environments.

References

Behra, A. and M. Thirunavoukkarasu. 2006. *In-vitro* micropropagation of medicinally important *Desmodium gangeticum* (L.) DC. through nodal explant. Indian J. Plant Physiol. 11: 83-88.

Bohidar, S., M. Thirunavoukkarasu and T.V. Rao. 2008. Propagation of *Ruta graveolens* L. by *In-vitro* culture of nodal explants. Indian J. Plant Physiol. 13 (2): 125-129.

Gangopadhyay, G., S. Das, S.K. Mitra, R. Poddar, B.K. Modak and K.K. Mukherjee. 2002. Enhanced rate of multiplication and rooting through the use of coir in aseptic liquid culture media. Plant cell tiss. Organ cult. 68: 301-310.

ICUN. 2004. Red List of Threatened Species. www.iucnredlist.org.

Kawale, M.V. and A.D. Choudhary. 2009. *In-vitro* multiple shoot induction in *Trichosanthes cucumeriana* L. Indian J. Plant Physiol. 14 (2): 116-123.

Mahato, S., Anita Mehta and R.K. Pandey. 2009. *In-vitro* regeneration and callus formation from different parts of seedling of *Plantago ovata* Forsk. The Bioscan 4 (1): 131-134.

Murashige, T. and F. Skoog. 1962. A revised medium for rapid growth and bioassay with tobacco tissue culture. Physiol. Plant. 15: 473-497.

Narayanaswamy, S. 1994. Plant cell and tissue culture. Tata McGraw Hill Publishing Company Limited, New Delhi.

Reddy, M.T., S.C. Kumar and V. Kumar. 2009. Efficient *In-vitro* plant regeneration in *Ruta graveolens*. In. J. Bioscan 4(4): 703-706.

Rout, G. 2006. Effect of auxins on adventitious root development from single node cuttings of *Camellia sinensis* (L.) Kuntze and associated biochemical changes. Plant Growth Regulators 48: 111-117.

Sahoo, S. and B.K. Debata. 1998. Micropropagation of *Plumbago zeylanica* Linn. J. Herbs Spices and Med. Plants 5: 87-94.

Saradvathi, J., A. Kedarnath Reddy and N. Savithramma. 2009. High frequency *In-vitro* shoot multiplication of *Talinum cuneifolium* (Vahl.) Willd. The Bioscan 4(3): 511-514.

Singh, G. and S.K. Guru, 2007. Multiple shoots induction in intact shoot tip, excised shoot tip and nodal segment explants of *Rauwolfia serpentina*. Indian J. Plant Physiol. 12: 360-365.

Zebakhan, Mohammad Anis and Iram Siddique. 2009. *In-vitro* regeneration of plantlets of *Solanum melongena* L. using cotyledonary node explants. J. Indian Bot. Soc. 88 (3 & 4): 141-145.

Chapter 9

Pearl Millet Tissue Culture and its Genetics: Current Status

T.S.S. Mohan Dev[1], P.V. Arjuna Rao[2], D. Sanjeeva Rao[3], Y.V. Rao[1], B.V. Rao[1] and M.V. Subba Rao[1]

[1]*Department of Botany, Andhra University,Vishakhapatnam – 530 003*
[2]*M.V.R. P.G. College, Gajuwaka, Visakhapatnam – 530 026*
[3]*Department of Microbiology, Dr. L.B.P.G. College, Visakhapatnam*

Introduction

Pearl millet [*Pennisetum glaucum* (L.) R. Br. = *P. americanum* (L.) Leeke = *P. typhoides* S&H; 2n = 14] is an important cereal of traditional farming systems in tropical and subtropical Asia and Sub-Saharan Africa. It is the sixth most important cereal crop after wheat, rice, maize, barley and sorghum in terms of annual global production (FAO, 1992). It is a staple food grain and a source of feed, fodder, fuel and construction material, grown on 29 million hectares (FAO, 2005) supporting millions of poor rural families in the hottest and driest drought-prone semi-arid regions of Africa and the Indian sub-continent, where rainfed agriculture is practiced. It belongs to the subfamily Panicoideae of family Poaceae.

Pearl millet originated in Africa where it was domesticated on the southern margins of the Saharan central highlands at the onset of the present dry phase some 4000-5000 years ago and was introduced into India about 4000 B.C (Anand Kumar, 1989). It is the fourth most important cereal crop in India after rice, wheat and sorghum. In India, the average cultivated area under pearl millet amounts to 9.5 M ha with an average annual grain production of 8.3 M tons and average grain yield of 880 kg/ha (FAO, 2005).

Pearl millet tolerates drought, low soil fertility and low soil pH better than other cereals, yet responds well to favorable weather and soil conditions. It has a higher level of heat tolerance than sorghum or maize. These facts make pearl millet an important food crop in semi-arid areas where other crops tend to fail because of inadequate rainfall and poor soil conditions (FAO and ICRISAT, 1996). Therefore pearl millet is considered as one of the most drought tolerant of all domesticated cereals (Bidinger and Hash, 2004). A very wide range of genetic variability is available in the primary germplasm pool of *Pennisetum* for improvement of this species.

Besides the agronomically useful characters, pearl millet also has the following desirable features, which make it suitable for basic genetic, cytogenetic and molecular genetic studies (Jauhar, 1981; Armuganathan and Earle, 1991):

1. Low haploid number of chromosomes (n=7) of large size,
2. Protogynous nature which facilitates cross-pollination (both natural and controlled),
3. Large number of seeds per panicle.
4. Short span of life cycle (approx 90 days), and
5. Large genome size (about 2300 Mbp) that is nearly equal to that of maize (2400 Mbp).

With in the first decade of report of successful tissue culture work in cereals by LaRue (1949) in maize, pearl millet drew attention in the field of *in vitro* culture. The earliest work on *in vitro* culture of this crop plant was that of Narayanaswamy(1959) who reported proliferation of scutellar tissue from cultured immature embryos and formation of non-morphogenic callus. He also showed that it is essentially the epidermal layer which became meristimatic although the entire scutellum showed callus formation. Subsequently a number of workers tried various media, hormones, explants and genotypes of pearl millet and considerable wealth of information on tissue, cell and protoplast culture of this important agricultural crop (Table 9.1). The formation of two different morphotypes of calli *viz.*, embryogenic and nonembryogenic, a distinct feature of callus in cereals and grasses, was also noticed in pearl millet. Another general feature reported in several graminaceous members *viz.*, regeneration following either complete withdrawal (Gamborg *et al.*, 1970) or reduction of the 2,4-D level (Sears and Deckard, 1982) was also noticed in pearl millet system.

(*a*) Nature of Explant and Identification of the 'Competence Window'

In addition to the scutellar tissue used by Narayanaswamy (1959), a variety of plant parts were tried for their response to tissue culture conditions. These include immature embryos (Vasil and Vasil, 1981b), mature embryos (Botti and Vasil, 1983), seeds and root tips (Nagarathna, *et al.*, 1991), mesocotyl (Rangan, 1976; Subba Rao and Nitzsche 1984; Krishna Rao *et al.*, 1988) and immature inflorescences (Talwar and Rashid, 1990; Vasil and Vasil, 1982a; Mythili *et al*, 1997).

Morrish *et al.* (1987) suggested the existence of a 'selective phase' or 'competence window' to which the production of embryogenic callus is mainly confined. Immature inflorescences of pearl millet measuring 2 – 25 mm (Vasil and Vasil, 1981a), 10±2 mm long inflorescences (Talwar and Rashid, 1990), 3 – 4 cm long inflorescences referred to as a stage – I(Mythili, 1993) and 0.5 – 1 cm long immature inflorescence(Botti and Vasil, 1984) represented the suitable developmental stages required to obtain optimum callus induction and regeneration in various accessions of pearl millet. Mythili (1993) made a detailed study to identify the growth stage of immature inflorescence that represented the best 'selective phase'. She used inflorescences representing three developmental stages which were named as stage – I, stage – II and stage – III. Identification of the three stages was based on the relative positions of the flag and boot leaves, inflorescence length and extent of spikelet development as described below.

Stage – I

Flag leaf was longitudinally folded or curled and was very close(< 0.5 cm) to that of the boot leaf; inflorescence measured 3-4 cm with pinhead sized spikelet primordia all along its length.

Stage – II

Lamina of flag leaf was completely expanded and was approximately 0.5 to 1 cm above the boot leaf; inflorescence measured 6 – 8 cm but still remained enclosed with in the sheaths of the flag and boot leaves; small spikelets were visible only in the upper half of the inflorescence.

Stage – III

Apical 1.5 – 2 cm of the inflorescence emerged out of the boot leaf while the remaining portion was still enclosed with in the sheath of the flag leaf; inflorescence measured about 12- 15 cm; small spikelets were clearly visible all along the length of the inflorescence and pre-meiotic mitotic divisions could be observed in the pollen mother cells of the primary spikelets present in the upper one-third portion of the inflorescence.

All the three stages of the inflorescence segments in the five inbreds studied produced calli. However, the embryogenic calli formation was observed in all the 2,4-D concentrations from stage – I explants, but only in lower concentration from stage – II and stage – III explants. Though the quantity of total callus produced by the explants of second and third stages was known than that of stage – I, the quantity of embryogenic callus from these older stages was less than that produced from explants of stage – I. Such decline or loss of competence in the morphogenic potential observed in relatively older or developmentally advanced stages was suggested to result on several factors such as qualitative or quantitative changes in DNA, cell cycle status, genotype and endogenous levels of hormones in relation to senescence and aging and also on the exogenous hormones (Morrish *et al.*, 1987). Thus in general its needs that immature inflorescences ranging in size between 0.5 – 4 cm length might represent the competence window. However the different genotypes might vary in their characteristic and more precise lengths.

(b) Media and Phytohormones

Various media *viz.*, MS, B₅ (Subba Rao and Nitzsche, 1984), LS (Ketchum *et al*,

1987) and N$_6$ (Talwar and Rashid, 1990; Mythili, 1993) have been tried but MS has been the most frequently used medium. Talwar and Rashid (1990) observed formation of somatic embryos directly from the inflorescences when the N$_6$ medium was supplemented with higher concentrations of 2,4-D. Though the same concentrations of 2,4-D for used in combination with MS as well as N$_6$ media by Mythili (1993), embryogenic callus formation was observed only on MS. Thus the N$_6$ medium neither supported to the direct formation of embryoids nor callus in the genotypes used by Mythili (1993).

A variety of phytohormones such as IAA, 2,4-D, 2,4,5-T, ABA, BAP, Kinetin etc. were tried in various concentrations and in combinations. The responses to these phytohormones varied depending on the genotypes used suggesting the role of a strong genotypic component.

(c) Genotypic Differences

Various genotypes were used by different workers but a concerted effort to bring out the significance of genotypic differences was made by Subba Rao and Nitzsche (1984). These authors used eight inbred lines for their callus forming ability from mesocotyl explants, callus growth rate and shoot bud production. Subsequently, the work by Mythili *et al.* (1997) and Satyavathi *et al.* (2006) not only confirmed the existence of genotypic differences but also led to a detailed analysis of these genotypic differences. Existence of such genotype differences was attributed to the inherent levels of endogenous growth factors (Abe and Futsuhara, 1986).

Callus Histology

Vasil and Vasil (1981a) reported the occurrence of three morphotypes of calli *viz.,*

1. An yellow to white ridge of foam and opaque embryogenic callus,
2. A soft, translucent, friable and unorganized callus, and
3. Soft, watery and musilagenous callus

Mythili (1993) could identify two types of calli based on their colour, texture and nature (friability or compactness). The relatively more compact, nodular and white or pale yellow callus was more embryogenic and probably corresponded to the E callus of Vasil and Vasil (1981a). Where as the more friable, pale yellow or light brown, crystalline or watery callus corresponded to the NE (non embyrogenic) type. These two types of calli showed changes in their colour and texture with increasing subcultures. The E callus became yellowish or dark brown, compact and less nodulated. Even after becoming brown it showed proliferation. In contract to this the NE callus turned brown and became more watery or mucilaginous.

Marked differences were also observed between the E and NE types from scanning electron microscopic studies (Mythili, 1993).The E callus was characterized by the presence of regular and densely arranged cells while the NE callus had loosely arranged cells with irregular shapes. The nodulated regions of the callus showed a distinct organization where the cells were compactly arranged in rows. Somatic embryogenesis involved the formation of rounded and elongated structures called pro-embryoids (Vasil and Vasil, 1982b).

Mythili *et al*. (1995) reported the presence of three types of cells in 6-12 weeks old E calli and these were:

1. Small cells with an average diameter of 11.53±.013 mM which were observed in high frequencies in the original explants and also in all the E calli irrespective of the 2,4-D concentration. The chromocenters representing the heterochromatin were less prominent in the nuclei of these cells. In view of their abundance and relatively higher frequencies of mitotic division these cell populations were regarded as the normal embryogenic type.

2. Medium sized cells having an average diameter of 17.5±0.31 mM were observed in calli of only one accession(IP 3128) cultured on 2.5 mg/L 2,4-D. Nuclei of these cells also appeared relatively larger than in the normal embryogenic cells.

3. Large cells with an average diameter of 32.56±1.19 mM showed relatively large and densely stained nucleus and increased number of chromocenters at interphase. These cells were observed in 6 week old calli of accessions CO5, IP 3128, VG 587 and IP 1346. In spite of an increase in the cell and nuclear volume as well as number of chromocenters, the large and medium size cells (types b and c) showed normal diploid chromosome number (2n = 14) only.

Regeneration

In all the genotypes of pearl millet studied so far, regeneration from the callus was through somatic embryogenesis. Vasil and Vasil (1982 b) and Mythili (1995) reported the formation of somatic embryos either directly from single cell or after the formation of a mass of pro-embryonic cells. Genotypic differences were also observed for the frequency of shoot bud formation (Subba Rao and Nitzsche, 1984). Though the gramenaceous trend of initiation of somatic embryos following withdrawl of auxins was generally observed, the various genotypes differed marginally in the optimum concentrations and combinations of hormones required for maximum frequency regeneration. Rangan(1976) reported that maximum shoot bud formation required Coconut milk (CM), IAA and Kinetin. Interestingly, Vasil and Vasil (1982a) reported that 2,4-D is also required for regeneration in addition to BAP, NAA and CM. Talwar and Rashid (1990) observed that somatic embryo formation was dependent on the concentration of 2,4-D and minerals in the medium, especially ammonium levels. Mythili (1993) noticed that the concentration of 2,4-D used for callus induction influenced the hormonal requirement for regeneration. The optimum hormonal concentrations for the maximum frequency of regeneration in the five accessions used by her were 0.25 mg / L. Kinetin and 0.25 mg / L BAP.

Cell Culture

Calli derived from embryos (Vasil and Vasil, 1981b), mesocotyl (Subba Rao, 1984) and inflorescence (Vasil and Vasil, 1982a) were used to establish cell suspension cultures. Irrespective of the source of the cell culture, two morphotypes of cells – i) large vacuolated irregularly shaped cells with scanty cytoplasm containing little or no starch, and ii) small, more or less spherical cells with prominent nucleus with abundant cells. Another common feature of the cell cultures is the presence of mucilage that generally appears by the end of first or early second week of culture. Vasil and Vasil (1982a) reported that embryogenesis from cell cultures is favored by a 1:1 dilution with MS medium containing lower concentration of 2,4-D.

Protoplast Culture

Cell suspension cultures originating from immature embryos were used to isolate protoplast (Vasil and Vasil, 1979) which were of two kinds – those with vacuolated cytoplasm which did not divide and those with non-vacuolated cytoplasm which underwent sustained cell divisions. Mesophyll derived protoplasts were isolated and cultured by Subba Rao (1981) but regeneration was not reported.

Anther Culture

Nitsch *et al.* (1982) produced androgenic haploids and observed that anthers containing uninucleate microspores with the nucleus positioned away from the pole were ideal for androgenesis. Because of the small size of anthers at this stage of development, culture of old florets rather than individual anthers was suggested to be more convenient. The recommended culture conditions included keeping the anthers/florets at 14°C in the dark for seven days and at 20°C under red light for seven days followed by continued culture under red light (16 hrs.) at 25-27°C, use of L-proline(100 mg / L), 2,3,5-Triiodobenzoic acid (TIBA), (0.1mg / L), sucrose (60mg / L) and activated charcoal (5 mg / L) in combination with N6 or Yu-Pei medium.

Somaclonal Variation

By analyzing the variation in the regenerated plants using protein, Alcohol dehydrogenase (ADH) and Malate dehydrogenase (MDH) patterns, Swedlund and Vasil (1985) found no significant somaclonal variability. On the basis of chromosomal studies in the immature inflorescence – derived calli and their regenerants, Mythili (1993), suggested the presence of a selection mechanism operating in the callus that favours the regeneration of only normal diploid plants. Morrish *et al.* (1990) used quantitative characters to study somaclonal variation and suggested the role of epigenetic factors.

In addition to pearl millet which is a cultivated species of the genus *Pennisetum*, the following information is also available on callus culture of some of the other species (*P. villosum, P. mezinum, P. clandestinum, P. orientale, P. setaceum* and *P. hohenackeri*) and interspecific hybrids [(*P. glaucum* X *P. purpureum*; apomicitic derivatives of (*P. glaucum* X *P. purpureum*) X (*P. glaucum* X *P. squamulatum*)] (Ozias-Akins *et al.*, 1989). While no regeneration was reported in *P.clandestinum, P.setaceum* and *P.hohenackeri*, regeneration potential of the callus in *P.purpureum* was shown to be related to the endogenous levels of Abscisic acid and IAA.

Genetics of Tissue Culture

The first study on the genetic basis of tissue culture related characters in this crop plant was by Mythili *et al.* (1997). These authors used five inbred lines IP14254, CO 5, IP 3128, VG 587 and IP 1346 and four quantitative characters to represent the *in vitro* response.

**Table 9.1: Types of Explants, Media and
Accessions Used for Tissue Culture in Pearl Millet**

Type of Explant	Medium Used	Accessions	Reference
Immature embryos	MS	Ghai-3	Vasil and Vasil (1981b)
Mature embryos	MS	-	Botti and Vasil (1983)
Seeds	LS	HMP-559 and SB	Ketchum *et al,* (1987)
	MS	Ghai, HMP-559, 7203, Senegal bulk and WC-C75	Nabors *et al.* (1983)
Scutellar tissue	MS	-	Narayanaswamy (1959)
Root tips	MS	-	Nagarathna *et al.,*1991
Mesocotyl	MS	23A X J 993	Rangan (1976)
	MS	IP 3128, IP 3676,IP 4122,	Subba Rao and
	B5	IP 5313, Tift 28, Tift 23 DB, Tift 18 DB and Tift 23 B	Nitzsche, (1984)
Immature inflorescence	N6	-	Talwar and Rashid (1990)
		IP 14254, CO5, IP3128, VG 587 and IP 1346	Mythili (1997)
	MS	Ghai-3	Vasil and Vasil (1982a)
		IP 14254, CO5, IP3128, VG 587 and IP 1346	Mythili (1997)

These included:

1. *Total callus quantity:* the quantity of embryogenic(E) and non-embryogenic(NE) portions of the callus together was measured using a thin transparent polythene graph paper drawn in milli meters and was expressed as the area (in cm²) occupied.

2. *E callus quantity:* The proportion of embryogenic callus area in the total callus and expressed in percentage was taken as E callus quantity.

3. *Callus growth rate:* This was expressed in terms of percentage increase in fresh weight during 4th-6th week.

4. *Regeneration frequency:* This indicated the number of plantlets produced per explant at the end of 9th week after initial inoculation (*i.e.* 3 weeks after transfer of calli to regeneration medium)

With the exception of E callus quantity the other three characters showed high degree of heritability (53-82 per cent) indicating the existence of definite genetic basis. Such high heritability values suggest the possibility for improving these characters through breeding and selection. Diallel analysis showed that callus quantity was the only character showing epistatic gene action (non-allelic interaction) and the alleles responsible for this were predominantly contributed by IP 3128. The authors tried to explain this interaction by proposing two models.

1. The volume of NE callus might have been under the influence of epistatic genetic effects and since NE type forms a major component of the total callus, the epistatic gene action of NE callus itself might have been reflected in the genetic control of the total callus quantity also.

2. The genes responsible for NE callus and those for E callus might be acting additively and might be interacting with each other in an epistatic manner. The implication of both these models is that the genes controlling E callus and NE callus quantities might be non-allelic.

VG 587 showed complete dominance for E callus quantity while the other four showed partial dominance. For callus growth rate and percentage regeneration, all five inbreds showed partial dominance. Thus in these five accessions additive gene action was predominant for total callus quantity, callus growth rate and regeneration indicating their good selection potential.

The diallel analysis also indicated equal gene frequencies with positive (u) and negative(v) effects, suggesting that there was no selection against the favourable alleles in these accessions. The direction of dominance was towards decrease in the total callus quantity, E callus quantity and percentage regeneration.

Heterosis both in terms of mid parental values and better parental values was inferred to be equally frequent in the hybrids which offers the possibility to exploit the phenomenon for improving these *in vitro* characters especially quantity as well as growth rate of E callus and the percentage of regeneration.

Mythili *et al.* (1997) also carried out combining ability analysis in these five inbreds which indicated that IP 1346 was the best general combiner for total callus quantity, E callus quantity, callus growth rate and regeneration frequency. Therefore it might be possible to improve these *in vitro* characters genetically by using lines with high GCA values. SCA effects resulting in increase of these callus characters were also observed in certain hybrid combinations. The overall conclusion from the combining ability analysis was that IP 1346 (among the five) is a suitable parent for transferring the *in vitro* traits to other cultivars.

The general trend in the quantitative genetic basis of these *in vitro* characters was further confirmed by Satyavathi (2006) using a different set of inbreds VG 272, D2 dwarf developed, IP 3128, and AR 4209. The availability of a distinct dwarf phenotype was used to an advantage by Satyavathi *et al.* (2006) who analysed the association with each of the four *in vitro* characters with the d_2 locus. This was possible because the callus produced by the dwarf plants (named as D type) was highly compact, nodular, friable and slightly tanned and thus could be visually differentiated from the callus (ND type, that was soft, translucent and slightly mucilagenous) of the tall parental lines. Polygenic control was inferred for each of these *in vitro* characters since the estimated number of effective genetic factors was greater than two.

Based on the the joint segregation patterns of the dwarf nature and each of the *in vitro* characters together with correlation analysis between plant height and the *in vitro* characters, the following hypothetical model was suggested by Satyavathi *et al.,* in 2006.

1. Dwarf gene d2 might be relatively closely linked (so as to give significant correlation) with majority of the loci governing E callus production,

2. Loci governing E callus production might be closer (than d2) to atleast some of the loci controlling regeneration frequency and

3. If the very loci controlling E callus production are also assumed to be governing regeneration frequency, a significant correlation of the latter with d2 is also expected. Since such significant correlation has not been observed between plant height and regeneration frequency, it might be suggested that at least some of the loci governing E callus production and regeneration frequency are different. In such a case, the d2 locus is expected to be more closely linked to loci for E callus production (in view of the significant correlation) than those responsible for regeneration.

A model was also provided by Satyavathi *et al.* (2006) to indicate the probable arrangement of the loci :

$$d_2 \quad Ec_1 \ldots \ldots Ec_4 \quad Rg_1 \ldots \ldots Rg_5 \quad Cgr_1 \ldots Cgr_3$$

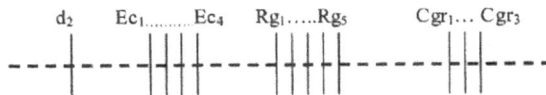

Applications of Tissue Culture

In vitro Selection

Callus cultures of pearl millet were successfully used for the selection of ergot resistance (Bajaj *et al.*, 1980, Sharma and Chahal, 1989), and S-(2-amino ethyl)-cysteine (AEC) tolerance (Boyes and Vasil, 1987), where as cell suspension cultures of both pearl millet (Rangan and Vasil, 1983) and *P. purpureum* (Bajaj and Gupta, 1987) were used for the selection of NaCl tolerance lines.

Somatic Hybridization

Protoplast derived from AEC resistant cell suspension cultures of pearl millet were used in somatic hybridization experiments with *Saccharim officinarum* (Tabaeizadech *et al.*, 1986), *Panicum maxium* (Ozias – Akins *et al.*, 1986) and *Triticum monococcum* (Vasil *et al.*, 1988). In the case of *P.maxium* + *P.glaucum*, inactivation of pearl millet protoplast by Iodo-acetate treatment was also combined with AEC tolerance. A common identification marker for the hybrids in all these cases was 6-phospho gluconate dehydrogenase (6-PGD).

Cryopreservation

Cryopreservation of immature embryo-derived cell suspension cultures was attempted by Gnanapragasam and Vasil (1992). A survival rate of up to 29 per cent was reported in cultures pre-condioned in media supplemented with 0.33M mannitol for 3 days and cryoprotection by gradual addition of 0.5M sorbitol and 0.7M Dimethyl sulfoxide (DMSO). Increase in the plating efficiency of cryoprotected cells and protoplast yield from them was also noticed.

Genetic Transformation

Tissue or cell cultures are increasingly used as the target material in genetic transformation experiments in various plant systems. In pearl millet also embryonic callus and cell suspension cultures were used (Lambe, 1995, 2000; Latha *et al.*, 2006) for genetic transformation using biolistic method.

References

Abe, T and Futsuhara, Y. 1986. Genotypic variability for callus formation and plant regeneration in rice (*Oryza sativa* L). Theoritical and Applied genetics., 72: 3-10.

Anand kumar, K. 1989. Pearl millet: current status and future potential. Outlook on Agriculture. 18: 46-53.

Armuganathan, K. and Earle, E.D. 1991. Nucelar DNA content of some important plant species. Plant Molecular Biology Reports. 9:208-218.

Bajaj,Y.P.S. and Gupta, R. 1987. Plants from salt tolerant cell lines of napier grass, *Pennisetum purpureum* Schum. Indian J. Exp. Biol. 25: 58-60.

Bidinger, F.R., and Hash, C.T. 2004. Pearl millet. Chapter 5: pp 225-270 in Physiology and Biotechnology Integration for plant breeding. (H.T.Nguyen and A.Blum, eds.). Marcel Dekker, New York.

Botti, C. and Vasil, I.K. 1983. Plant regeneration by somatic embryogenesis from parts of cultured mature embryos of *Pennisetum americanum* (L.) K Schum. Zeitschrift fur Pflanzenphysiologie 111: 319-325.

Botti, C. and Vasil, I.K. 1984. Ontogeny of somatic embryos of *Pennisetum americanum* L. II. In cultured immature inflorescences. Can. J. Bot. 62: 1629¬1935.

Boyes, C.J. and Vasil, I.K. 1987. *In vitro* selection for tolerance to S-(2-aminoethyl)-L-Cysteine and overproduction of lysine by embryogenic calli and regenerated plants of *Pennisetum americanum* (L.). K. Schum. Plant Science 50: 195-203.

FAO. 1992. Crop production year book., FAO, Rome, Italy.

FAO. 2005. Global network on Integrated soil management for sustainable use of salt-affected soils. Rome, Italy: FAO land and plant nutrition management service. http://www.fao.org/ag/agl/gall/spush.

Gnanapragasam, S. and Vasil, I.K. 1992. Cryopreservation of immature embryos, embryogenic callus and cell suspension cultures of graminaeceous species. Plant Science 83: 205-215.

Jauhar, P.P. 1981. Cytogenetics and breeding of Peral millet and related species. Progress and topics in Cytogenetics I (Ed.Avery A.Sandberg) Alan R.Liss Inc., New York.

Ketchum, J.L.F., Gamborg, O.L., Hanning, G.E. and Nabors, M.W. 1987. Millet tissue culture for crops-project progress report. Department of Botany, Colarado State University, Fort Collins, Colarado, pp.38-39.

Krishna Rao M., Rajyalakshmi, T.V. and Surya Kumari, T.V. 1988. Induction of callus from different explants of pearl millet. Page 319. In: International Conference on Research in Plant Sciences and its Relevance to Future. New Delhi, India.

LaRue, C.D., 1949. Cultures on the endosperm of maize. American J. Bot. 34: 585-586.

Lambe, P., Dinant, M., Matagne, R.F.. 1995. Differential long-term expression and methylation of the hygromycin phosphotransferase(hph) and b-glucuronidase(GUS) genes in transgenic pearl millet (*Pennisetum americanum*) callus. Plant Sciences. 108:51-62.

Lambe, P., Dinant, M., Deltour, R..2000.Transgenic pearl millet(*Pennisetum glaucum*). In Bajaj, Y.P.S. (ed) Biotechnology in agriculture and forestry, transgenic crops I, Vol. 46, Springer, Berlin, pp 84-108.

Latha, M.A., Rao, K.V., Reddy, T.P., Reddy, V.D..2006. Development of transgenic pearl millet (*Pennisetum glaucum* (L)R.Br.) plants resistant to downy mildew. Plant cell Reports. 25-927-935.

Morrish, F.M., Hanna W.W. and Vasil, I.K. 1990. The expression and perpetuation of inherent somatic variation in regenerants from embryogenic cultures of *Pennisetum glaucum* (L.) R Br. (pearl millet). Theoretical and Applied Genetics 80: 409-416.

Morrish, F.M., Vasil, V. and Vasil, I.K. 1987. Development, morphogenesis and genetic manipulation in tissue and cell cultures of the Gramineae. Advances in Genetics 24: 431-482.

Mythili, P.K 1993. Callus culture and *in vitro* morphogenesis: Some cytological and genetic aspects in five inbreds of pearl millet, *Pennisitum glaucum* (L.) R. Br. Ph. D. Thesis, Andhra University, Visakhapatnam, India.

Mythili, P.K., Subba Rao, M.V. and Manga, V.1995. Cytology of Explants, Calli and Regenerants in five inbred lines of Pearl millet, *Pennisetum glaucum* (L) R.Br. Cytologia. 60: 23-29.

Mythili, P.K, Satyavathi, V., Pavankumar, G., Rao, M.V.S. and Manga, V. 1997. Genetic analysis of short term callus culture and morphogenesis in pearl millet, *Pennisetum glaucum*. Plant cell, tiss. Organ cult.. 50:171-178.

Nabors, M.W., Heyser, J.W., Dykes T.A. and.Demott, K.J. 1983. Long duration, high frequency plant regeneration from cereal tissue cultures. Planta, 157 : 139-142.

Nagarathna, KC., Shetty, S.A., Harinarayana, G. and Shetty, H.S. 1993. Selection for downy mildew resistance from the regenerants of pearl millet. Plant Science 90: 53-61.

Narayanaswamy, S. 1959. Experimental studies on growth of excised grass embryos *in vitro*. 1. Overgrowth of the scutellum of *Pennisetum* embryos. Phytomorphology 9: 358-267.

Nitsch, C., Andersen, S. Godard, M. Neuffer M.G. and Sheridan, W.F. 1982. Production of haploid plants of *Zea mays* and *Pennisetum* through androgenesis. Pages 69–91.

In: Variability in Plants Regenerated from Tissue Culture. (Elizabeth D. Earle and Yues Demanly, eds.). Praeger Publisher, New York, USA.

Ozias-Akins, P., M. Dujardin, Hanna, W.W. and Vasil, I.K., 1989. Quantitative variation recovered from tissue cultures of an apomictic interspecific *Pennisetum* hybrid. Maydica 34: 123-132.

Ozias-Akins, P., Ferl, R.J. and Vasil, I.K. 1986. Somatic hybridization in Gramineae: *Pennisetum americanum* (L.) K. Schum. (pearl millet) + *Panicum maximum* Jacq. (Guinea grass). Molecular and General Genetics 203: 365-370.

Rangan, T.S. 1976. Growth and plantlet regeneration in tissue cultures of some Indian millets: *Paspalum scrobiculatum* L., *Eleusine coracana* Gaertner and *Pennisetum typhoideum*. Zeitschrift für Pflanzenphysiologie 78: 208-216.

Rangan, T.S. and Vasil, I.K. 1983. Sodium chloride tolerant embryogenic cell lines of *Pennisetum americanum* (L.) K. Schum. Annals of Botany 52: 59-64.

Satyavathi.V., Subba Rao, M.V., Manga, V and Chitti babu, V. 2006. Genetics of some *in vitro* characters in pearl millet. Euphytica 148 : 243-249.

Sharma, S.B. and Chahal, S.S.1989. Effect of culture filtrate of the ergot pathogen on pearl millet seedlings and callus cultures. Indian J. Exp. Biol. 27: 187-188.

Subba Rao, M.V. 1981. Studies of some factors in the isolation of mesophyll protoplasts, *Pennisetum americanum*. In: Proceedings International Symposium on Plant Cell Culture in Crop Improvement, Calcutta, India.pp.48-49.

Subba Rao, M.V. and Nitzsche, W. 1984. Genotypic differences in callus growth and organogenesis of eight pearl millet lines. Euphytica 33: 923-928.

Swedlund, B. and Vasil, I.K. 1985. Cytogenetic characterization of embryogenic callus and regenerated plants of *Pennisetum americanum* (L.) K. Schum. Theoretical and Applied Genetics 69: 575-581.

Tabaeizadeh, Z., Ferl, R.J. and Vasil, I.K. 1986. Somatic hybridization in the Gramineae: *Saccharum officinarum* L. (sugarcane) and *Pennisetum americanum* (L.) K. Schum (pearl millet.). In: Proceedings of the National Academy of Science, USA 83: 5616-5619.

Talwar, M. and Rashid, A. 1990. Factors affecting formation of somatic embryos and embryogenic callus from unemerged inflorescences of a graminaceous crop, *Pennisetum*. Ann. Bot. 66: 17-21.

Vasil, V., Ferl, R.J. and Vasil, I.K. 1988. Somatic hybridization in the Gramineae: *Triticum monococcum* L. (einkorn) + *Pennisetum americanum* (L.) K. Schum. (pearl millet). J. Plant Physiol. 132: 160-163.

Vasil, V. and Vasil, I.K. 1979. Isolation and culture of cereal protoplasts. 1. Callus formation from pearl millet (*Pennisetum americanum*). Theoretical and Applied Genetics 56: 97-99.

Vasil, V. and Vasil, I.K. 1981a. Somatic embryogenesis and plant regeneration from tissue cultures of *Pennisetum americanum* and *P. americanum* x *P. purpureum* hybrid. Amer. J. Bot. 68: 864-872.

Vasil, V. and Vasil, I.K. 1981b. Somatic embryogenesis and plant regeneration from suspension cultures of pearl millet (*Pennisetum americanum*). Amer. J. Bot. 47: 679-686.

Vasil, V. and Vasil, I.K. 1982a. Characterization of an embryogenic cell suspension culture derived from cultured inflorescences of *Pennisetum americanum* (pearl millet, Gramineae). Amer. J. Bot. 69: 1441-1449.

Vasil, V. and Vasil, I.K. 1982b. The ontogeny of somatic embryos of *Pennisetum americanum* (L.) K. Schum. 1. In cultured immature embryos. Bot. Gaz. 143: 454-565.

Chapter 10

Molecular Characterization of Somaclones Using RAPD Markers in Patchouli

M.C. Gayathri* and Asmita Behera

Department of Botany, Plant Biotechnology Unit, Bangalore University, Bangalore – 560 056

ABSTRACT

In the present investigation, the plants regenerated through indirect organogenesis of leaf explant of *Pogostemon cablin* Benth. an essential oil yielding plant showed some variation with regard to morphological characters. Hence, Randomly Amplified Polymorphic DNA (RAPD) analysis was employed to assess genetic instability among regenerated patchouli plants (*Pogostemon cablin* Benth.). Sixteen plants of SC_5 (first generation after an *in vitro* phase followed by SC_2–SC_5) generation were subjected to RAPD analysis. Out of 8 random 10-mer primers, 5 primers generated polymorphism. A total of 649 bands ranging from 100 bp to 3 kbp were produced. Five plants produced polymorphic bands out of 16 regenerated plants. Our result demonstrates that RAPD analysis can be used successfully to determine the genetic instability among regenerated plants and confirms that they are morphologically and genetically different from that of control.

Keywords: Pogostemon cablin, Regenerated plants, Morphological variants, RAPD analysis.

---

* Corresponding Author: E-mail: gayatrimc@ hotmail.com

Introduction

Patchouli (*Pogostemon cablin* Benth.), belonging to the family Lamiaceae, is the source of patchouli oil. The commercial oil of patchouli is obtained by steam distillation of the shade dried leaves, and is one of the most important naturally occurring oil used in the perfumery industry. Due to rare flowering and failure of seed setting, propagation through seeds is not possible. Hence, patchouli plant generally propagated asexually through shoot tip cuttings. Because of vegetative propagation, genetic variation is not much in the cultivated plants. Therefore, regeneration of plants by *in vitro* cell and tissue is a fundamental step for the genetic manipulation and improvement of crops.

Occurence of somaclonal variation may be due to pre-existing genetic variation within the explants or during the tissue culture phase (induced variation) (Evans *et al.*, 1984). Different approaches have been applied for identifying variants among regenerated plants such as phenotypic variation (Vuylsteke *et al.*, 1988), Karyotypic analysis of metaphase chromosomes (Jha *et al.*,1992) and biochemical analysis (Damasco *et al.*, 1996). But the above approaches are not fully suitable for detecting DNA sequence polymorphism of *in vitro* raised plants. However, Randomly Amplified Polymorphic DNA (RAPD) analysis is often preferred because of reduced complexity. Many investigators have successfully employed RAPD analysis to find the genetic diversity among micropropagated plants (Isabel *et al.*, 1993; Valles *et al.*, 1993; Rani *et al.*, 1995; Damasco *et al.*, 1996).

Material and Methods

Callus Initiation and Plant Regeneration

Callus cultures were initiated from leaf explants of patchouli (*Pogostemon cablin* Benth.), variety Johor on Murashige and Skoog's (1962) (MS) Basal Medium supplemented with NAA and BAP. The pH was adjusted to 5.8. Agar was used at 0.8 per cent as a gelling agent. Cultures were kept at $25\pm2°C$ with photon flux density of $30-50$ $mEm^{-2}s^{-1}$ under a photoperiodic regime of 16h light and 8h dark cycles. After 183 days, the callus was subcultured on MS basal medium supplemented with BAP to regenerate multiple shoots and then transferred to the rooting medium MS basal medium supplemented with IBA.

Plants regenerated through callus phase were acclimatized and successfully transferred to green house following the standard method. After 30 days of hardening, plants were transferred to field. Somaclones were isolated from six month old hardened and acclimatized 100 plants after screening for morphological traits such as plant height, number of branches per plant, leaf area, leaf size, petiole length, herb yield.

DNA Extraction and PCR Reaction

In the present study, young leaf was chosen from the somaclonal variants for DNA extraction. DNA was extracted from 100 mg of *in vivo* leaf tissue by CTAB method and amplified with 8 random 10 mere decamer primers (Table 10.1). RAPD reaction was carried out in 25 µl volume containing 200 ng/µl genomic DNA, 1 unit

Taq polymerase, 4 µl dNTP's, 2.5 µl of 10x reaction buffer with 3.5 µl random primer. The volume was made up using sterile water. Amplification was carried out for 30 cycles with initial heat-denaturation of the DNA at 95°C for 3 minutes. The thermal cycling was performed with the following temperature regimes- 95°C for 45 seconds, 40°C for 2 minutes and 72°C for 3 minutes. The final extension step was performed at 72°C for 5 minutes followed by cooling to 4°C for completion of the programme.

Table 10.1: Primer Sequences Showing Amplification in Somaclone and the Control Plant of Patchouli

Primer	Sequence 5'-3'	Somaclones Showing Amplification	Number of Bands Produced	Number of polymorphic Bands
OPA 03	AGTCAGCCAC	1,2,3,4,5,6,7,8,9,10,11,12,13,15	95	-
OPA 09	GGGTAACGCC	1,2,3,4,5,6,7,8,9,10,11,12,13,14,15,16	78	12
OPA 18	AGGTGACCGT	1,2,3,4,5,6,7,8,9,10,11,12,13,14,15,16	87	15
OPD 08	GTGTGCCCCA	1,2,3,4,5,6,7,8,9,10,11,12,13,14,15,16	66	6
OPG 01	CTACGGAGGA	1,2,3,5,6,7,9,10,11,12,13,14	41	-
OPG 09	CTGACGTCAC	1,2,3,4,5,6,7,8,9,10,11,12,13,14,15,16	165	4
OPG 11	TGCCCGTCGT	1,2,3,4,5,6,7,8,10,11,12,13,14,15	30	-
OPZ 13	GACTAAGCCC	1,2,3,5,6,7,9,10,11,12,13,14,15,16	87	1

The PCR product was separated on 1 per cent agarose gel 0.5 X TBE buffer. The gel was visualized using ethidium bromide stain and photographed in gel documentation system. The bands were identified based on intensity and molecular weight marker (100 bp ladder).The various morphological characters such as plant height, number of branches per plant, leaf area, herb yield per plant and DNA content of the somaclones were recorded (Table 10.2).

All reactions were repeated 3-4 times and the bands that were bright and reproducible were only scored for the analysis.

Results and Discussion

Callus cultures were initiated and maintained on MS basal medium containing 10.76 µM NAA and 2.22 µM BAP and then transferred to medium containing 4.44 µM BAP to induce multiple shoot formation and further growth (Figure 10.1). This agrees with the findings of Mishra (1996). Then the plants were transferred to MS medium containing IBA 2.0µM for root formation. Plants regenerated from leaf callus were acclimatized (Figure 10.2) and successfully transferred to green house with 85 per cent survival frequency (Hembrom *et al.*, 2006, Behera and Thirunavoukkarasu, 2006, Sharma, 1999). After 30 days of hardening, plants were transferred to field. Six month old field grown plants were compared with that of the same age control with regard to morphological characters such as plant height, number of branches, leaf area and herb yield per plant in the first generation SC$_1$ (first generation after an *in vitro* phase).

Table 10.2: Morphological and Biochemical Parameters in Control and Somaclones of *Pogostemon cablin* Johor Variety

Somaclones	Plant height (Inch)	No. of Branches	Leaf Area (cm²)	Yield/Plant (gm)	Moisture (per cent)	DNA Content (ng)
Control	26.6± 1.01	5.6±0.8	70±3.4	185±7.0	76.5±0.5	103±8.3
Tall plant with few branches and thick leaf	35±3.2	5.6±1.01	104.9±11.3	228±11.3	79.8±2	100.4±2.8
Tall plant with more branches and broad leaf	33.6±2.4	5.4±1	113.6±7.7	268±8.6	78.6±1.04	112.4±3.2
Dwarf plant with more branches and normal leaf	21.4±3.8	9.2±0.4	79.2±6.4	227±9.9	80±0.5	97.4±1.01

Figure 10.1: Multiple Shoot Differentiation from Callus

Figure 10.2: Acclimatized *in vitro* Plant

Figure 10.3: Tall Plant with Few Branches and Thick Leaf

Figure 10.4: Tall Plant with More Branches and Broad Leaf

Figure 10.5: Dwarf Plant with More Branches and Normal Leaf

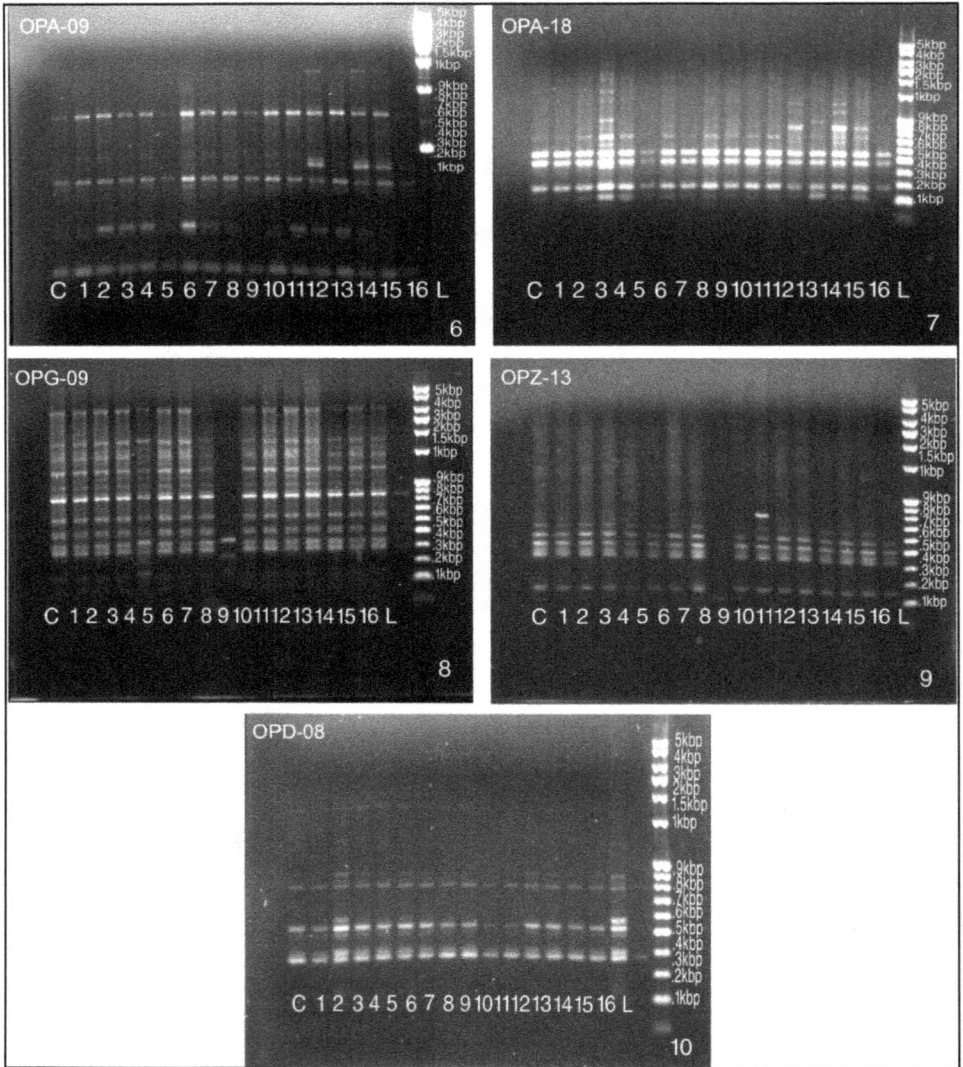

Figure 10.6: RAPD with Primer OPA-09

Figure 10.7: RAPD with Primer OPA-18

Figure 10.8: RAPD with Primer OPG-09

Figure 10.9: RAPD with Primer OPZ-13

Figure 10.10: RAPD with Primer OPD-08

Lane: 1: Control plant; 2–6: Tall plant with more branches and broad leaf; 7–11: Tall plant with few branches and thick leaf; 12–17: Dwarf plant with more branches and normal leaf; 18: Ladder (100 bp)

Three different types of somaclones were isolated out of 100 regenerated plants. These are (1) tall plant with few branches and thick leaf (Figure 10.3), (2) tall plant with more branches and broad leaf (Figure 10.4) and (3) dwarf plant with more branches and normal leaf (Figure 10.5). The present result is in agreement with the findings of Kavyasree *et al.* (2005), Ravindra *et al.* (2004), Gayatri and Kiran (2003), Seeta *et al.* (2000)

The number of bands produced by each primer in the somaclones and control is depicted in Table 10.1. Among the 8 operon primers used for amplification, 5 primers such as OPA-09, OPA-18, OPG-09, OPZ-13 and OPD-08 showed polymorphic bands. In the other 3 primers such as OPA-03, OPA-01, OPG-11, amplification was found to be monomorphic. A total of 649 bands were scored, of which 38 bands were polymorphic for the somaclones. The band size ranged between 100 bp to 3 kbp. The maximum and minimum size of the bands was ranged between 3 kbp in OPG-09 primer to 100 bp in OPA-09 primer respectively.

In primer OPA-09, 7 unique bands were noticed in dwarf plant with more branches and normal leaf, 3 unique bands in tall plant with more branches and broad leaf and 2 unique bands in tall plant with few branches and thick leaf (Figure 10.6). In primer OPA-18, 7 unique bands were observed and intensity was more in 2 bands of molecular weight 300 bp and 450 bp in tall plant with more branches and broad leaf, 8 unique bands were found in case of dwarf plant with more branches and normal leaf (Figure 10.7). In primer OPG-09, 2 polymorphic bands were in tall plant with more branches and broad leaf and 2 polymorphic bands in tall plant with few branches and thick leaf (Figure 10.8). In primer OPZ-13 only 1 unique band were noticed in tall plant with few branches and thick leaf (Figure 10.9). In primer OPD-08, 3 unique bands were noticed in tall plant with more branches and broad leaf and 3 unique bands in dwarf plant with more branches and normal leaf (Figure 10.10). The present finding confirmed the fact that variation occurred as a result of mutation during micropropagation as reported by Rani *et al.* (1995).

The aim of the present study was to provide polymorphic RAPD markers for detection of somaclonal variations in tissue culture derived plants. Various studies indicate that *in vitro* culture produces abundant cytological anomalies affecting both structural and numerical chromosome constitutions which can be correlated with many phenotypic abnormalities as reported by Evans and Sharp (1986). The occurrence of variation in morphological characters was mostly because of regeneration from callus phase thus confirms the influence of undifferentiated cell phase on promoting somaclonal variation.

The analysis of banding pattern revealed sufficient information to estimate the genetic variability among different genotypes. Polymorphism among the somaclones showed diffeent banding pattern suggesting the occurrence of a single mutation. The polymorphism in the amplification products may be either from changes from in the sequence of the primer binding site (*e.g.* point mutation) or changes which alter the size or prevent the successful amplification of the target DNA (*e.g.* insertion, deletions; inversions) as suggested by Rani *at al.* (1995). The better understanding of genetic variation at the intraspecific level help in identifying superior genotypes for crop improvement.

Acknowledgement

Financial assistance provided by Council of Scientific and Industrial Research (CSIR), New Delhi to one of the author, Asmita Behera is gratefully acknowledged.

References

Behera, A. and M.Thirunavoukkarasu. 2006 *In vitro* micropropagation of *Desmodium gangeticum* (L.) DC. through nodal explants. Indian J. Plant Physiol.11(1): 83-88.

Damasco, O.P., Godwin, I.D., Smith, M.K., Adkins, S.W.1996. Gibberlic acid detection of dwarf off –types in micropropagated Cavendish bananas. Aust.J.Expt.Agric. 36:237-241.

Evans,D.A. and Sharp, W.R., Medina-Filho, H.P. 1984. Somaclonal and gametoclonal Variation. Am J Bot 71: 759-794.

Evans,D.A. and Sharp, W.R.1986. Somaclonal and gametoclonal variation. In:Evans, D.A.,Sharp, W.R. and Ammirato, P.V. (eds.), Handbook of plant cell culture, techniques and applications, Macmillan, New York, 4:97-132.

Gayatri, M.C. and Kiran, S. 2003. Somaclonal variation in Ginger- A new report under *in vitro*. J.Cytol Genet 4(NS): 75-78.

Hembrom, M.E., Martin, K.P; Suresh Kumar Patchathundikandi and Joseph Madassery. 2006 Rapid *in vitro* production of true-to type plants of *Pogostemon heyneanus* through dedifferintiated axillary buds. *In Vitro* Cell. Dev.Biol-Plant. 42: 283–286.

Isabel, N.I; Tremblay, M.M, ; F.M; Bousquet, J. 1993. RAPD as an aid to evaluate the genetic integrity of somatic embryogenesis derived population of *Picea mariana* (Mill)   B.S.P. Theor. Appl.Genet. 86: 81-87.

Jha, T.B., Jha, S. and Sen, S.K. Somatic embryogenesis from immature cotyledons of an elite Darjeeling tea clone. Plant Sci. 84: 209-213.

Kavyashree, R. 2008. Molecular and Morphological characterization of seven varieties of *Zingiber officinale*. J.Cytol.Genet. 9 (NS): 101-107

Mishra, M. 1996. Regeneration of patchouli (*Pogostemon cablin* Benth.) plants from leaf and node callus, and evaluation after growth in the field. Plant Cell Reports. 15: 991-994.

Murashige, T. and Skoog, F. 1962. A revised medium for rapid growth and bioassays with tissue cultures. Physiol. Plant. 15: 473-497.

Rani, V; Ajay, P; Raina, S.N.1995. Random amplified polymorphic DNA (RAPD) markers for genetic analysis in micropropagated plants of *Populus deltoides* Marsh. Plant Cell Rep. 14: 459-462.

Ravindra, N.S., Kulkarni, R.N., Gayatri, M.C.and Ramesh, S. 2004. Somaclonal variation for some morphological traits, herb yield, essential oil content and essential oil composition in an Indian cultivar of rose -scented geranium. Plant Breeding. 123: 1-5.

Seeta, P., Talat, K. and Anwar, S.Y. 2000. Somaclonal variation- An alternative source of genetic variability in Safflower. J.Cytol.Genet. 1 (NS): 127-135.

Sharma, N. 1999. Conservation of patchouli (*Pogostemon patchouli*) through *in vitro* methods. Indian Perfumer 43 (1): 19-22.

Valles, M.P., Wang, Z.Y., Montana, P., Portykos, I.and Spangenberg, G.1993. Analysis of genetic stability of plants regenerated from suspension cultures and protoplasts of meadow fescue (*Festuca pratensis* Huds). Plant Cell Rep. 12:101-106.

Vuylsteke, D., Swennen, R., Wilson, G. F., Langhe, E.D. 1988. Phenotypic variation among *in vitro* propagated plantain (*Musa* sp. Cultivar ABB). Sci. Hort. 36:79-80.

Chapter 11

Somatic Embryogenesis in *Sauropus androgynous* (L.) Merr.

D.H. Tejavathi and S. Padma

Department of Botany, Jnanabharathi, Bangalore University,
Bangalore – 560 056, Karnataka

ABSTRACT

A successful protocol for *in vitro* regeneration of *Sauropus androgynous via* somatic embryogenesis using different explants has been developed. Somatic embryos were derived from the embryogenic callus derived from shoot tip, node and leaf cultures on Phillips and Collins medium supplemented with NAA at various concentrations ranging from 5.3µM to 26.85µM within two months of inoculation. Thus obtained embryos were subcultured to modified Phillips and Collins media supplemented with increased vitamin level for further growth. Somatic embryos with well-developed cotyledons were transferred to normal and modified L_2 basal medium for conversion. The plantlets thus obtained were subjected to brief acclimatization before transferring them to land. About 95 per cent of survival was recorded. Effect of induced stress on somatic embryogenesis is discussed.

Keywords: Multivitamin, Modified L_2 medium, Somatic embryos and Sauropus androgynous.

Introduction

Somatic embryogenesis is a remarkable illustration of the dictum of plant totipotency. *Sauropus androgynous*(L.) Merr. commonly known as "Star Goose Berry" is a member of family Euphorbiaceae. It is popularized as a "Multivitamin plant" and consumed as a leafy green vegetable due to its high nutritional value (Bender and Ismail, 1975; Padmavathi and Rao, 1990; Hemalatha *et al.*, 1999; Asmah-Rahmat *et al.*, 2003). It is also a promising medicinal plant which has many therapeutic values including effective for weight reduction, in controlling hypertension, gynecological problems, gallstones, diabetes and constipation (Sai and Srividhya, 2002).The leaf extract is effective towards some breast cancer cell lines (Asmah-Rahmat *et al.*, 2003). Inspite of being nutritious and medicinal it is also cautioned of being toxic by causing "Bronchiolitis Obliterans" on excessive consumption (Lin-Tzengjih *et al.*, 1996; Liao-Xuekun *et al.*, 1996 and Luo Ping Ge *et al.*, 1997). The damage was so severe that there were even cases of lung transplantation. The toxicity is believed to be associated with an alkaloid called "papaverine" (Guo-Juo Xian *et al.*, 2005). Due to these drawbacks the consumption level has decreased which in turn has affected the commercial cultivation making the *Sauropus androgynous* commercially and regionally threatened.

Conventionally the plant is propagated vegetatively using cuttings but recombination breeding for the occurrence of genetic variation and production of better variety of *Sauropus* is handicapped by negligible amount of fruit set, poor seed longevity and short viable period of seed. Hence an attempt for *in vitro* propagation via somatic embryogenesis has been carried out as the process affords high multiplication rates and results in propagules which possess both root and shoot axes. Further, this can be combined with other techniques like genetic transformation, protoplast fusion, mutagen treatment etc., which can open new ways for production of safer and innovative variety of *S. androgynous*.

In this paper, we describe a comparative study of both direct and indirect somatic embryogenesis of *Sauropus androgynous*. Though there are reports of organogenesis by Philomena in 1993 and Tejavathi *et al.*, in 2010, there are no reports of somatic embryogenesis. Hence, this present research work was aimed to standardize the simple reliable protocol for direct and indirect somatic embryogenesis in *S. androgynous*. Further, NAA supplemented medium was modified by increasing the vitamin contents and the effect on induction and maturation of somatic embryos was studied.

Material and Methods

Explant Sterilization and Inoculation

Explants (Shoot tips: 0.75-1.0 cm, leaves: 1.5-2.0 sq cms and nodal segments: 1.-1.5 cms) were selected from healthy plants which were procured from GKVK Campus, Bangalore and grown in polyhouse in Department of Botany, J.B. Campus, Bangalore University. Third to sixth branches from the tip of the plant were excised for explants and were surface sterilized using Tween -20 for 15 min followed by freshly prepared 0.1 per cent (w/v) Bavistin for 10 min. These surface sterilants were removed by repeated washing for 2-3 times with tap water and distilled water respectively under lab condition. The explants were further sterilized using 0.1 per cent (w/v) Mercuric chloride for 10 min followed by repeated washings using sterile distilled water for 2-3 times to remove traces of mercuric chloride under aseptic condition. The explants were trimmed and were inoculated on nutrient medium.

Growth Medium

Callus Induction Medium (CIM):

The nutrient media selected for the present studies were Murashige and Skoog's (MS) medium (1962) and Phillips and Collins (L_2) medium (1979) supplemented with various auxins like NAA, 2,4-D, IAA and IBA at various concentrations.

Embryo Induction and Maturation Medium (EIMM)

L_2 medium supplemented with NAA at concentration ranging from 5.37 µM-26.85 µM and IAA ranging from 5.71 µM -28.55 µM were found to be optimal for induction of embryogenic callus and direct embryos respectively.

The L_2 medium supplemented with NAA 26.85 µM was further modified by increasing the amount of vitamins *i.e.*, stock –D by 5,10,15, 20 and 25 times from the normal level and termed as modified L_2 media *i.e.*, M L_2(5)- 5 times increased vitamin level, M L_2 (10)-10 times increased vitamin level, M L_2(15)-15 times increased vitamin level, M L_2(20)- 20 times increased vitamin level and M L_2(25)-25 times increased vitamin level were used for the development of somatic embryos.

Embryo Conversion Medium (ECM)

For conversion of somatic embryos into plantlets the media used were normal L_2-Basal medium and modified L_2-Basal medium with 10 times increased level of vitamin *i.e.*, M L_2 (10)-Basal medium.

All the above media were fortified with 3.0 per cent (w/v) sucrose as carbon source and 0.9 per cent (w/v) agar or 0.2 per cent (w/v) clarigel as gelling agent. pH of the media was adjusted to 5.6-5.8 before autoclaving at 108 kpa for 15 min. The cultures were incubated at 25±2°C under cool fluorescent tube lights with 16hrs light and 8 hrs dark cycles at light intensity of 25 µmol $m^{-1}s^{-1}$.

Results and Discussion

In the present study a reliable method to induce somatic embryogenesis from the cultures of *S.androgynous* was established. To our knowledge, this is the first report on somatic embryogenesis in *S.androgynous* wherein the key factors influencing the embryogenic potential of explants is analyzed. The important factor that determines the success for establishing the embryogenic protocol in *S.androgynous* is choice of explant. The choice of explant is analyzed on the basis of explant type (shoot tip, leaf or nodal explants), texture of explants, season and developmental phase of explant donor plant and orientation of explant.

Primarily, the season and developmental phase (vegetative/ reproductive) of explant source plant/ donor plant played a vital role in explant contribution for induction of somatic embryogenesis. The explants selected during the month of April, May, June, November and December showed better response for somatic embryogenesis which may be attributed to the non-flowering or vegetative phase of donor plant. Explants selected for inoculation did not respond or poorly responded for somatic embryogenesis in the month of August, September when the flowering is at its peak. Instead they produced rhizogenic callus which proved to be the major obstacle in the development of somatic embryos. Such results, where the low frequency of somatic embryogenesis coincided with flowering was reported by Santana *et al.*, in 2004 in coffee clones.

In order to select the appropriate explant as a potential initial material for induction of embryogenic callus in *S.androgynous,* the explants were always selected from 3-6th branch from the tip of donor plant. Nearly mature leaves responded well for callusing compared to young and old leaves. Further, the texture of nodal explants should be soft and flexible for embryonic calli induction.

Orientation of explant specifically nodal explants on the nutrient media also plays a role in induction of callus. A vertical orientation in which the basal half of the stem was inserted into the medium demonstrated much reduced stem necrosis and better survival leading to callus formation. While the explants which were placed horizontally on the medium showed stem necrosis and very poor survival. Such effect of orientation of explants on the morphogenetic potential was studied earlier in *Euphorbia pulcherrima* and *Euphorbia lathyris* by Yogesh *et al.* (2003) and Ripley and Preece (1986) respectively.

Above all, the nutrient medium formulation and growth regulators act as key ingredients for induction of somatic embryogenesis. Generally auxins are known to be essential for the induction of somatic embryogenesis and 2,4-D is the most commonly used auxin (Ammirato, 1983). Further, a combination of 2,4-D or NAA with cytokinin was also reported to be essential for the induction of somatic embryos in several taxa (Gingas and Lineberger, 1986; William and Maheshwaran, 1986). Certain cells may need simple MS medium for the induction and further development of somatic embryos (Jasrai *et al.*, 1999). But the results of the present study showed that L_2 medium supplemented with NAA alone at various concentrations is sufficient for yielding embryogenic calli compared to MS medium. Further, the increased vitamin level in modified L_2 medium to particular level also enhanced the yield of somatic embryos to maximum. All the above mentioned factors with extremely wide range of manipulations in selection of explants and media formulations are must for somatic embryogeneic process to happen.

Induction of Callus Tissue

On the callus induction medium (CIM) callus initiation occurred on sides of the cut ends of explants on both MS and L_2 medium supplemented with different concentrations of auxins like NAA, 2,4-D, IAA and IBA. Callus initiation was observed within 10 days in the presence of NAA and 2,4-D. Whereas, on IAA it took more than 15 days, while on IBA induction of callus was delayed for a month. Frequency of callus was better on L_2 media compared to MS media. L_2 media favoring for enhanced callus growth in comparison with MS media was earlier reported by Tejavathi *et al.* (2010) in *Macrotyloma uniflorum*. Profuse callus was formed on NAA supplemented medium followed by 2,4-D and IAA supplemented media but callus induction was comparatively low or nil on IBA supplemented media. Leaf explants promoted maximum amount of callus followed by node and shoot tip explants.

Colour and Texture of Callus

The quality and type of callus produced on different growth regulators (auxins) varied. The callus induced from explants on media containing 2,4-D showed compact hard with rough velvety surfaced creamish to green callus. While explants on IAA and NAA supplemented media produced predominantly white to creamish callus and also green coloured soft, translucent granulated to nearly nodulated callus (Figure 11.1). Intermingling with this there also occurred much softer cottony friable patches of callus. Whereas, explants on IBA supplemented media produced no or little browner callus which was moderately harder compared to callus raised on 2,4-D supplemented medium.

Rhizogenic Callus

Rhizogenic callus was observed from all explants at all auxins levels but maximum rhizogenic callus was observed on medium supplemented with NAA followed by 2,4-D and IAA respectively. Similar callus induction was observed in *Papaver somniferum album* by Kassem in 2001. Leaf explants promoted higher percentage of rhizogenic callus followed by nodal and shoot tip explants mainly on medium supplemented with NAA, particularly at concentrations 21.48µM and 16.11µM. Further this rhizogenic callus did not contribute to plantlet regeneration instead, on subculture, continued growth for period of time and ultimately turned brown. To our observation it was noticed that about 90 per cent of explants on inoculation promoted for rhizogenesis when they were collected during reproductive phase of explant donor plant or mother plant.

Indirect Somatic Embryogenesis

Embryo Induction and Maturation

The callus induced on CIM was allowed to grow on the same media for 40-45 days. Further the same cultures were subjected to stress by keeping them without subculturing for 50-60 days for induction of somatic embryos. Same was favored for the induction of somatic embryos in Ginger (Lincy *et al.*, 2009).

In vitro culture conditions represent an unusual combination of stress factors that plant cell encounters. The stress associated with *in vitro* induction of somatic embryo may result in chromatin reorganization. Increasing number of studies (Steward *et al.*, 2002; Williams *et al.*, 2003; Law *et al.*, 2005) demonstrated that release of the somatic embryo is highly organized by patterns of DNA methylation/ demethylation and histone acetylation/ deacetylation during cellular dedifferentiation/ differentiation stages. All these pronounced changes in the cellular environment may generate stress effects, such as exposing wounded cells or tissues to sub optimal nutrient or Plant Growth Regulators (PGRs).

PGRs act as central signals to reprogram somatic cells towards embryogenic pathways by controlled chromatin remodeling and gene expression (Dudits *et al.*, 1995; Pasternak *et al.*, 2002; Feher *et al.*, 2003; Gaj, 2004; Feher, 2005, 2008; Jimenez and Thomas. 2005; Thomas and Jimenez. 2005). Auxins are considered to be the most important PGRs that regulate somatic embryo induction. Mainly 2,4-D and higher levels of other auxins, such as 1- naphthalene acetic acid (NAA) and Indole-3-acetic acid (IAA) are found to bring about chromatin remodeling for induction of somatic embryos. However, most importantly the endogenous content and the supplemented exogenous auxins are both determining factors during the somatic embryo induction

(Thomas and Jimenez. 2005). This is because, for the couple of reasons, the endogenous auxin level appears to be an inadequate marker to possess embryogenic potential. Thus exogenous supplementation of auxins is necessary to apply stress for somatic embryo induction process.

Further, there should be critical interaction or specific balance between the endogenous and supplemented hormones (auxins) level for a cell to undergo stress and to possess totipotency for induction of somatic embryos. If this is disturbed the cell may express for only caulogenesis or rhizogenesis and may sometimes shows other abnormalities or deviations. In the present condition, the flowering phase of the donor plant may disturb this specific balance of endogenous and supplemented auxins and trigger most of the explant cells for rhizogenesis.

The induction of embryo was observed only on L_2 medium supplemented with NAA (Figures 11.2 and 11.3). Though no reports of somatic embryo development in *S.androgynous* is available so far but somatic embryogenesis has been reported in other plants mainly on MS medium supplemented with NAA either alone or in combination with BAP (Chand and Singh, 2001; Sahrawat and Chand, 2001; Jasrai *et al.*, 1999; Ikram-Ul-Haq, 2005; Gliozeris *et al.*, 2006; Sharry *et al.*, 2006 and Naz *et al.*, 2005).

The mass of translucent callus produced on NAA at all concentrations produced few cotyledonary notches which were white to green in colour. The frequency and average number of somatic embryo induction on different explants at various concentrations of NAA on analysis showed better somatic embryo induction from leaf explants (Table 11.1).

Table 11.1: Effect of Various Concentrations of NAA in L_2 Media on the Induction of Somatic Embryos from Different Cultures (Data after 12 subcultures)

Sl.No.	Concentration of NAA on L_2 Medium (µM)	Leaf Explant Mean ± S.E.	Nodal Explant Mean ± S.E.	Shoot Tip Explant Mean ± S.E.
1.	5.37	64.8 ± 0.3976	52.2 ± 0.3556	19.6 ± 0.3376
2.	10.74	139.4 ± 0.4262	84.2 ± 0.3976	31.8 ± 0.3610
3.	16.11	375.6 ± 0.4385	272.3 ± 0.4053	156.2 ± 0.3712
4.	21.48	439.2 ± 0.4842	333.4 ± 0.4553	202.2 ± 0.4443
5.	26.85	568.3 ± 0.5087	437.2 ± 0.4949	281.2 ± 0.4907

Further this embryogenic callus was subcultured (approximately 1-1.10 gm.) on to different modified L_2 media with varied levels of vitamins as the vitamins play a vital role in somatic embryogenesis (Pullman *et al.*, 2006.). The modification criteria were solely based on increasing vitamin levels and the increased vitamins help in carbohydrate metabolism and biosynthesis of aminoacids which are beneficial for the growth and quality of somatic embryo. This is in conformation with Robichand *et al.* (2004) who elucidated the significance of amino acids in maturation and germination of somatic embryos in American chestnut. Visual observation showed comparatively enhanced growth of embryogenic callus with more number of somatic embryos on ML_2(10) media. Further, there was gradual detoriation in the growth of embryogenic callus and the number of embryos when vitamin level was increased. At ML_2 (25) *i.e.*, 25 times increased level of vitamin media lead to development of more dense, necrotic and less embryogenic callus. 26.85 µM concentration of NAA in the media was found to be conducive for the increased number of somatic embryo formation (Table 11.2). Same concentration of NAA was found to be optimum for induction of somatic embryos in *Pimpinella pruatjan* (Roostika *et al.*, 2007).

Table 11.2: Effect of L_2 Media and Different Modified L_2 Media Supplemented with NAA at 26.85 µM on Biomass of Embryogenic Callus and Number of Somatic Embryos from Leaf Explants

Sl.No.	Medium	Biomass of Fresh Callus at the End of 40 Days (gms) Mean ± S.E.	Number of Somatic Embryos Formed Mean ± S.E.
1.	L_2-NAA(26.85 µM)	5.78 ± 0.1249	12.7 ± 0.3164
2.	L_2(5)-NAA(26.85 µM)	6.27 ± 0.1768	22.7 ± 0.3189
3.	L_2(10)-NAA(26.85 µM)	9.28 ± 0.1942	40.8 ± 0.4196
4.	L_2(15)-NAA(26.85 µM)	8.87 ± 0.1377	16.8 ± 0.2845
5.	L_2(20)-NAA(26.85 µM)	8.80 ± 0.1705	10.8 ± 0.1985
6.	L_2(25)-NAA(26.85 µM)	8.97 ± 0.1524	5.4 ± 0.2597

Addition of vitamins to particular level to the medium significantly enhanced the rapid formation of somatic embryos, from the callus. The enhanced response of somatic embryos induction was maximum when medium was co-supplemented with 10 times increased level of vitamins along with NAA. The rate of induction of somatic embryos increased linearly with internal vitamin concentration thus confirming that the induction response by vitamin was manifested as an osmotic rather than a nutrient factor. Vitamins serve as a nitrogen source. When it is supplied up to 10 times increased level exerts mild osmotic stress responsible for driving embryogenesis. Further, increase in vitamins beyond desirable level was responsible for high stress causing cells to become necrotic and ultimately death of cells by not tolerating the extreme stress level.

Further, some somatic embryos were produced with wider hypocotyls and cotyledons and developed asynchronously. These structures further failed to transform into complete plantlet. Such embryos were observed in all media with maximum number in ML_2 (15) and ML_2 (20). Similar types of embryos were reported by Faure *et al.* (1991) in Muscadine grape cultivar triumph.

Direct Somatic Embryogenesis

Very low frequency *i.e.*, about 3 per cent of leaf explants inoculated on L_2 medium supplemented with IAA 17.13µM, 22.84µM and 28.55µM showed direct somatic embryogenesis with dark green cotyledons and hypocotyls developed within 40-45 days of culture. Ikram-Ul-Haq in 2005 reported direct somatic embryogenesis on IAA supplemented media in *Gossypium constarricense*.

Figure 11.1: Embryogenic Callus on L$_2$-NAA (26.85 µM)

Figures 11.2–11.3: Induction and Maturation of Embryo
on L$_2$-NAA (26.85 µM) and L$_2$(10)-NAA (26.85 µM) medium

Figure 11.4: Rhizogenic Callus on L$_2$-NAA (26.85 µM)

Figure 11.5: Heart Shaped Embryo

Figure 11.6: Torpedo Shaped Embryo

Figure 11.7: Conversion of Embryo into Plantlet ML$_2$ (10)–Basal Medium

Figure 11.8: Hardened Plantlets in Different Hardening Mixtures: A: Soilrite, B: Soilrite+vermiculate mixture in 3:1 ratio

Figure 11.9: Hardened Plants in Pots Containing Soil, Sand and Manure Mixture

Embryo Conversion

Well-developed somatic embryos were selected and were transferred to conversion media (Figures 11.5 and 11.6). The somatic embryos on ML_2 (10)-Basal medium showed complete development by hypocotyl elongation followed by rooting within 7-10 days of culture. Further, the embryos showed vigorous rooting with tufts of roots and healthier plantlets with branched well developed shoots with greener leaves on the same medium (Figure 11.7). However, in comparison with ML_2 (10)-Basal medium, normal L_2-Basal medium developed no or few roots with weak shoots.

Hardening and Acclimatization

Fully grown plantlets were separated and were transferred to plastic cups containing different hardening mixture consisting of soilrite, vermiculate, vermicompost, cocopeat, sand and garden soil either alone or mixed in different ratio (Figure 11.8). Further these cups were either kept open or were covered with perforated polythene cover and kept moist by regular watering and maintained at room temperature for a month.

The plantlets in soilrite showed better survival rate followed by soilrite + vermiculate mixture in 3:1 ratio and soilrite+vermiculate+cocopeat in 1:1:1 ratio (Figure 11.9). These plants after attaining hard textured stem were then transferred to pots containing 1:2:1 ratio sand, soil and manure. The pots were further maintained in polyhouse for 30-40 days and then transferred to field. The survival rate was about 95 per cent.

Somatic embryogenesis is the preferred *in vitro* regeneration route for plants, as this morphogenetic pathway may increase the number of regenerated plants in comparison with organogenesis. The production of somatic embryos capitalizes upon the totipotency of plant cells and involves the development of bipolar structures resembling zygotic embryos (Dodeman *et al.*, 1997; Tejavathi *et al.*, 2000).

Somatic embryogenesis is the proper way to regenerate plant from single somatic cell and opens up possibility to understand process of cell cycle reprogramming from somatic to embryogenic type, cloning and characterization involving in wounding, hormone activation, cell division, differentiation and developmental processes. The sequence of events for somatic embryogenesis as a morphogenic phenomenon is frequently expressed as discrete phases or steps. These phases are characterized by distinct biochemical and molecular events (Suprasanna and Bapat. 2005). The first phase of somatic embryogenesis is the induction stage in which differentiated somatic cells acquire embryogenic competence either directly (without a dedifferentiation step) or indirectly (by dedifferentiation and usually involving callus phase). After the appropriate stimulus, this phase is followed by the expression or imitation of somatic embryo in which competent cells or proembryos start developing. Finally, during maturation, somatic embryos anticipate germination by desiccation and reserve accumulation (Jimnez. 2001).

Somatic embryo must therefore consist of replacement of the existing pattern of gene expression in the explant tissue for a new embryogenic gene expression program (Chugh and Khurana. 2002; Zeng *et al.*, 2007). This is only possible if the cells are both competent and receive the appropriate inductor stimuli. For all these to happen several factors should be imposed by *in vitro* conditions. Thus by the drastic changes in the cellular environment of the *in vitro* culture, induced by a 'stressor' in the culture medium or the physical environment of the culture responsible for reprogramming of gene expression resulted in significance success in obtaining embryo formation, regeneration and plantlet formation in *S.androgynous*. Based on the present work it could be said that somatic embryo as a cell response to exogenously applied stressors.

Though, further work is required in specific areas as genomics, proteomics and metabolomics to clarify the role of stress in somatic embryo induction in *S.androgynous*. But the present studies can be exploited for regeneration of plantlets from somatic embryos on simple medium which might be used for the production of somaclones of plants and for the storage and maintenance of germplasm. Further, it could be of practical application for raising hybrid seedlings of difficult crosses and mutagenesis *in vitro*. It can also be applied for plant transformation through biotechnological approaches.

References

Ammirato, P.V. 1983. Embryogenesis. In: Handbook of Plant Cell Culture. Vol. 1, Evans, Sharp, W.R., Ammirato and Yamada (Eds.), Mac Millan Publishing Co., New York, pp. 82-123.

Andre's, M. Gatica, Griselda Arrieta, Ana, M. Espinoza. 2008. Direct somatic embryogenesis in *Coffea arabica* L. Cvs. *Caturra* and *Catuai*: Effect of Triacontanol, light condition and medium consistency. Agronomia constarricense. 32(1): 139-149.

Asmah-Rahmat, Vijay kumar, Loo-Mei Fong, Susi-Endrini and Huzaimah-abdullah-Sani. 2003. Determination of total antioxidant activity in three types of local vegetables shoots and the cytotoxic effect of their ethanolic extracts against different cancer cell lines. Asia pacific journal of Clinical nutrition. 12(3): 292-295.

Bender, A.E and Ismail, K.S. 1975. Nutrititive value and toxicity of a Malaysian food, *Sauropus androgynous*. Plant Foods Man. 139-143.

Chand, S., and Ajay Kumar Singh. 2001 Direct somatic embryogenesis from zygotic embryos of a timber yielding leguminous tree. *Hardwickia binata* Roxb. Curr. Sci. 8(7).

Chugh, Archana and Khurana Paramjit. 2002. Gene expression during somatic embryogenesis- recent advances. Curr. Sci. 83(6): 715-739.

Dodeman, Valrie Laurence, Ducreux, Georges, Kreis and Martin. 1997. Zygotic embryogenesis versus somatic embryogenesis. J. Exp. Bot. 48(7): 1493-1509.

Dudits, Denes, Gyorgyey, Janos, Bogre, Laszlo and Bako. 1995. Molecular biology of somatic embryogenesis. In: Thorpe, Trevor A. (ed.) *In vitro* embryogenesis in Plants. Kluwer Academic publishers. Dordrecht. The Netherlands. pp. 267-308.

Feher, Attila, Pasternak, Taras,P. and Dudits Denes. 2003. Transition of somatic plant cells to an embryogenic state. Plant Cell Tiss. Org. Cult. 74(3): 201-228.

Feher, Attila. 2005. Why somatic plant cells start to form embryos? In: Mujid, Abdul and Samaj, Josef. (eds.) Somatic Embryogenesis. Plant Cell Monographs, Springer; Berlin/Heidelberg. 2: 85-101.

Feher, Attila. 2008. The initiation phase of somatic embryogenesis. What we know and what we don't? Acta Biologica Szegediensis. 52(1): 53-56.

Faure, O., M. Gengoli., A.Nougarede and N.Bagni. 1991. Polyamine pattern and biosynthesis in zygotic and somatic embryo stages of *Vitis vinifera*. J.Plant physiology. 138 :545-549.

Gaj, Malgorzata D. 2004. Factors influencing somatic embryogenesis induction and plant regeneration with particular reference to *Arabidopsis thaliana* (L.) Heynh. Plant Growth Regulation. 43(1): 27-47.

Gingas, V.M and Lineberger, R.D. 1986. Asexual embryogenesis and plant regeneration in *Quercus*. Plant Cell Tiss. Org. Cult. 17: 191-203.

Gliozeri, S., Egidija venskutoniene, Alfonsas tamosiunas and Laimute stuopyte. 2006. Somatic embryogenesis of *Asarina erubescens* (A. Gray) Pennel. Biologiya. 4: 92-95.

Guo-Ju Xian, Yang-Xian and Guo-Lang Liang. 2005. Studies on the toxicology of *Sauropus*, a wild vegetable in South China. J. South-China-Agric. Univ. 26(4): 10-14.

Hemalatha, G., Sundharaiya, K. and Ponnuswamy, V. 1999. Comparitive analysis of nutritive value in some leafy vegetables. South-Indian-Horticulture. 47(1/6): 295.

Ikram-Ul-Haq. 2005. Callus proliferation and somatic embryogenesis in Cotton. (*Gossypium hirsutum L.*). African J. Biotech. 4(2): 206-209.

Jasrai, Y.T., Chauhan, V.A. and Palmer, J.P. 1999. Plant somatic embryogenesis. In: Recent Trends in Developmental Biology. Gakhar, S.K and Mishra, S.N.(Eds.). Himalayan Publishing House. Mumbai. pp. 193-203.

Jimenez, Victor, M. 2001. Regulation of *in vitro* somatic embryogenesis with emphasis on the role of endogenous hormones. Revista Brasileira de fisiologia Vegetal. 13(2): 196-223.

Jimenez, Victor M. and Thomas, C. 2005. Participation of plant hormones in determination and progression of somatic embryogenesis. In: Mujid, A. and Samaj, J. (eds.) Somatic Embryogenesis: Plant Cell Monographs. Springer; Berlin/Heidelberg. 2: 103-118.

Kassem, A.M. and Annie Jacquin. 2001. Somatic embryogenesis, rhizogenesis and morphian alkaloids production in two species of Opium poppy. Journal of Biomedicine and Biotechnology. 1:2: 70-78.

Law, R., David, Suttle, and Jeffry, C. 2005. Chromatin remodeling in plant cell culture: patterns of DNA methylation and histone H3w and H4 acetylation vary during growth of asynchronous potato cell suspensions. Plant Physiology and Biochemistry. 43(6): 527-534.

Liao-Xuekun and Li-Yong Hua. 1996. Fatty acid composition of the seed oil from *Sauropus androgynous* (L.) Merr. Journal of Tropical and Subtropical Botany. 4(3): 70-71.

Lin-Tzengjih, Lu-Chong Chen, Chen-Kuan Wen and Deng-Jou fang. 1996. Outbreak of obstructive ventilator impairement associated with consumption of *Sauropus androgynous* vegetable. Journal of Toxicology. Clinical Toxicology. 34(1): 1-8.

Lincy, A.K, Azhimala, B. Remashree and Bhaskaran Sasikumar. 2009. Indirect and Direct somatic embryogenesis from aerial stem explants of ginger (*Zingiber officinale* Rosc.) Acta. Bot. Croat. 68(1): 93-103.

Luo-Ping Ger, Ambrose, A., Chiang Ruay-Sheng Li, Su-Meichen and Ching-Jiunn Tseng. 1997. Association of *Sauropus androgynous* and Bronchiolitis Obliterans Syndrome; A hospital based case control study. American Journal of Epidemology.145 (9): 842-849.

Murashige, T and Skoog, F. 1962. A revised medium for rapid growth and bioassays with tobacco tissue cultures. Physiol. Planta. 15: 473-497.

Naz, S., Amir Ali, Fayyaz Ahmed Siddique and Javed, Chang-Sheng Wang. 2005 Optimization of somatic embryogenesis in suspension cultures of Horse gram (*Macrotyloma uniflorum* (Lam.) Verdc.) – A hardy grain legume. Sci. Hort., 106: 427-439.

Padmavathi, P. and Prabhakara Rao. 1990. Nutritive value of *Sauropus androgynous* leaves. Plant food for human nutrition. 40(2): 107-113.

Phillips, G.C. and Collins, G.B. 1979. *In vitro* tissue cultures of selected legumes and plant regeneration from callus cultures of *red cloves*. Crop.Sci. 19:59-64.

Philomena, P.A. 1993. Micropropagation of *Sauropus androgynous* (L.) Merr. Multivitamin plant. Advances in plant sciences. 6(2): 273-278.

Pullman, G.S., Chopra, P. and Chase, K.M. 2006. Loblolly Pine (*Pinus taeda* L.) Somatic embryogenesis: Improvements in embryogenic tissue initiation by supplementation of medium with organic acids, vitamins B_{12} and E. Scientia Horticulturae. 170: 648-658.

Ripley, K.P. and Preece, J.E. 1986. Micropropagation of *E. lathyris* Linn. Plant cell tissue and organ culture. 5: 213-218.

Robichand, R.L., Lessard, V.C and Merckle, S.A. 2004. Treatments affecting maturation and germination of American chestnut somatic embryos. Journal of Plant Physiology. 161: 957- 969.

Roostika, R., Purnamaning Sih., Darwati, I. and Mariska, I. 2007. Regeneration of *Pimpinella pruatjan* through somatic embryogenesis. Indonesian journal of Agricultural sciencies 8 (2): 60-66.

Sai, K.S and Srividya, N. 2002. Blood glucose lowering effect of the leaves of *Tinospora cordifolia* and *Sauropus androgynous* in diabetic subjects. Journal of Natural remedies. 2(1): 28-32.

Sahrawat, A.K. and Suresh Chand. 2001. Continuous somatic embryogenesis and plant regeneration from hypocotyl segments of *Psoralea corylifolia* Linn., an endangered and medicinally important Fabaceae plant. Curr. Sci. 81(10): 1328-1331.

Sharry, S.E., Jaime, A. and Teixeira da silva. 2006. Effective organogenesis, somatic embryogenesis and salt tolerance induction. *In vitro* in the Persian Lilac Tree (*Melia azedarach* L.). Floriculture, Ornamental and plant Biotechnology. 2: 317-324.

Santana, N., Conzalez, M.E., Valcarcel, M., Canto-Flick, A., Hernandez, M.M., Fuentes-Cerda, C.F.J., Barahona, F., Mijangos-Cortes, J., and Loyola-Vargas, V.M. 2004. Somatic Embryogenesis: A valuable alternative for propagating selected Robusta Coffee (*Coffea canephora*) clones. *In vitro* Cell. Dev. Biol.-Plant. 40: 95-101.

Steward, Nicolas, ITO, MIkako, Yamaguchi, Yube, Koizumu, Nozumo, Sano and Hiroshi. 2002. Periodic DNA methylation in maize nucleosomes and demethylation by environmental stress. Journal of Biological Chemistry. 226(6): 1449-1458.

Suprasanna, P. and Bapat, V.A. 2005. Differential gene expression during somatic embryogenesis. In: Mujib, A. and Samaj, J. (eds.) Somatic embryogenesis, Plant Cell Monographs. Berlin; Springer-Verlag. 2: 305-320.

Tejavathi, D.H., Devraj, V.R., Savitha, Murthy, M., Anitha, P. and Nijagunaiah, R. 2010. Regeneration of multiple shoots from the callus cultures of *Macrotyloma uniflorum* (Lam.) Verdc. Indian J. Biotech.. 9: 101-105.

Tejavathi, D.H., Padma, S., Gayatramma, K. and Pushpavathi, B. 2010. *In vitro* studies in *Sauropus androgynous* (L.) Merr. ISHS Acta Horticulturae. 865: 371-375.

Thomas, Clement and Jimenez Victor, M. 2005. Mode of action of plant hormones and plant growth regulators during induction of somatic embryogenesis: molecular aspects. In: Mujib, A. and Samaj, J. (eds.) Somatic Embryogenesis, Plant cell Monographs. Berlin; Springer-Verlag. 2:157-175.

Whilliam, E.G. and Maheshwaran, G. 1986. Somatic embryogenesis factors influencing co-ordinate behavior of cells as an embryogenic group. Ann. Bot. 57: 442-462.

Williams, Leor, Zhao, Jing, Morozava, Nadya, Li, Yan, Avivi, Yigal, Grafi and Gideon. 2003. Chromatin reorganization accompanying cellular differentiation is associated with modifications of histone H3, redistribution of HP1 and activation of E2F-target genes. Develpomental Dynamics. 228(1): 113-120.

Xu, X., Lu, J., Lamikanra, O. and Schell, I. 1995. Somatic embryogenesis and plant regeneration in Muscadine grape cultivar triumph. Proc. Fla. State Hort. Soc. 108: 358-359.

Yogesh, T., Jasrai, K.N., Thaker and M.C. D'Souza. 2003. *In vitro* propagation of *Euphorbia pulcherrima* Willd. through somatic embryogenesis. Plant tissue culture. 13(1): 31-36.

Zeng, Fanchang, Zhang, Xianglong, Cheng, Lei, Hu, Lisong, Zhu, Longfu, Cao, JInglin, Guo and Xiaoping. 2007. A draft gene regulatory network for cellular totipotency reprogramming during plant somatic embryogenesis. Genomics. 90(5): 620-628.

Chapter 12

Callusing Efficiency and Regeneration in *Solanum surattense* Burm f.: A Medicinal Herb

T. Ugandhar, A. Lakshman, M. Praveen, M. Rambabu,
D. Sharada, M. Upender, K. Madhusudhan
and N. Rama Swamy*

Plant Biotechnology Group, Department of Biotechnology
Kakatiya University, Warangal – 506 009, A.P.

ABSTRACT

Solanum surattense Burm f. is a medicinally important species used in Ayurveda. Callusing ability of different explants such as cotyledon, hypocotyls and leaf was observed on MS medium supplemented with various concentrations of auxins IAA, and 2,4-D (1-5 mg/l) individually. Callusing efficiency was found to be higher in leaf cultures followed by cotyledon and hypocotyl explants. Among the auxins, NAA induced high amount of callus followed by IAA and 2, 4-D in all the explants tested. Multiple shoots were induced from leaf derived callus on MS medium fortified with different concentrations of cytokinoins BAP/Kn (1-8 mg/l) alone and also in combination with 0.5 mg/l IAA/NAA. A high frequency of shoots were regenerated per explant on MS medium amended with 0.5 mg/l IAA+ 3.0 mg/l BAP compared to all other hormonal combinations and concentrations. The *in vitro* regenerated micro-shoots were rooted on ½ strength

---

* Corresponding Author: E-mail: swamynr_dr@gmail.com

MS medium supplemented with 1.0 mg/l IAA. The plantlets were successfully hardened and maintained in the field. The protocol can be used to multiply and propagate the species.

Keywords: *Callus induction, Regeneration, Multiple shoots, Solanum surattense, Medicinal herb.*

Introduction

Solanum surattense Burm f. (syn. *S. xanthocarpum* Shrad & Wendl.) (Sanskrit: Kantakri), Indian solanum is used in ayurvedic medicine as Dasamula for chest pains, cough, asthma and fever (Nayar *et al.*, 1989). It is an important therapeutic agent and is also used against difficult urination, bladder stones, rheumatism, sore-throat, enlargement of liver and spleen, vomiting and skin diseases (Sivarajan and Balachandran, 1999). It is also used in preparation of kamavirya, a wonderful ayurvedic massage oil and also as one of the ingredients in bronchicyl.

Although the species has much pharmaceutical importance, it was not exploited *in vitro*. The literature shows a limited work on *in vitro* studies in S. *surattense* (Baburaj and Thamizhchelvan, 1991; Prasad *et al.*, 1998, Pawar *et al*; 2002; Ugandhar, 2002). Hence, an attempt was made in the present investigations to establish the efficeient protocol using various growth regulators in inducing the callus from different explants and also plant regeneration in S. *surattense*.

Methodology

Seeds of *S. surattense* were soaked in sterile distilled water for 24 hrs and were surface sterilized with 0.1 per cent (w/v) mercuric chloride solution for 3 – 5 minutes followed by three rinses with sterile distilled water. These were germinated aseptically on MS basal medium in 250 ml Erlenmeyer flasks.

The explants *viz.*, hypocoty1 (0.5-0.8 cm long) and cotyledon (0.6-0.8 cm²) from 3-week-old and leaf (0.8 – 1.0cm²) from 6 week-old axenic seedling were excised for callus induction. These explants were inoculated on MS (Murashige and Skoog, 1962) medium supplemented with various concentrations (1.0 – 5.0 mg/l) of auxins such as 2, 4-D, IAA and NAA.

For *in vitro* organogenesis the leaf explants from six-week-old axenic seedlings were implanted on MS basal medium supplemented with 4.0 mg/l NAA to induce white-friable callus. Calli obtained after 4 weeks of culture were regularly sub-cultured on the same medium for further proliferation. After three passages of subculture, 0.5 – 1.0 cm² fresh callus was transferred to regeneration medium containing MS salts, 3 per cent (w/v) sucrose and various concentrations of growth regulators *viz.* BAP or Kn (1, 2, 3,4,6,8 mg/l) individually and also in combination with auxins IAA and NAA (0.5 mg/l). The pH of the media was adjusted to 5.8 either with 0.1 NaOH or 0.1 N HC1. The medium was solidified with 0.8 per cent (w/v) difco-bacto agar and autoclaved at 121°C for 15-20 minutes. The cultures were incubated at 25°C with 16/8 hours of photoperiod under white florescent light of 40-60 µmol m⁻²S⁻¹ intensity.

For *in vitro* rooting, micro-shoots (3-4 cms) were excised and cultured on ½ strength MS medium containing 1.0 mg/lIAA. The plantlets were washed to remove remains of agar and transferred to plastic pots containing sterlite vermiculite: garden soil (1:1). The plantlets were covered with polythene bag in order to maintain the RH (70-80 per cent). After 3 weeks of hardening, these were shifted to earthenware pots containing garden soil. These plants were maintained under shady conditions.

Results

Callus induction ability of different explants *viz.*, cotyledon, hypocotyl and leaf was investigated by using varying concentrations of different auxins individually. Callus proliferation was initiated at the cut surfaces of the explants studied and later it covered the entire surface. Both color and texture of the callus also varied with growth regulators supplemented. The explants *viz.*, cotyledon, hypocoty1 and leaf cultured on MS medium supplemented with different concentrations (1.0 to 5.0mg/l) of auxins such as 2,4-D, IAA and NAA individually exhibited initiation of callus after 10 days of incubation while it took 12-15 days in cotyledon cultures.

On 2, 4-D supplemented medium, an early induction of callus was observed except at 1.0 mg/l in all the explants tested. Callus induction was observed from all the explants cultured and in all the concentrations of 2, 4-D (Fugre 12.1a). Maximum percentage of response was observed in cotyledon cultures followed by leaf and hypocotyl cultures at 1.0 mg/l 2, 4-D. As the concentration increased, there was a gradual decrease in the response in hypocotyl and cotyledon cultures but it was interesting to note that 100 per cent response was observed in all the concentrations of 2, 4-D in cotyledon explants. High amount of callus was induced on 2, 4-D in all the explants except at 4.0 and 5.0 mg/l 2, 4-D. Morphology of callus varied at different levels of 2, 4-D in leaf and hypocotyl cultures whereas white-compact calli were induced in almost all the concentrations of 2, 4-D.

To know the callusing efficiency of cotyledon, hopocotyl and leaf explants cultured on MS medium containing various concentrations of IAA. High percentage of response was observed at low levels of auxins used in all the explants. At 5.0 mg/l IAA the callus induction was inhibited in cotyledon and leaf explants, while less percentage of response was observed in hypocotyl cultures. As the concentration of IAA was increased to 2–4 mg/l, the percentage of responding cultures decreased. High amount of callus was induced at 3.0 mg/l IAA in hypocotyl cultures compared to all other concentrations of explants too. Embryogenic callus was found at 2.0 and 3.0 mg/l IAA in leaf cultures. Friable callus was observed at all other concentrations of IAA in cotyledon, hypocotyls and leaf cultures, except at 1.0 mg/l in cotyledon and 4.0mg/l in leaf explants in which white and brown compact calli were induced respectively.

Effect of NAA on Callusing Ability of Cotyledon, Hypocotyls and Leaf Explants

Highest percentage (100 per cent) of response was observed at 4.0 mg/l NAA in leaf cultures whereas less percentage of responding cultures were noted in all the concentrations of NAA used in hypocotyls explants (Figure 12.1b). Cotyledon and leaf explants responded well at all levels of NAA. Very high amount of callus was produced from cotyledon and leaf cultures on MS medium supplemented with 3.0 and 4.0 mg/l NAA while at 3.0mg/l NAA more amount of callus was induced in hypocotyl explants. The callus induction was moderate at high concentrations of NAA in leaf and hypocotyl explants. Induction of nodular callus was observed in both cotyledon and leaf explants on MS medium containing different concentrations of NAA whereas white – compact calli were found at 1.0 mg/l NAA in cotyledon and hypocotyl explants. Friable callus was produced on 3.0 – 5.0 mg/l NAA in hypocotyl and cotyledon cultures and at 4.0 and 5.0 mg/l from leaf explants.

In vitro Organogenesis

During the present investigation, freshly isolated callus showed greater morphogenetic response for induction of shoots. The morphogenetic callus continued to generate shoots upon subculture. The number of shoots produced from callus cultures varied with the hormones added to MS basal medium in S. *surattense*. Auxin in combination with a cytokinin produced maximum number of shoots with a greater frequency when compared to cytokinin as a sole growth regulator (Figure 12.4).

Callus derived from leaf explants when cultured on cytokinin alone or in combination with auxins showed the shoot bud induction after one week of culture (Figure 12.1c-f).

A frequency of 50 per cent cultures responded at 1 mg/l BAP with an average number of 12.4+0.3 shoots per explant (Figure 12.2). At 3.0 mg/l BAP, 63.0 per cent cultures responded with a maximum number of shoots 14.9±0.7 per explant. Higher levels of BAP *i.e.*, at 4.0 mg/l and 6.0 mg/l produced less number of shoots 11.7 and 11.2 with a lower frequency of 47.0 and 26.0 response respectively. Shoot bud suppression was observed at 8.0 mg/l BAP. Kinetin was found to be less responsive compared to BAP. A high frequency of 56.2 per cent cultures responded with a maximum number of shoots/explant (13.9±0.5) at 3.0 mg/l Kn. Low and high levels of Kn showed less response as observed in the case of BAP. As the concentration of Kn increased to 8.0 mg/l, shoot bud initiation was completely suppressed.

After the study of cytokinin as a sole growth regulator on shoot regeneration from callus cultures, auxin was taken in combination with cytokinin to study the effect on regeneration ability (Figures 12.3 and 12.4). The concentration of auxin IAA was kept constant at 0.5 mg/l and different concentrations of BAP (1, 2, 3, 4, 6, 8.0 mg/l) were added to the MS medium. Lower level of BAP 1.0 mg/l induced less number of shoots (16.5±0.54) with a frequency of 46.5 per cent cultures responding (Figure 12.3). Whereas at 3mg/l BAP, 60.5 per cent cultures responded and maximum number of shoots/explant (20.5±0.53) were recorded. As the concentration of BAP was increased to 4.0 and 6.0 mg/l the percentage of responding cultures reduced to 58.0 and 45.0 respectively and there was also a decrease in the number of shoot bud proliferation 15.6 and 13.5 respectively. Only green nodular callus was produced without shoot bud initiation at 8mg/l BAP (Figure 12.3).

The calli cultured on MS medium supplemented with 0.5mg/l IAA in combination with various concentrations of Kn showed less response compared to the combination of BAP (Figure 12.1f). Different concentrations of Kn (1, 2, 3, 4, 6, 8mg/l) were added with 0.5mg/l IAA. At 1 and 2 mg/l Kn 46 per cent and 60 per cent cultures responded with an average 14.5 and 15.6 shoots/explant respectively. A

Figures 12.1: Callus Induction and Multiple Shoot Regeneration in *S. surarttense*

a: Induction of white-compact callus from cotyledon explant on MS + 3.0 mg/L 2, 4-D; b: Induction of friable callus from hypocotyls explant on MS + 3.0 mg/L NAA; c: Callus derived on MS + 4.0 mg/L NAA from leaf explant cultured on 3.0 mg/L BAP (Note the organogenesis after 10 days); d: Multiple shoot bud induction from the same explant after three weeks of culture; e: Multiple shoot regeneration after six weeks of the same; f: Profusely developing multiple shoots from leaf derived callus cultures on MS + 0.5 mg/L IAA + 3.0 mg/L BAP.

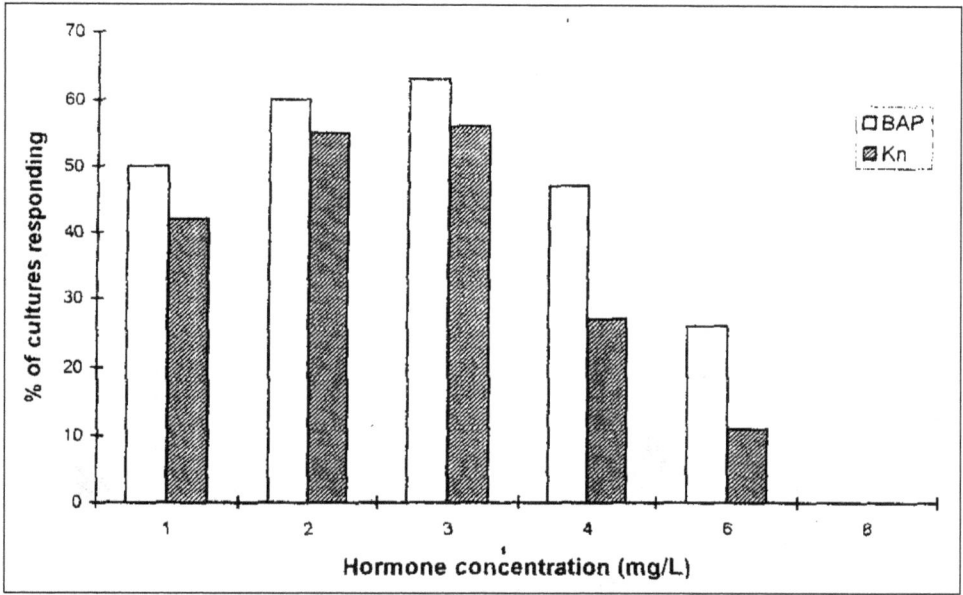

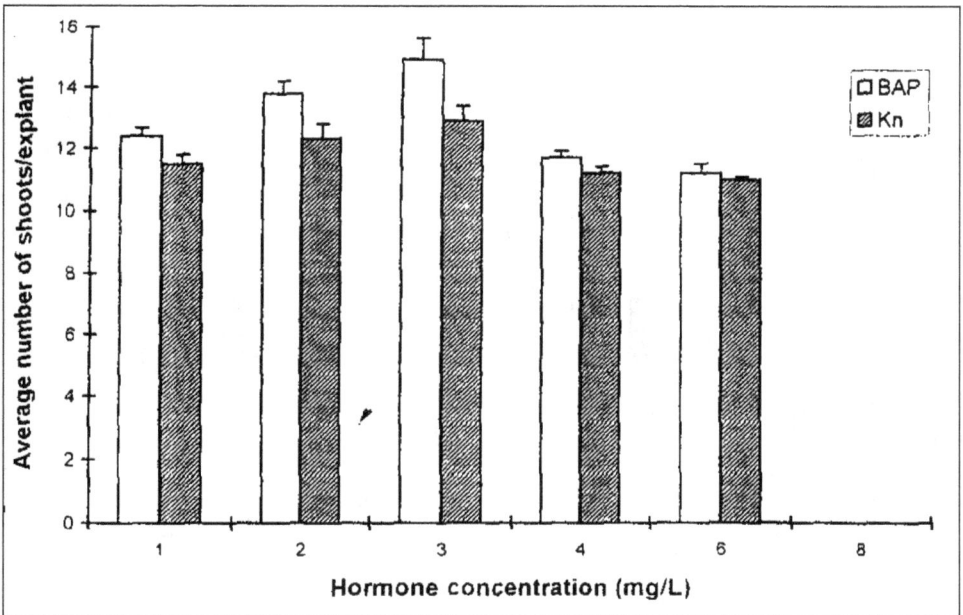

Figures 12.2: Effect of BAP and Kn on Induction of Multiple Shoots from Leaf Derived Callus of S. *surarttense*

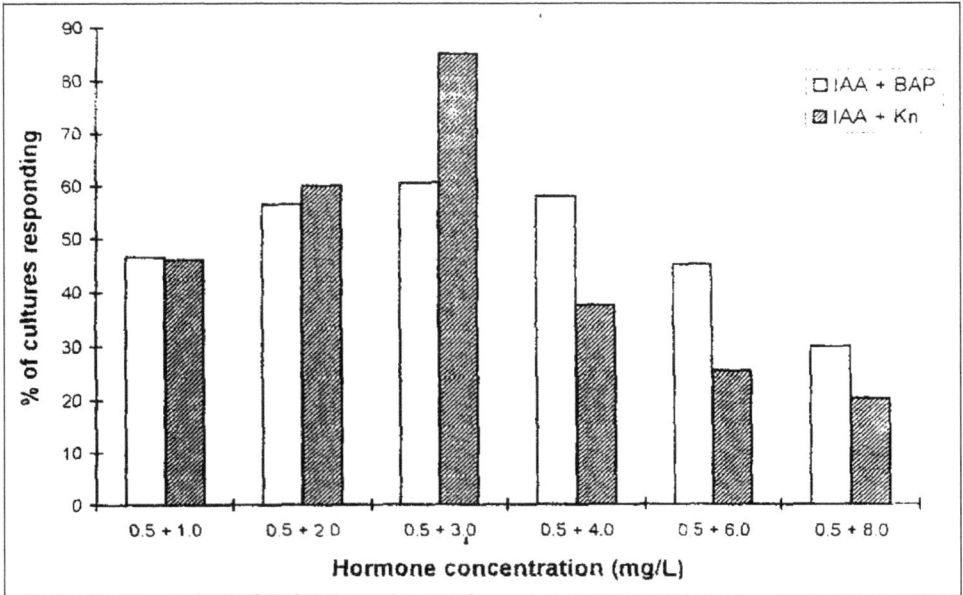

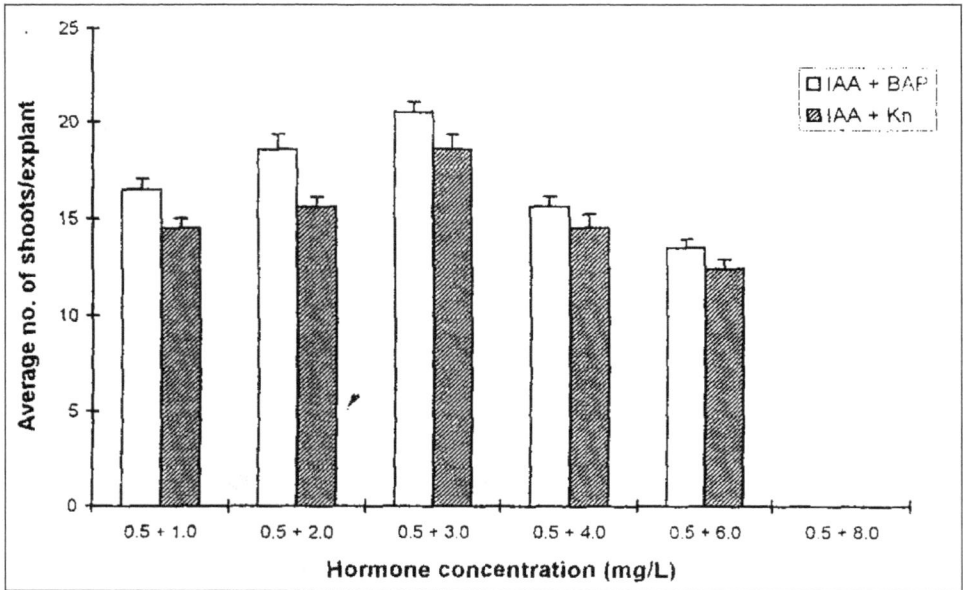

Figures 12.3: Effect of IAA in Combination with BAP and Kn on Induction
of Multiple Shoots from Leaf Derived Callus Cultures of S. *surarttense*

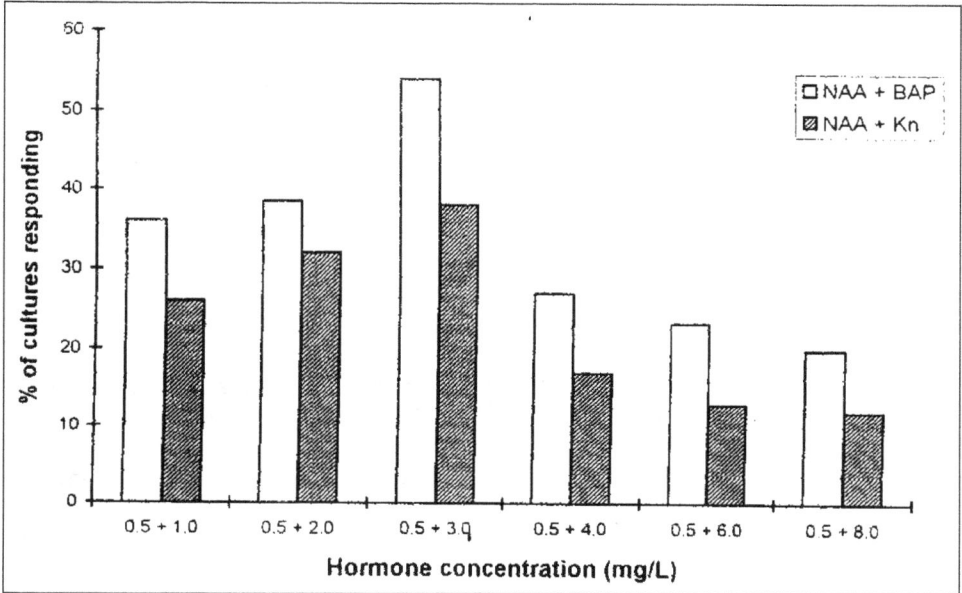

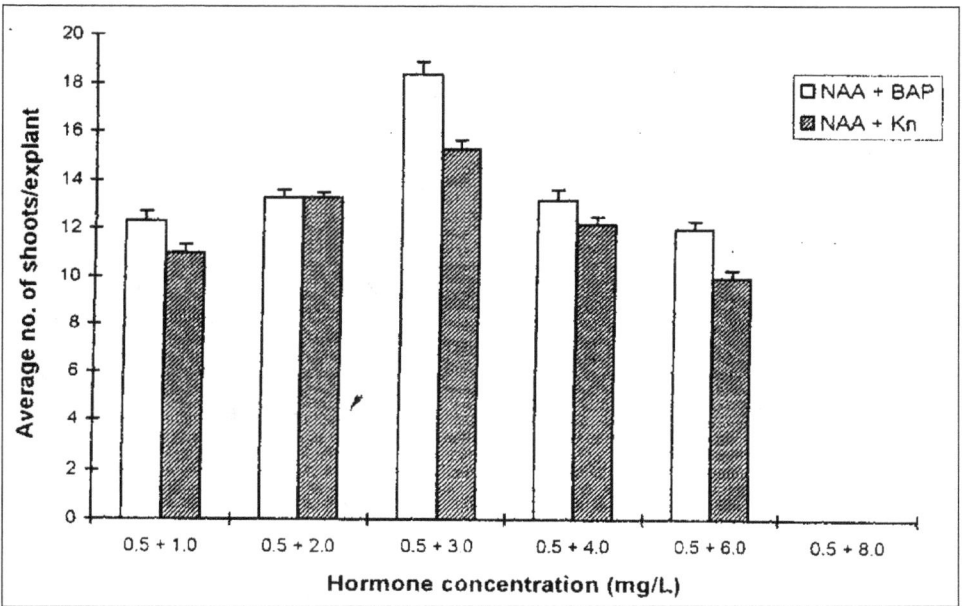

**Figures 12.4: Effect of NAA in Combination with BAP and Kn on Induction
of Multiple Shoots from Leaf Derived Callus Cultures of S. *surarttense***

maximum of 18.5 shoots/explant was recorded at 3mg/l Kn with a frequency of 85 per cent cultures responding. Number of shoots considerably was reduced as the concentration of Kn was increased. Responding cultures were also reduced after 3.0 mg/l Kn as observed on BAP. Complete suppression of shoot bud induction was found at high concentration of Kn.

Similarly, the callus derived from leaf explants was cultured on 0.5 mg/l NAA in combination with different concentrations of BAP or Kn (Figure 12.4). Low level of BAP (1.0, 2.0 mg/l) induced less number of shoots/ explant 12.3 and 13.3 with less percentage of responding cultures. At 3.0 mg/l BAP, 54 per cent cultures responded and a maximum number of shoots (18.4) per explant was recorded. It was also noted that as the concentration of BAP was increased to 4 and 6 mg/l the percentage of responding cultures reduced to 27 per cent and 23 per cent and also a decrease in number of shoots 13.2 and 12.0 per explant respectively. Only green nodular callus was induced without any shoot bud initiation at 8 mg/l BAP (Figure 12.4). Regeneration of shoot buds from callus cultures on 0.5 mg/l NAA in combination with different concentrations of Kn was found to be less responsive compared to BAP. Low levels of Kn induced less number of shoots/explant showing less percentage of cultures response. At 3.0mg/l Kn maximum number of shoots were produced. Higher levels of Kn (4.0 and 6.0 mg/l) showed less response, producing less number of shoots/explant. At 8.0mg/l only callus proliferation was observed without any shoot bud initiation as observed on BAP (Figure 12.4).

Micro-shoots were rooted on ½ strength MS medium augmented with 1.0 mg/l IAA. The *in vitro* regenerated plants were successfully hardened and maintained in the field under shady conditions. The survival rate was found to be 53 per cent. Morphology of plants was similar to the donor plants.

Discussion

In the present investigations the callus was induced from all the explants *viz.*, hypocotyls, cotyledon and leaf in all the concentrations of auxins 2, 4-D, IAA and NAA with the exception of 5.0 mg/l IAA. At this high concentration of IAA the morphogenetic response was inhibited in cotyledon and leaf explants. Callusing efficiency was found to be higher in leaf explants followed by cotyledon and hopocotyl explants of *S. surattense*. This difference in callusing ability suggests the presence of different levels of endogenous hormones in the tissues. The auxins such as NAA, IAA and 2, 4-D alone suppressed shoot bud formation in all the explants studied but promoted callusing.

Among the auxins tested, NAA induced the high yield of callus followed by IAA and 2, 4-D. Similarly, Omar (1988) observed the same findings with NAA in *Rhyazya stricta*, a medicinal plant.

The auxin 2, 4-D has been determined as a potent callus inducing phytohormone in studies with many plant species such as *Capsicum* (Gunay and Rao, 1978: Philips and Hubstenberger, 1985), *Cucumis sativus* (Rajasekharan *et al.*, 1983), *Solanum melongena* (Sreenivasa Swamy *et al.*, 1988), whereas in the present investigations 2, 4-D induced less amount of callus proliferation compared to all other auxins used in all the explants studied. Praveen *et al.* (2001) have studied the callusing ability of different explants in *Strychnos potatorum* – a medicinal plant on MS medium supplemented with various growth substances *viz.*, IAA, NAA and 2,4-D. They observed the maximum callus growth on medium containing 2, 4-D in contrast to present findings.

Tejavathi and Bhuvana (1998) have also observed the callusing ability of different explants in *Solanum viarum*, using auxins NAA and 2, 4-D. Among these auxins NAA (2 mg/l) was found to be the best to induce callus from hypocotyls, while 2, 4-D (3 mg/l) either alone or with CM (10 per cent) elicited callus formation from root, stem and leaf explants. Kumari and Kumar (1995) have observed the friable callusing and rhizogenesis in the explants cultured on a medium containing IAA, NAA and 2, 4-D at 1-25 µM range of concentration in *Thevetia peruviana*. Shahzad *et al.* (1999) have observed the callus induction on MS medium supplemented with NAA (2 mg/l) and 2, 4-D (2 mg/l) in leaf cultures of *Solanum nigrum*. They found the faster proliferation and very high yield of callus on 2, 4-D compared to NAA in contrast to our present observations.

Callus produced from different explants showed variability in texture, form and coloration. This difference is dependent upon the responses of plant tissues to various growth promoting substances. Thus, successful callus induction depends upon various factors such as composition of the nutrient medium, hormonal balance besides the type, age and genotype of the explant (Huang and Murashige, 1976; Narayana Swamy, 1977).

Thus it is evident that the auxin NAA was found to be potent for callus proliferation followed by IAA and 2, 4-D. Proliferation in callus was faster and a very high yield of callus mass was achieved within 4 weeks on NAA followed by IAA and 2, 4-D. Maximum amount of callus production was observed on MS medium supplemented with 3.0 and 4.0 mg/l NAA.

The callus produced from 2, 4-D was white and brown/gray compact whereas from IAA and NAA, the callus was friable embryogenic in almost all the concentrations and explants used indicating the capability for regeneration.

Since Guilietti *et al.* (1991) have used the cell suspension culture technique for isolating Solasodine in *Solanum elaeagnifolium* from callus induced on different auxins such as IAA, 2, 4-D, NAA, IBA and 2, 4, 5-T, the same technology can also be used to isolate the different glycoalkaloids (Solasonine, Solamargine, Solasodine) from *in vitro* cultures of *S. surattense*.

We were successful in regenerating plants from callus cultures on MS medium fortified with different concentrations of cytokinins *i.e.*, BAP/Kn individually and also in combination with 0.5 mg/l IAA / 0.5 mg/l NAA. Maximum number of shoot buds were induced at 3.0 mg/l BAP in comparison to Kn as a sole growth regulator. When low level of auxins (0.5 mg/l) were added to the medium containing BAP/Kn, it was interesting to find out that the shoots induction was enhanced in all the concentrations of cytokinins tested. However, the shoot bud proliferation was found to be more on 0.5 mg/l IAA in combination with BAP/Kn compared to 0.5 mg/l

NAA. Probably, IAA might have triggered the action of BAP/Kn in a proper way for inducing more number of shoots/explant. But the combination of IAA + BAP induced highest number of plantlet regeneration among all hormonal combinations and concentrations used.

De Langhe and De Bruijne (1976) have observed the maximum shoot regeneration on IAA + BA in comparison to NAA + BA in leaf callus cultures of tomato as in *S. surattense*. Similar observations were reported in callus cultures of *Theobroma cacao* (King and Rao, 1981), *Brassica campestris* (Singh *et al.*, 1985) and *Piper longum* (Bhat *et al.*, 1992) the maximum proliferation of shoot buds on auxin and cytokinin combination. Azad and Amin (1998) also found maximum number of shoot bud proliferation on the medium containing NAA + BA in internodal explants of *Adhatoda vasica*, a medicinal plant.

Similarly, Hoque *et al.* (2000) have reported the high frequency of plant regeneration of MS medium containing 2.0 mg/l BA in combination with 0.5 mg/l NAA from cotyledon derived callus in *Momordica dioica*. They have also found the maximum number of shoots per explant on BA compared to Kn and induction was higher on NAA + BA than Kn as it was found in the present investigations. The essentiality of both auxin-cytokinin combination for inducing shoot organogenesis has been reported in leaf callus cultures of *Cicer arietinum* by Arockiasamy *et al.* (2000). The callus was produced on MS medium containing NAA + BA and regenerated on BA + GA_3 in *C. arietinum*. They have also found that BA/GA_3 alone was ineffective in eliciting shoot regeneration.

Though more number of shoots were formed per explant in auxin-cytokinin combination but the combination of IAA + BAP / Kn was more effective then NAA + BAP / Kn. Whereas Shahzad *et al.* (1999) have found the efficacy of auxin-cytokinin combination NAA + BAP in inducing shoot organogenesis from leaf callus cultures of *Solanum nigrum*. We have achieved the multiple shoot induction from leaf derived callus cultures in *S. surattense*. Thus *in vitro* micropropagation through callus cultures is an important technique to multiply and propagate *S. surattense*, a medicinally important herb in large numbers.

Acknowledgements

We thank Prof. K. Subhash, Head, Department of Biotechnology, Kakatiya University for his encouragement. NRS is grateful to TWAS (Italy) and UNESCO (France) for the financial assistance.

References

Arockiasamy, S., Varghese, G. and Ignacimuthu, S. 2000. High frequency regeneration of chick pea (*Cicer arietinum* L.), plantlets from leaf callus. Phytomorphology 50: 297-302.

Azad, M.A.K. and Amin, M.N. 1998. *In vitro* regeneration of plantlets from internode explants of *Adhatoda vasica* Nees. Plant cell Tiss. Org. Cult. 8: 27-34.

Baburaj, S. and Thamizhchelvan, P. 1991. Plant regeneration from leaf callus of *Solanum surattense*. Indian J. Exp. Biol., 29: 391-392.

Bhat, S.K. Kachar, A. and Chandel, K.P.S. 1992. Plant regeration from callus cultures of *Piper longum* L. by organogenesis. Plant Cell Rep., 11: 525-528.

De Langhe, E. and Bruijne, A. 1976. Continuous propagation of tomato plants by means of callus culture. Sci. Hortic., 4: 2221-227.

Guilietti, A.M. Negra, H.M. and Caso, O. 1991. *Solanum eleaegnifolium* (silver leaf night shade). *In vitro* culture and the production of Solasodine In: Bajaj Y.P.S. (ed.), Biotechnology in agriculture and foresty. Vol. 15 Medicinal and Aromatic Plants III. Springer-Verlag, Berlin, pp. 432-450.

Gunay, A.L. and Rao, P.S 1978. *In vitro* plant regeneration from hypocotyls and cotyledon explants of red pepper (*Capsicum*). Plant Sci. Lett., 11: 365-372.

Hoque, A., Islam, R. and Arima, S. 2000. High frequency plant regeneration from cotyledon derived callus of *Momordica dioeica* (Roxb.) Willd. Phytomorphology, 50: 267-272.

Huang, H,C. and Murashige, T. 1976. Plant tissue culture media, major constituents, their preparation and some applications. Tissue Culture Assn. Manual, 3: 539-549

King, L.S. and Rao, A.N. 1981. Induction of callus and organogenesis in Cocoa tissues. In Proc. Casted symp. on Tiss. Cult. of Economically Important Plants, pp. 107-112.

Kumari, A.S. and Kumar, A. 1995. Plant regeneration from cultured embryogenic axis *Thevetia peruviana* L. Indian J. Exptl. Biol., 33: 190-193.

Murashige, T. and Skoog, F. 1962. A revised medium for rapid growth and bioassay with tobacco tissue culture. Physiol. Plant, 159: 473-497.

Narayanaswamy, S. 1977. Regeneration of plants from tissue cultures. In: Reinert, J. and Bajaj, Y.P.S. (eds.), *Applied and fundamental aspects of plant cell, tissue and organ culture*. Springer-Verlag, Berlin, pp. 179-206.

Nayar, M.P., Ramamurthy, K. and Agarwal, V.S. 1989. Economic plants of India, Vols. I & II, B.S.I., Calcutta.

Omar, M.S. 1988. *Rhazya stricta* Decaisne: *In vitro* culture and the production of indole alkaloids. In: Bajaj, Y.P.S. (ed.) *Biotechnology* in *Agriculture* and *forestry*. Springer Verlag Berlin, pp. 529-540.

Pawar, K., Pawar, C.S., Narkhede, B.A., Teli, N.P., Bhalsing, S.R. and Maheshwari, V.L. 2002. A technique for rapid micropropagation of *Solanum surattense* Burm. f. Indian J. Biotech, 1: 201-204.

Philips, G.C. and Hubstenberger, J.F. 1985. Organogenesis in pepper tissue culture. Plant Cell Tissue Org. Cult., 4: 262-269.

Prasad, R.N., Sharma, M., Sharma, A.K. and Chaturvedi, H.C. 1998. Androgenic stable somaclonal variant of *Solanum surattense* Burm.f. Indian J. Exptl. Biol., 36: 1097-1012.

Praveen, M., Lakshman, A., Ugandhar, T. and Ramaswamy, N. 2001 Callusing efficiency and plant regeneration from different explants of *Strychnos potatorum*- A medicinally important forest tree. In: Sadanandam, A., Reddy, K.J.M., Ram Reddy, S. and Ramaswamy, N. (eds), Frontiers of Plant Biotechnology, pp. 193-197.

Rajasekharan, K., Mullins, M. and Nair, Y. 1983. Flower formation *in vitro by* hypocotyl explants of cucumber (*Cucumis sativus*). Annals Bot., 52: 417-420.

Shahzad, A., Hasan, H. and Siddiqui A.S. 1999. Callus induction and regeneration in *Solanum nigrum* L. *in vitro*. Phytomorphology, 49: 215-220.

Singh, S., Garg, K. and Chandra, N. 1985. Growth and differentiation in internode culture of *Brassica campestris* var. Yellow sarson. Acta Bot. Indica 13: 45-50.

Sivarajan, V.V. and Balachandran, I. 1999. Ayurvedic drugs and their plant sources. Oxford & IBH publishing Co. Pvt. Ltd., New Delhi.

Srinivasaswamy, M., Charistopher, T. and Subhash, K. 1988. Multiple shoot formation in embryo culture of *Solanum melongena*. Curr. Sci. 57: 197-198.

Tejavathi, D.H. and Bhuvana, B. 1998. *In vitro* morphogenetic studies in *Solanum viarum* Dunal. J. Swamy Bot. Cl. 15: 27-30.

Ugandhar, T. 2002. Tissue culture studies in *Solanum surattense* Burm. f. Ph.D. Thesis, Kakatiya University, Warangal.

Index

www.ingramcontent.com/pod-product-compliance
Lightning Source LLC
Chambersburg PA
CBHW050515190326
41458CB00005B/1544